FATE OF POLLUTANTS IN THE AIR AND WATER ENVIRONMENTS

Part 2

FATE OF POLLUTANTS IN THE AIR AND WATER ENVIRONMENTS

Part 2

Chemical and Biological Fate of Pollutants in the Environment

Edited by
I. H. SUFFET
Drexel University
Department of Chemistry
Environmental Engineering and Science
Philadelphia, Pennsylvania

From paper presented in part at the symposium on "Fate of Pollutants in the Air and Water Environments" at the 165th National American Chemical Society Meeting in the Environmental Chemistry Division held in April 1975 in Philadelphia, Pennsylvania

A Wiley-Interscience Publication
JOHN WILEY & SONS
New York / London / Sydney / Toronto

Library of Congress Cataloging in Publication Data:
Main entry under title:

Fate of pollutants in the air and water environments.

(Advances in environmental science and technology; v. 8)
"A Wiley-Interscience publication."
Based on "an American Chemical Society symposium in the Environmental Chemistry Division, 'Fate of pollutants in the air and water environments,' held at the 165th national meeting of the American Chemical Society in Philadelphia, Pennsylvania, on April 6-11, 1975."
Includes indexes.
CONTENTS: Pt. 1. Mechanisms of interaction between environments. Mathematical modeling and the physical fate of pollutants.—Pt. 2. Chemical and biological fate of pollutants in the enviroment.
1. Pollution—Congresses. 2. Environmental chemistry—Congresses. I. Suffet, Irwin H. II. American Chemical Society. III. Title. IV. Series.

TD180.A38 vol. 8 [TD172.5] 363.6 76-58408
ISBN 0-471-83539-0 (v. 1)
ISBN 0-471-01803-1 (v. 2)

Printed in the United States of America

10 9 8 7 6 5 4 3 2 1

Dedication

To
my loving family:

my wife, Eileen; my children,
Alison Michelle and Jeffrey Hugh;
and my parents

Contributors

DR. A. P. ALTSHULLER
Director of Environmental Science Laboratory
U. S. Environmental Protection Agency
Research Triangle Park, North Carolina 27711

MR. EDMOND ARSENAULT
Environment Canada, Fisheries and Marine Service
Biological Station
St. Andrews, New Brunswick, Canada EOG 2XO

DR. BERT BOLIN
Department of Meteorology
University of Stockholm
Arrhenius Laboratory
S-104 05 Stockholm, Sweden

DR. JOACHIM BORNEFF
Department of Hygiene
University of Mainz
Mainz, Germany

DR. KENNETH L. DEMERJIAN
Calspan Corporation
Buffalo, New York 14221
Present Address:
U.S.E.P.A.
Environmental Science Research Labs
Research Triangle Park, North Carolina 27711

DR. SAMUEL D. FAUST
Department of Environmental Science
Rutgers University
New Brunswick, New Jersey 08903

DR. STEVEN M. GERTZ
Porter-Gertz Consultants, Inc.
Radiological Protection and Environmental Services
Ardmore, Pennsylvania 19003

DR. ERNEST F. GLOYNA
Dean, College of Engineering
The University of Texas
Austin, Texas 78712

DR. JERRY L. HAMELINK
Lilly Research Laboratories
Division of Eli Lilly and Co.
Box 708
Greenfield, Indiana 46140

DR. JULIAN HEICKLEN
Dept. of Chemistry and Center for Air Environment Studies
The Pennsylvania State University
University Park, Pennsylvania 16802

DR. GEORGE M. HIDY
Rockwell International Science Center
Thousand Oaks, California 91360
Present Address:
Environmental Research and Technology, Inc.
Western Technical Center
741 Lakefield Road
Westlake Village, California 91361

DR. I. C. HISATSUNE
Dept. of Chemistry and Center for Air Environment Studies
The Pennsylvania State University
University Park, Pennsylvania 16802

DR. JAMES W. HOGAN
Fish-Pesticide Research Lab.
U. S. Department of the Interior
Fish and Wildlife Service
Columbia, Missouri 65201

DR. B. THOMAS JOHNSON
Fish-Pesticide Research Lab.
U. S. Department of the Interior
Fish and Wildlife Service
Columbia, Missouri 65201

DR. MORRIS KATZ
Department of Chemistry
York University
4700 Keele Street
Toronto, Ontario, Canada
M3J 1P3

DR. SHAHAMAT U. KHAN
Chemistry and Biology Research Institute, Canada Agriculture
Ottawa, Ontario, Canada
KIA OC6

DR. DAVID B. KITTELSON
Department of Mechanical Engineering
University of Minnesota
Minneapolis, Minnesota 55455

DR. WARREN C. KOCMOND
Calspan Corporation
Buffalo, New York 14221
Present Address:
Desert Research Institute
University of Nevada
Reno, Nevada 89507

DR. FREDERICK C. KOPFLER
U. S. Environmental Protection Agency
Health Effects Research Lab.
Cincinnati, Ohio 45268

MR. DWIGHT A. LANDIS
Rockwell International Science Center
Thousand Oaks, California 91360
Present Address:
Environmental Research and Technology, Inc.
Western Technical Center
741 Lakefield Road
Westlake Village, California 91361

DR. DOUGLAS A. LANE
Department of Chemistry
York University
4700 Keele Street
Toronto, Ontario, Canada
M3J 1P3
Present Address:
Sciex Limited
Thornhill, Ontario, Canada
L3T 1P2

DR. MARTIN LIPELES
Rockwell International Science Center
Thousand Oaks, California 91360

DR. FOSTER L. MAYER
Fish-Pesticide Research Lab.
U. S. Department of the Interior
Fish and Wildlife Service
Columbia, Missouri 65201

DR. PAUL M. MEHRLE
Fish-Pesticide Research Lab.
U. S. Department of the Interior
Fish and Wildlife Service
Columbia, Missouri 65201

DR. ROBERT G. MELTON
U. S. Environmental Protection Agency
Health Effects Research Lab.
Cincinnati, Ohio 45268

DR. ROBERT L. METCALF
Environmental Studies
Institute and Departments
of Biology and Entomology
University of Illinois
Urbana-Champaign, Illinois 61801

MS. JUDITH L. MULLANEY
U. S. Environmental Protection
Agency
Health Effects Research Lab.
Cincinnati, Ohio 45268

DR. EUGENIO SANHUEZA
Dept. of Chemistry and Center
for Air Environment Studies
The Pennsylvania State University
University Park, Pennsylvania
16802
Present Address:
Instituto Venezolano de Inves-
tigaciones Cientificas
Centro de Ingenieria
Apartado 1827
Caracas 101, Venezuela

DR. RICHARD A. SCHOETTGER
Fish-Pesticide Research Lab.
U. S. Department of the Interior
Fish and Wildlife Service
Columbia, Missouri 65201

DR. EDWARD J. SOWINSKI
Occupational Health & Safety
Western Electric Company
Allentown, Pennsylvania 18103
Present Address:
Environmental Health and Safety
Uniroyal Chemical
Nagatuck, Connecticut 06770

DR. DAVID L. STALLING
Fish-Pesticide Research Lab.
U. S. Department of the Interior
Fish and Wildlife Service
Columbia, Missouri 65201

DR. IRWIN H. SUFFET
Department of Chemistry
Environmental Studies
Institute
Drexel University
Philadelphia, Pennsylvania 19104

DR. ROBERT G. TARDIFF
U. S. Environmental Protection
Agency
Health Effects Research Lab.
Cincinnati, Ohio 45268

DR. RONALD C. WAYBRANT
Michigan Department of Natural
Resources
Stevens T. Mason Building
Lansing, Michigan 48926

DR. KENNETH T. WHITBY
Department of Mechanical
Engineering
University of Minnesota
Minneapolis, Minnesota 55455

DR. J. Y. YANG
Calspan Corporation
Buffalo, New York 14221

MR. PHILLIP R. YANT
Department of Zoology
University of Michigan
Ann Arbor, Michigan 48107

DR. YOUSEF A. YOUSEF
College of Engineering
Florida Technological Univ.
Orlando, Florida 32816

DR. VLADIMIR ZITKO
Environment Canada, Fisheries
and Marine Service
Biological Station
St. Andrews, New Brushwick,
Canada EOG 2XO

Introduction to the Series

Advances in Environmental Science and Technology is a series of multiauthored books devoted to the study of the quality of the environment and to the technology of its conservation. Environmental sciences relate, therefore, to the chemical, physical, and biological changes in the environment through contamination or modification; to the physical nature and biological behavior of air, water, soil, food, and waste as they are affected by man's agricultural, industrial, and social activities; and to the application of science and technology to the control and improvement of environmental quality.

The deterioration of environmental quality, which began when man first assembled into villages and utilized fire, has existed as a serious problem since the industrial revolution. In the second half of the twentieth century, under the ever-increasing impacts of exponentially growing population and of industrializing society, environmental contamination of air, water, soil, and food has become a threat to the continued existence of many plant and animal communities of the ecosystem and may ultimately threaten the very survival of the human race.

It seems clear that if we are to preserve for future generations some semblance of the existing biological order and if we hope to improve on the deteriorating standards of urban public health, environmental sciences and technology must quickly come to play a dominant role in designing our social and industrial structure for tomorrow. Scientifically rigorous criteria of environmental quality must be developed and, based in part on these, realistic standards must be established, so that our technological progress can be tailored to meet such standards. Civilization will continue to require increasing amounts of fuel, transportation, industrial chemicals, fertilizers, pesticides, and countless other products, as well as to produce waste products of all descriptions. What is urgently needed is a total systems approach to modern civilization through which the pooled talents of scientists and engineers, in cooperation with social scientists and the medical profession, can be focused on the development of order and equilibrium among the presently disparate segments of the human environment. Most of the skills and tools that are needed already exist. Surely a technology that has created manifold environmental problems is also capable of solving them. It is our hope that the series in Environmental Science and Technology will not only serve to make this challenge more explicit to the established professional but will also help to stimulate the student toward the career opportunities in this vital area.

The chapters in this series of Advances are written by experts in their respective disciplines, who also are involved with the broad scope of environmental science. As editors, we asked the authors to give their "points of view" on key questions; we were not concerned simply with literature surveys. They have responded in a gratifying manner with thoughtful and challenging statements on critical environmental problems.

JAMES N. PITTS, JR., Editor
Statewide Air Pollution Research Center and Department of Chemistry
University of California
Riverside, CA 92521
Telephone: (714) 787-4584

ROBERT L. METCALF, Editor
Environmental Studies Institute
Departments of Biology and Entomology
University of Illinois
Urbana-Champaign, IL 61801
Telephone: (217) 333-3649

DANIEL GROSJEAN, Associate Editor
Statewide Air Pollution Research Center
University of California
Riverside, CA 92521
Telephone: (714) 787-3629

Preface

The "environmental fate" of specific chemical species is becoming a main criterion for assessing pollutant impact. The fate of pollutants is presented as a frame of reference to relate the work of scientists and engineers interested in the natural environment and environmental protection. It is hoped that a new consciousness to interpret scientific data will develop from the concept of "environmental fate."

To describe this impact Part 2 of this treatise elucidates the chemical and biological influences to which compounds are subjected. In the present context chemical influences are defined as physiochemical reactions of parent compounds in the environment encompassing adsorption, coagulation, equilibrium, and so on. Biological influences are defined as physiochemical interactions or metabolic changes associated with receptor organisms. In relation to these influences, physical influences are defined as parameters that transport or disperse chemicals throughout the lithosphere, hydrosphere, and atmosphere. That the primary chemical, physical, and biological influences of any pollutant can overlap environmental interfaces and that specific sinks fail to constitute segregated entities are embodied in the concept of fate.

Part 1 of this treatise illustrates how pollutants become interconnected within the confines of "Planet Earth," the atmosphere, lithosphere, and mesospheres. In pursuit of the fate of one pollutant, we observe how one pollutant's fate can become another pollutant's source. Groups of persistent pollutants can disappear from one area, but they later show up in some unsuspected locale.

Many chapters deal with pesticides, because they have been a widely studied group of environmentally troublesome organic compounds. The fate of pesticides was discussed in Part 1 of this treatise as a global phenomenon (pp. 7-26) and as it exchanges between environments (pp. 49-126). In this volume pesticides provide a focus for both biological (pp. 195-222 and 261-282) and chemical fate (pp. 317-366). The fate of other classes of organic compounds discussed includes phthlates, hydrocarbons (especially the polynuclear type), heavy metal complexes, chlorinated organics, and humic acids. Other chapters treat the environmental concerns of nutrients, heavy metals, acid gases, and radioactive materials within the context of their environmental fate.

This volume on the biological and chemical fate of pollutants in the environment (Part III to V) compares approaches to the study of pollutants in the air and water environments, as well as presents specific information on pollution problems. Two primary approaches are the development of model experimental environments and evaluation of data collected from the environment. These approaches must be coordinated so that an evaluation can be made of the dynamics and the knowns and unknowns of the system.

The editor gratefully acknowledges the assistance of the Environmental Protection Agency, which financed in part with Federal funds under Grant No. R8035801-0, an American Chemical Society Symposium in the Environmental Chemistry Division, "Fate of Pollutants in the Air and Water Environments," held at the 165th National Meeting of the American Chemical Society in Philadelphia, Pennsylvania on April 6-11, 1975, on which this work is in large part based. The contents of this work do not necessarily reflect the views and policies of the Environmental Protection Agency nor its mention of trade names or commercial products constitute endorsement or recommendation for use.

The preparation of this book has been a process in which many have participated. Each chapter has been technically reviewed by at least two outside referees and the editor to assure that the material meets the scientific rigor of a technical or review journal and that the combination of papers are within the framework of the goal of the treatise. Each reviewer is commended for his candor. Special thanks is due to Linda Kosmin, Engineering Librarian at Drexel University, who helped me technically edit selected chapters. The editor is deeply indebted to the speakers and session chairmen for their contributions and patience. I also acknowledge that this treatise includes Dr. A. P. Altshuller's address upon receiving the 1975 ACS Award for Pollution Control sponsored by the Monsanto Co.

At Drexel University, I have been privileged to be in a truly interdisciplinary atmosphere where engineers, physical scientists, and social scientists work together in an environmental program housed in its own facility. The idea for this treatise evolved from 1970 to 1975 as a "green" environmental analytical chemist taught and completed research with colleagues and students in the areas of air pollution, water pollution, solid waste, and industrial hygiene. Primary focus was on problems relating to environmental quality and toxic substances, especially pesticides. Professional growth led me across many disciplines. During this evolutionary process, I undertook the creation of Drexel Graduate Course F208 "Fate of Pollutants," as suggested by Dr. Henry Wohlers. The graduate students who participated in this course at Drexel are commended for openly exchanging ideas from whence evolved the refinement of ideas for this treatise. Also, useful suggestions were made by a number of colleagues regarding selection of topics in order to make this book an effective working reference. Particularly helpful were brainstorming sessions with Drexel colleagues Dr. Edward Glaser, Dr. James Friend, Dr. P. Walton Purdom, Dr. Robert Schoenberger, and Dr. Henry Wohlers. I must also praise the influence of Dr. Samuel D. Faust of Rutgers University, who showed me how a creative and industrious atmosphere can stimulate ideas. This I have strived to maintain.

Finally, I must convey my deepest affection and appreciation to my wife Eileen and to my children, Alison Michelle and Jeffrey Hugh, for their keen sense of understanding during the preparation of this treatise; and to my mother, Lee, whose early encouragement and patience helped bring me to this apex.

Irwin H. Suffet

Philadelphia, Pennsylvania
October, 1976

The editors are deeply indebted to Mrs. Dolores V. Tanno and Mrs. Glenna Paschal for their painstaking effort in final typing this volume of the Series.

Contents

PART 1. FATE OF POLLUTANTS IN THE AIR AND WATER ENVIRONMENTS
SECTION I. MECHANISMS OF INTERACTION BETWEEN ENVIRONMENTS

SECTION II. METHODS OF MATHEMATICAL MODELING, THE FATE OF POLLUTANTS, AND THE PHYSICAL FATE OF POLLUTANTS

FATE OF POLLUTANTS IN THE AIR AND WATER ENVIRONMENTS

Part 2

Fate of Pollutants in the Air and Water Environment: Chemical and Biological Fate

IRWIN H. SUFFET
Department of Chemistry
Environmental Studies Institute
Drexel University
Philadelphia, Pennsylvania

The quality of life hinges on understanding of the "environmental fate" of a plethora of potential pollutants that could be detrimental to our planet. How and where, in what form, and in what concentration pollutants are globally and locally distributed also has been considered a barometer for evaluating Man's tomorrows. These concerns for an understanding of pollutant impact is evident on every page of this text.

The research, wisdom, and insight of respected scientists are brought together to crystallize their concepts about the chemical and biological fate of pollutants. All the scientists contributing to this treatise have an interest in relating their work to the natural environment or environmental protection. Solving environmental problems requires these cooperative efforts of groups of scientists and engineers representing various subject specialties.

This treatise crosses between relevant disciplines to narrow the communication gap between principal investigators by revealing parallel approaches of investigation into the chemical and biological fate of pollutants in air and water. Similar oxidative and hydrolytic considerations are offered to follow the fate of a pollutant in the litho-, atmo-, hydro-, and biospheres; comparable surface mechanisms as phase partitioning processes between lipid and water interfaces in the biosphere and at clay-organic surfaces in the aquatic environment are proposed to understand biological fate and aquatic fate; and, parallel free radical mechanisms are used to explain both atmospheric smog formation and photochemical degradation of pollutants in rivers.

The environmental fate of specific chemical species is becoming a main criterion for assessing pollutant impact. This treatise elucidates the chemical, physical, and biological influences acting on compounds so that such impact can be described.

In the present context, chemical influences are defined as physiochemical reactions of parent compounds in the environment encompassing adsorption, photolysis, and oxidation. Physical influences are defined as processes that transport or disperse chemicals throughout the lithosphere, hydrosphere, and atmosphere. Biological influences are defined as physiochemical interactions or metabolic changes associated with receptor organisms.

A frame of reference for various aspects of environmental fate of specific chemical species is displayed in Table 1. Embodied is the idea that the chemical, physical, and biological

TABLE 1. Fate of Pollutants in the Air and Water Environments-A Frame of Reference

---

Mechanisms of Interaction Between Environments

A. The Air-Water Interface
B. The Solid-Water Interface
C. The Solid-Air Interface

Methods of Mathematical Modeling the Fate of Pollutants

Physical Fate

A. Transport
B. Mixing

Biological Fate

A. Food chain relationships
B. Metabolic changes

Chemical Fate

A. Adsorption
B. Coagulation
C. Complexation
D. Photolysis
E. Hydrolysis
F. Precipitation
G. Oxidation
H. Isomerization

---

fate of any pollutant can overlap environmental interfaces and that specific sinks fail to constitute segregated entities. Mathematical modeling readily lends itself to describing interactions within environmental territories or across environmental boundaries. Figure 1 represents "Planet Earth" and uses multi-directional arrows to emphasize that one should focus on the total environment when studying even a minute segment. This treatise is presented within the frame of reference of Table 1 and Figure 1. A holistic view is stressed to develop insight.

The concentration of a specific pollutant changes with time by the following equation in the usual format:

$$\frac{dC_i}{dt} = \text{Turbulent diffusion} + \text{Advection} + \text{Emission rate} + \text{Sink} + \text{Reactivity}$$

Or in terms of the fate of pollutants, this equation can be expressed as:

$$\frac{dC_i}{dt} = \underset{\text{(Physical fate)}}{\text{Mixing \& Transport}} + \underset{\substack{\text{(Biological and chemical fate)} \\ \text{(Physical removal processes)}}}{\text{Input-Output \& Reactions}}$$

Continuity of mass and heat must also be satisfied.

Models are used to depict simplified segments of the world on a scale that can be conceptualized and understood. Models are not a complete representation of the real world; they only present a segment of the world in a simplified manner. Models provide a conceptual framework - a simplification of nature - to enable man to comprehend, respect, and (we hope) help predict the complex behavior of nature.

Two points of focus are possible within this frame of reference. How the system operates with chemicals in it or how the chemical operates in the system. Chemical and biological fate of a pollutant focuses interest on the chemical itself and how it interacts within the frame of reference of each environment or movement within and between environments. Laboratory and field studies look at the specific chemical. Focus can be on one mechanism, e.g., photochemical or multiple mechanisms, e.g., oxidation and hydrolysis acting in concert. Chemical structure and biological consequences of that chemical structure are not sufficiently understood at this time.

The fate of pollutants has been studied from both local (that is, point sources), regional and global distribution perspectives. Since "Planet Earth" formed, chemicals have been recycled in an overall steady-state manner throughout the environment. Yet, only within the last 30 years have scientists begun to grasp the significant nature of these cycles and how man may be affecting them. An awareness of the importance of

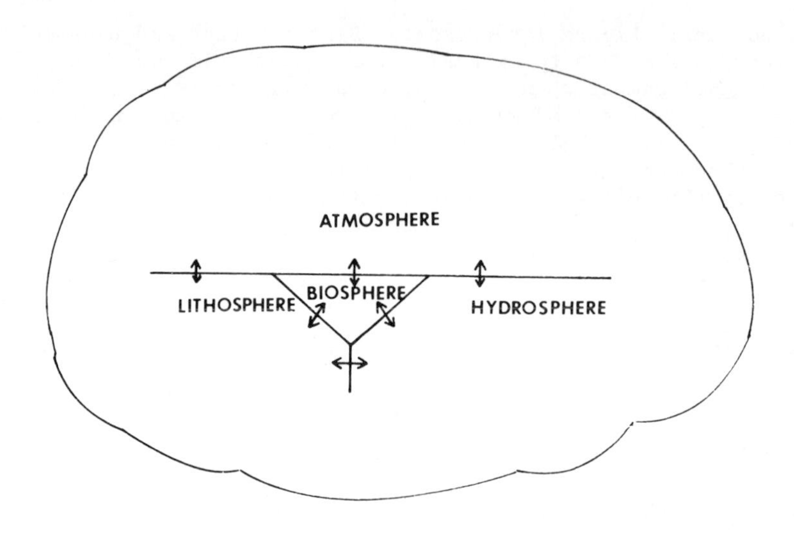

FIGURE 1. Planet Earth.

background concentrations and global cycles of pollutants now exists since the advent of new chemicals from man's industrialization and our attempts to understand and control environmental pollution.

It is reasonable to think that the study of a pollutant's fate and a model to represent it can be of great assistance in deciding strategy to control the pollutant. An example of this concept can be applied to a point source of pollution such as an industrial plant. A case in point is when an industrial chemical is put into the environment. Upon leaving the plant the chemical forms an environmental derivative. This derivative could be from an atmospheric reaction if it goes into the air (e.g., the photochemical smog mechanism), or from an aquatic reaction if it goes into a stream (e.g., the bioreaction of a mercury compound after it is placed in the water environment). Even after long dormant periods, an environmental impact can occur. This is a prime factor for studying the fate of pollutants. We strive to minimize environmental impact whenever possible. When an environmental derivative is formed, it then becomes a new chemical with a fate of its own. The factors that affect the fate of the pollutants are the result of the properties of the chemical, the method of use, and the environmental stresses put upon it. An industrial control strategy for studying fate could include analytical development and pilot-process monitoring. Moreover, laboratory modeling could be used to find out what can happen to the chemical within the industrial process in which it is used and within the environment. A control strategy, such as process modification, is the obvious best choice of abatement, because it can be designed into the process itself.

In summary, the fate of pollutants is a frame of reference to relate the work of scientists and engineers interested in the natural environment protection. It is hoped a new consciousness to interpret scientific data will develop from the concept of environmental fate because, at present, descriptions of the environmental fate of most pollutants are only partially known. Therefore, modeling and predictive efforts are limited.

SECTION III
CHEMICAL FATE OF POLLUTANTS IN THE AIR ENVIRONMENT

Formation and Removal of SO_2 and Oxidants from the Atmosphere*,†

A. P. ALTSHULLER
Director of the Environmental Sciences Laboratory
U. S. Environmental Protection Agency
Research Triangle Park, North Carolina

I. INTRODUCTION

A general concern about air pollution from the burning of dirty fuels has existed for centuries. Air pollution episodes in very large urban areas, such as London and New York, burning very large amounts of high-sulfur fuels focused concern on the sulfur oxide-particulate complex as a health factor in these urban communities.

By the 1950s, concern about large particulates and sulfur oxides had begun to extend to other pollutants such as carbon

---

*American Chemical Society Award Address given on April 9, 1975 for pollution control.

†Sponsored by Monsanto Company, 165th National ACS Meeting, Philadelphia, Pennsylvania.

monoxide and lead from the rapidly increasing population of motor vehicles. By the 1950s, it also had become apparent to the scientific community that photochemically initiated reactions of organic substances and nitrogen oxides from motor vehicles and stationary sources was causing a new problem with associated health, eye irritation, visibility, material, and plant damage effects. In this period, and more recently, concern also has been extended to biologically active particulates or vapors including beryllium, mercury, asbestos, polynuclear aromatic compounds, vinyl chloride, and certain other species.

Much of the concern about pollutants up to the 1960s focused on localized air pollution problems. Although the photochemical air pollution problem was clearly recognized as an area type of phenomena, even in California the problem appeared at first to be largely concentrated within Los Angeles County. Although lesser degrees of photochemical air pollution subsequently were identified in other United States communities, the problems were usually considered to be associated with the largest urban complexes, particularly in their central city areas with heavy vehicular traffic.

A particular set of circumstances eventually developed an awareness of the possibility for regional transport in the eastern United States. During the late 1960s, plant damage to Christmas trees grown in a rural area of West Virginia resulted in a field study to identify causes of the damage. In the course of measuring various pollutants, elevated oxidant concentration levels were observed in this rural area. Subsequent monitoring with ozone instruments clearly identified the oxidant as ozone. Extensive field studies at rural sites and with aircraft was conducted in 1973 and 1974 and elevated ozone levels were definitively associated with episodes during the summer over large rural areas in Ohio, Pennsylvania, and West Virginia. Similar results have been reported in new York State. By the 1960s, it was apparent that ozone concentration levels in the eastern portion of the southern California air basin were increasing rapidly. The ozone concentrations in the eastern basin (Upland, Redlands, San Bernardino, Riverside areas) approached or exceeded those in the western basin by the late 1960s and early 1970s.

Starting in the mid-1960s, monitoring for sulfur-containing particulates was increased in urban sites and at nonurban sites in both the eastern and western United States. Analysis of these results indicate elevated concentration levels of sulfur-containing particulates throughout large regions of the midwestern and northeastern United States. Sulfate concentrations at nonurban sites were one-third to two-thirds of the concentrations at various urban sites. Concentrations of sulfur-containing particulates west of the Mississippi are elevated in some urban centers with substantial power production from coal or oil but not in other communities. Analysis of rainwater from precipitation networks for sulfur compounds show the same distribution as for suspended sulfates. These results suggest transport of sulfur oxides throughout large regions of the eastern United States. Because ambient air, sulfur oxide concentrations are complexly associated with both near-ground level sources contributing to urban plumes and also with the elevated plumes of power plants located in both urban and rural sites, the characteristics and mechanism of transport is inadequately defined. The contribution of sulfur-containing particulates from distance plumes has been suggested as an important source of regional layer concentrations of sulfur-containing particulates.

Little effort has gone into determining the potential for regional transport of other species such as peroxyacyl nitrates, nitric acid, and particulate nitrates. Similarly, the fate of organic aerosols in the atmosphere have received very little attention.

The discussion to follow will concentrate on (a) the transport and chemical transformation of organics, nitrogen oxides, and ozone, and (b) the conversion of sulfur dioxide to sulfates and the transport of these species. This chapter will consider the chemical processes involved in and related to meteorological considerations.

II. FORMATION AND LOSS OF OZONE

A. Half Lives of Organic Substances in Atmosphere

Most organic substances in the presence of nitric oxide and nitrogen dioxide in the part-per-million range and sunlight will produce ozone as a major product. The rate of production and amount of ozone produced is related to the reactivity of the organic species. The production of ozone is also related to the ratio of organic species to nitrogen oxides. Olefins with internal double bonds react rapidly, converting nitric oxide to nitrogen dioxide within the first hour with subsequent rapid ozone formation. However, these types of olefins disappear within a few hours from actual atmospheric systems by reaction with reactive intermediate species and also with the ozone produced. With respect to ozone production the range of lifetimes of particular interest appear to be in the range from a few hours to several days. Such a range spans the scale of localized urban photochemical processes to the scale of large regional ozone formation, transport, and decay processes. All but the least reactive organics will eventually participate in the formation of substantial amounts of ozone.

Half lives can be estimated using various experimental results from irradiation studies (Kopczynski et al., 1972; Altshuller et al., 1967, 1968, 1969, 1970a, b; Kopczynski, et al., 1975; Altshuller and Cohen, 1963, 1964; Bufalini and Altshuller, 1963, 1967, 1969; Kopczynski et al., 1964; Gay et al., 1976) including the following reaction mixtures:
(a) actual atmospheric samples irradiated under ambient conditions of light intensity, temperature, and humidity; (b) simpler synthetic mixtures of hydrocarbons or other organics with nitrogen oxides irradiated under fixed laboratory conditions of light intensity, temperature, and humidity; (c) complex mixtures of hydrocarbons (and other organics) with nitrogen oxides and carbon monoxide irradiated under fixed laboratory conditions of light intensity, temperature, and humidity. Laboratory investigations indicate that half lives may vary by a factor of two or three over realistic ranges of concentration, temperature or light intensity. Therefore, $T_{1/2}$ values (continuous irradiation) obtained under varying experimental conditions in the foregoing types of study will not and should not be expected to be in close agreement. The results listed in Table 1 indicate such variability in $T_{1/2}$ values.

The $T_{1/2}$ values range from less than 1 hour for olefins such as _cis_- and _trans_-2-butene, 2-methyl-1-butene, 2-methyl-2-butene, and related olefins up to 2 days or more for acetylene and ethane. Consider the significance of this range for a large urban plume moving out of an urban area over a large body of water, arid lands, or lightly populated rural terrain. The most reactive olefins originally injected into the air mass will be depleted well before the end of the first day of transport. The slowest reacting olefin, ethylene, will be present at between one-quarter and one-half of its original concentration. The fastest reacting aromatics, the trimethylbenzene, also will be substantially depleted. However, about half of the slower reacting aromatics will remain. About half the fastest reacting paraffins will also remain, whereas most of the pentanes and lower molecular paraffinic hydrocarbons would be present in the air mass at the end of the first day of solar irradiation. By hydrocarbon class, the consumption of paraffins averages 3 percent/hour; aromatics, 7 percent/hour; and olefins, 16 percent/hour based on irradiations of atmospheric mixtures (Kopczynski et al., 1972). Experiments with paraffin, olefin, and aromatic mixtures simulating atmospheric compositions gave percent consumed in irradiated mixtures averaging 3 percent/hour for paraffins and 16 percent/hour for olefins with little effect of

TABLE 1. Estimated Half Lives of Organic Compounds in Atmospheric Photooxidations in the Presence of Nitrogen Oxides

	$T_{1/2}$ (hr) Solar Irradiations of Atmospheric Samples		$T_{1/2}$ (hr) Simulated Irradiation of Synthetic Mixtures	
Compound	I[a]	II[b]	III (multicomponent)[c]	IV (single organics)
Ethane	>100	ND	ND	>50[d]
n-Butane	20	25	20[d]	16-25[e]
Isopentane	12	25	11	25[e]
n-Pentane	12	25	10	ND
2-Methylpentane	ND	16	8	30[e]
2,4-Dimethylpentane	6	ND	8	ND
3-Methylhexane	8	ND	ND	20[e]
Methylcyclohexane	7	ND	ND	15[e]
Acetylene	>50	ND	>50	ND
Ethylene	5	ND	4	1.5-3.5[f]
Propylene	3	3	1.5	1-3[g]
1-Butene	4-5	3-4	1; 0.2-0.8[i]	1.5-6;[h] 0.5-1[i]
2-Butene	<4	<2	0.4-0.7	0.15-0.3[j]
2-Methyl-1-butene	<2	<3	0.7	ND
2-Methyl-2-butene	ND	<3	0.5	ND

(continued)

TABLE 1 (cont.)

	$T_{1/2}$ (hr)			
	Solar Irradiations of Atmospheric Samples		Simulated Irradiation of Synthetic Mixtures	
Compound	I[a]	II[b]	III (multicomponent)[c]	IV (single organics)
Toluene	9	25	7,9[d]	6-9[k]
O-Xylene	9	12	ND	4[k]
m-Xylene	6	12	3,5[j]	3-5[k]
1,2,4-Trimethylbenzene	ND	>3	2	ND
1,3,5-Trimethylbenzene	4	ND	0.8-2.5[i]	1-2,[h] 1-2,[i] 3[k]

[a] Kopczynski et al., 1972.
[b] Altshuller et al., 1970a.
[c] Kopczynski, Kuntz, and Bufalini, 1975.
[d] Altshuller et al., 1968.
[e] Altshuller et al., 1969.
[f] Altshuller and Cohn, 1963.
[g] Altshuller et al., 1967.
[h] Bufalini and Altshuller, 1969.
[i] Bufalini and Altshuller, 1967.
[j] Bufalini and Altshuller, 1963.
[k] Altshuller et al., 1970b.

aromatic composition (Kopczynski et al., 1975). Since overnight depletion of hydrocarbons by chemical processes will be minimal, the next morning within such an air mass these hydrocarbons are available for further reaction. The hydrocarbons remaining are capable of participating in generating substantial concentrations of ozone and other product species, provided that some nitrogen oxides still exist in the air mass or is available from local anthropogenic or biogenic sources. In the laboratory, substantial ozone production has been demonstrated from mixtures of low reactivity hydrocarbons such as *n*-butane (Altshuller et al., 1969) and paraffin mixtures (Kopczynski et al., 1975) at molar ratios of paraffins to nitrogen oxide of between 10 and 100. Therefore, very little nitrogen oxide is necessary to continue the reaction sequence.

The relatively low sensitivity of rates of conversion of organics to the ratio of organics to nitrogen oxides is demonstrated by the results in Figure 1. These curves of percent conversion of organics for 6-hour irradiation periods (or average irradiation period of 2 hours for dynamic runs) with change in organic species to nitrogen oxide ratio (or nitrogen oxide concentration) show some variations, but high rates of conversion occur over a wide range of ratios of organics to nitrogen oxides. These results are consistent with predictions by Calvert and McQuigg (1975) from a model for an olefin-paraffin-aldehyde-carbon monoxide-nitrogen oxide system. Their computed OH concentrations (the major species reacting with the organics) remained in the range $(2 \pm 1) \times 10^6$ molecules/cm^3 with variation in nitrogen oxide from 0.04 to 1.2 ppm (molar ratios of 0.3:1 to 9:1).

The results (Kopczynski et al., 1974; Altshuller and Cohen, 1963; Gay et al., 1976) given in Table 2 indicate some other systems which must be considered. A number of unsaturated chlorinated hydrocarbons are produced and utilized with substantial losses to the atmosphere (Gay et al., 1976). Some of the faster-reacting unsaturated species will be depleted at the rates similar to those of ethylene and propylene. Other chlorinated hydrocarbons such as perchloroethylene react very slowly over several days of transport.

The $T_{1/2}$ values for aldehydes (Kopczynski et al., 1974; Altshuller and Cohen, 1963) illustrate a complexity of the reaction system. Formaldehyde, acetaldehyde, propionaldehyde, and acrolein are important reaction products of the olefins (Kopczynski et al., 1975; Altshuller and Cohen, 1964; Altshuller et al., 1967), and the saturated aldehydes are formed in smaller amounts from the paraffins (Kopczynski et al., 1975; Altshuller

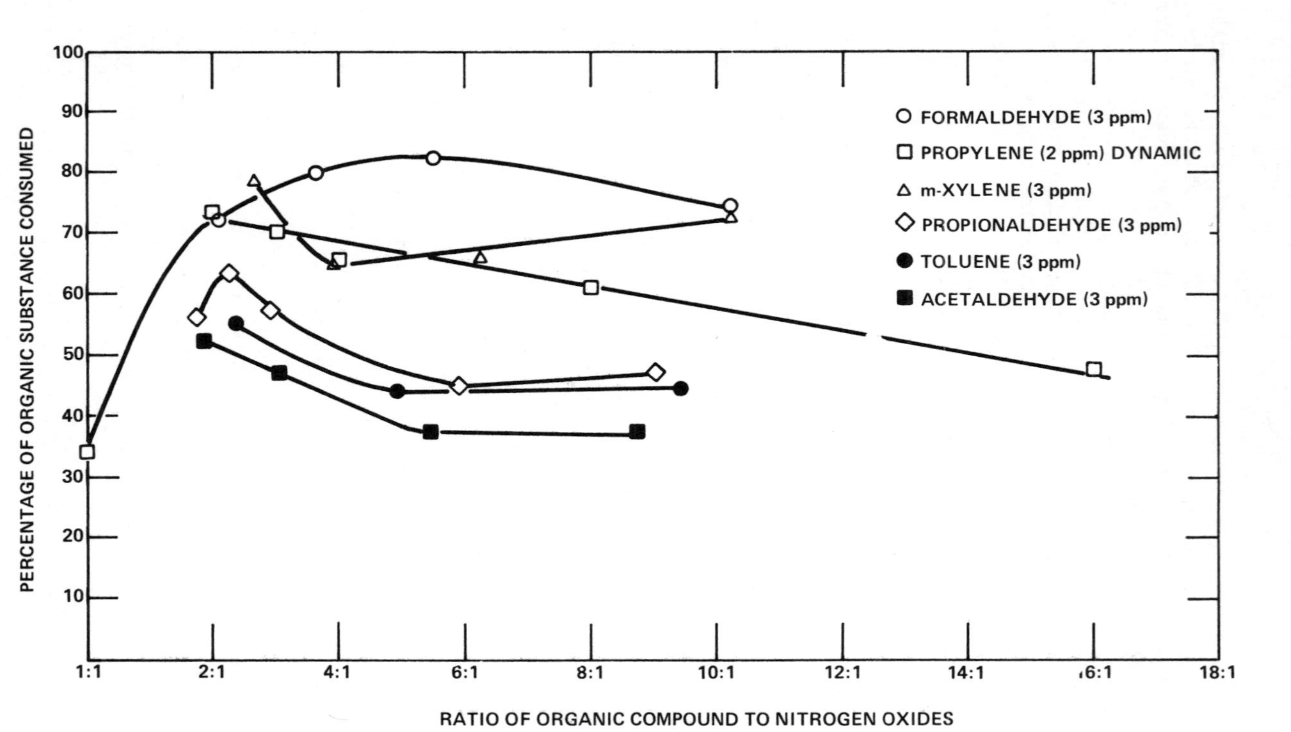

FIGURE 1. The rate of conversion of different organics as a function of the ratio of organics to nitrogen oxides.

TABLE 3. Estimated Half Lives, $T_{1/2}$ (hr) of Organic Compounds in Atmospheric Photooxidations in Presence of Nitrogen Oxides

Compound	Simple Mixtures
Formaldehyde	3[a]
Acetaldehyde	5-12[a]
Propionaldehyde	4-8[a]
Acrolein	4-8[b]
Vinyl Chloride	3[c]
1,1-Dichloroethylene	1[c]
1,2-Dichloroethylene	2[c]
Trichloroethylene	2[c]
Perchloroethylene	$\geq$20[c]

[a]Kopczynski, Altshuller, and Sutterfield, 1974.

[b]Altshuller and Cohen, 1963.

[c]Gay, Noonan, and Bufalini, 1976.

et al., 1968) and aromatics (Kopczynski et al., 1975; Altshuller et al., 1970b) reactions. These aldehydes usually will react more slowly than the olefins from which they are produced and will continue to be produced as the slower-reacting hydrocarbons continue to react. Each aldehyde participates in producing ozone, and all the aldehydes except formaldehyde will produce those aldehydes of lower molecular weight (Kopczynski et al., 1974). The atmospheric photolysis of these aldehydes also produces hydroperoxy radicals, HO_2, which attack hydrocarbons and react with nitric oxide to form nitrogen dioxide and OH radicals. The OH radicals appear to be a key intermediate species in consuming hydrocarbons. Therefore, aldehydes, as long as they continue to be generated in a transported air mass, have a very important function in maintaining the reaction sequences. Calvert and McQuigg (1975), in use of a hydrocarbon-aldehyde-carbon monoxide-nitrogen oxide system in model simulation computations, emphasize the importance of aldehydes in ozone formation, particularly when olefins are initially low or when olefins have been depleted.

Doyle and coworkers (1975) recently reported rates of aromatic hydrocarbons with hydroxyl radicals and used these values along with other hydrocarbon-OH rate constants to compute half lives in the atmosphere. Their half lives for n-butane, propylene, toluene, m-xylene, and 1,2,4-trimethylbenzene, which are one-half to one-third of those computed in Table 1 depend on an assumed atmospheric OH concentration of 10^7 radicals/cm^3. The results in Table 1 indicate that a more appropriate OH concentration would be 3×10^6 to 5×10^6 radicals/cm^3.

Some years ago, a series of experiments were reported that indicated a mechanism for consumption of hydrocarbons not involving nitrogen oxide photooxidations (Altshuller et al., 1967). Photolysis of formaldehyde, acetaldehyde, or propionaldehyde in the presence of olefins or alkylbenzenes resulted in a significant rate of consumption of the hydrocarbon with formation of hydrogen peroxide and alkyl hydroperoxides but not ozone. Some of these results are given in Table 3. The aspect of the overall mechanism involving reaction with the hydrocarbon suggest the following reaction steps for the simplest system—formaldehyde—hydrocarbon.

$$HCHO \rightarrow H + HCO \quad (1)$$

$$H + O_2 \rightarrow HO_2 \quad (2)$$

$$HCO + O_2 \rightarrow HO_2 + CO \quad (3)$$

$$HO_2 + HO_2 \rightarrow H_2O_2 + O_2 \quad (4)$$

$$H_2O_2 \rightarrow 2OH \quad (5)$$

$$RH + OH \rightarrow \text{Products} \quad (6)$$

$$RH + HO_2 \rightarrow \text{Products} \quad (7)$$

$$HCHO + OH \rightarrow HCO + H_2O \quad (8)$$

The formation of OH by Reaction 5 according to Calvert and McQuigg's computations (1975) accounts for only 5 percent of the overall rate at 1 hour into the reaction sequence in nitrogen oxide induced oxidations. Nitrous acid photolysis and the

TABLE 3. Aldehyde-Induced Photooxidations of Hydrocarbons

Hydrocarbon	Aldehyde	Light Source	Average Percent Reacted/Hour of Hydrocarbon
Ethylene	Propionaldehyde	SFL[a]	3
2-Methyl-1-butene	Formaldehyde	SFL	11
2-Methyl-1-butene	Acetaldehyde	SFL	13
2-Methyl-1-butene	Acetaldehyde	SR[b]	1
2-Methyl-1-butene	Propionaldehyde	SFL	13
2-Methyl-1-butene	Propionaldehyde	SRL	2
Trans-2-butene	Propionaldehyde	SFL	9
1,3,5-Trimethylbenzene	Formaldehyde	SFL	10
1,3,5-Trimethylbenzene	Propionaldehyde	SFL	11

[a]Sunlight fluorescent lamps.

[b]Solar radiation.

reaction of HO_2 with NO are much more important. An OH concentration at least an order of magnitude lower than in the nitrogen oxide-induced photooxidation is consistent with the nitrogen oxide-induced hydrocarbon photooxidations being a factor of 10 faster than aldehyde-induced hydrocarbon photooxidations. (Note the chain character of these reactions leading to regeneration of HO_2.) These aldehyde-induced hydrocarbon photooxidation reactions may possibly become significant under downwind transport conditions during which nitrogen oxides become greatly depleted relative to hydrocarbon and aldehydes.

1. Ozone Formation. Almost all of the organic species included in Tables 1 and 2 as well as many related organic compounds, will participate in generating ozone in the presence of nitrogen oxides and solar radiation. Many investigators have reported ozone or oxidant concentration-time profiles using a wide variety of simple and complex systems. The empirical relationships between hydrocarbon and nitrogen oxide concentrations and composition to ozone formation have received extensive laboratory investigation.

In air pollution research during the early 1960s, some systematic relationships began to be developed using vehicular exhaust mixtures under dynamic irradiation conditions under well-controlled laboratory conditions (Korth et al., 1964; Leach et al., 1964). Rates of hydrocarbon consumption, nitrogen dioxide formation, oxidant formation, and aldehyde and nitric formation, as well as eye irritation and plant damage, were investigated. Ozone formation was observed to increase almost in proportion to hydrocarbon concentration at fixed initial concentrations of hydrocarbons down to 0.3 ppm nitrogen oxide or to 18 to 1 hydrocarbon to nitrogen oxide ratio (Figure 2). Oxidant formation was suppressed at 3 to 1 and lower ratios of hydrocarbons to nitrogen oxides.

A detailed investigation was made under both dynamic and static irradiation conditions with the propylene-nitrogen oxide mixtures (Altshuller et al., 1967). At most propylene to nitrogen oxide ratios, oxidant concentration levels under dynamic irradiation conditions do not decrease in proportion to concurrent reductions in the concentrations of the propylene and of the nitrogen oxides. Oxidant concentrations were observed to decrease rapidly below certain propylene concentrations at initially fixed nitrogen oxide concentrations. At initial nitrogen oxide concentrations of 1.0 ppm of nitrogen oxide and under dynamic irradiating conditions, rapid decrease in oxidant concentrations occur only below a 2:1 molar ratio (6:1 carbon ppm to nitrogen oxide ratio). At an initial nitrogen oxide concentration of 0.25 ppm, decreases in oxidant concentrations

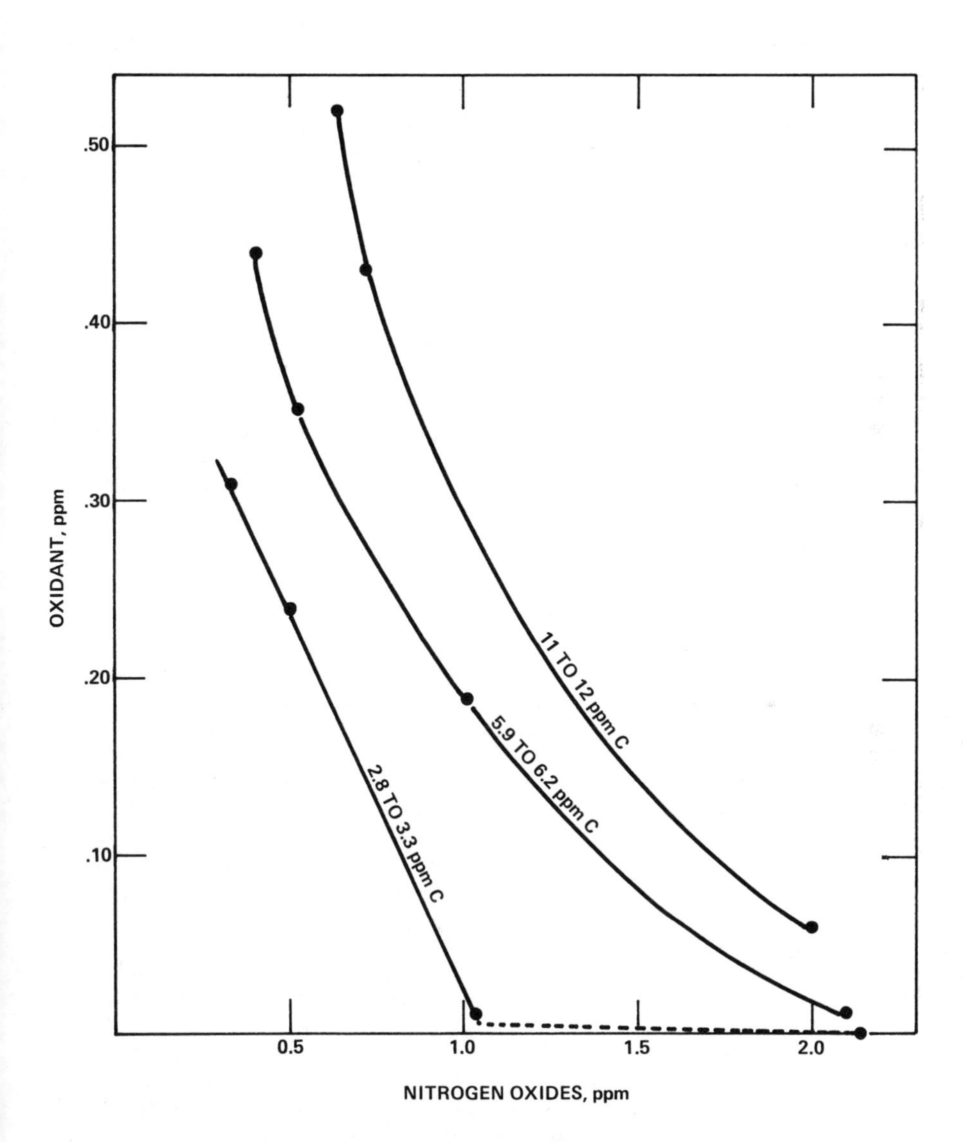

FIGURE 2. The effect of a fixed initial concentration of nitrogen oxide on the formation of oxidant at three different hydrocarbon levels.

occur on both sides of a 4:1 molar ratio of carbon ppm to nitrogen oxide. The ratios above 4:1 are at or above the average nonmethane hydrocarbon to nitrogen oxide ratios measured at urban monitoring sites (Altshuller, 1975). Unlike the dynamic irradiations of vehicular exhaust emissions (Korth et al., 1964), maximum oxidant concentrations were measured when nitrogen oxides were varied at initially fixed propylene concentrations. Peak oxidants occurred near 2:1 molar ratios. Oxidant concentrations above 0.1 ppm were measured down to the lowest nitrogen oxide concentrations used of 0.02 to 0.04 ppm. Oxidant concentrations below 0.1 ppm were consistently observed when sufficient nitrogen oxide was used to provide initial propylene to nitrogen oxide ratios of 1:1 or less (3:1 carbon ppm to nitrogen oxide ratios) (Altshuller et al., 1967). Under static 6-hour irradiation conditions, peak oxidant shifted from near a 2:1 molar ratio to a 1:1 molar ratio and oxidant concentrations were suppressed below 0.1 ppm at a propylene to nitrogen oxide ratio of 0.6:1 or less. Very few comparisons of such dynamic compared to static irradiation conditions exist. The results with the propylene-nitrogen oxide system, if they can be generalized, suggests that continuing emissions of reactants, particularly nitric oxide could result in lower oxidant concentrations than predicted from static irradiation experiments.

A study of the irradiation of lower reactivity paraffinic hydrocarbon-nitrogen oxide system has been conducted (Altshuller et al., 1969). With 6-hour static irradiations no oxidant was formed from the *n*-butane-nitrogen oxide mixtures at molar ratios below 3:1 (12:1 carbon ppm to nitrogen oxides). At molar ratios from 6:1 to 60:1 the oxidant concentrations formed from the *n*-butane-nitrogen oxide mixtures is in the same range of oxidant concentrations as formed from the propylene-nitrogen oxide mixtures at molar ratios from 1.5:1 to 6:1 (Altshuller et al., 1967). At nitrogen oxide concentrations of 0.05 or 0.2 ppm, oxidant concentrations rose rapidly as the *n*-butane concentration increased to the 0.5 or to 2 ppm *n*-butane concentration level (to a 10:1 molar ratio) and then the oxidant concentrations did not increase appreciably with further increases in the *n*-butane concentration.

Subsequent experiments were carried out to determine the effect of either addition of propylene or of toluene on oxidant formation from 6-hour static irradiations on either base *n*-butane-ethane-nitrogen oxide mixtures of ethane-nitrogen oxide mixtures (Altshuller, 1968). Neither addition of propylene up to 0.4 ppm or toluene at 1.5 ppm resulted in any increase in the already high oxidant concentrations produced from 3 ppm of *n*-butane and 3 ppm of ethane when irradiated with 0.3 ppm of nitrogen oxide. In contrast, addition of 0.4 ppm of propylene to 3 ppm of ethane and 0.3 ppm of nitrogen oxide increased the oxidant concentration of 0.2 to 0.7 ppm. Addition of 0.4 ppm of propylene to a 3 ppm n-butane and 3 ppm of ethane irradiated with 1.2 ppm of nitrogen oxide increased the oxidant concentration from 0.05 to 0.3 ppm. Addition of either 1.5 ppm of toluene or 0.5 ppm of propylene to the same base mixture produced 0.5 ppm of oxidant. Addition of 0.3 or 0.8 ppm of propylene to a 3 ppm of ethane irradiated with 1.2 ppm of nitrogen oxide produced no oxidant or increased the oxidant concentration from 0.0 to 0.35 ppm. Small additions of propylene to lower ratios of n-butane to nitrogen oxide mixtures definitely caused substantial increases in oxidant concentrations (Altshuller et al., 1968).

The prior lack of a systematic investigation of irradiated aromatic hydrocarbon-nitrogen oxide systems resulted in such a study of the toluene-nitrogen oxide and m-xylene-nitrogen oxide mixtures (Altshuller et al., 1970b). Oxidant dosages (ppm x time) during 6-hour irradiation periods associated with toluene-nitrogen oxide mixtures decreased very slowly either with decreasing nitrogen oxide concentration at fixed initial toluene concentrations or at constant toluene to nitrogen oxide ratios. Oxidant dosages decreased significantly with decreasing toluene concentration at or above 1 ppm of nitrogen oxide, but oxidant dosages were insensitive to toluene concentrations at nitrogen oxide concentrations below 1 ppm. The oxidant dosage curve for irradiated m-xylene-nitrogen oxide decreases with decreasing nitrogen oxide concentration below 1 ppm of nitrogen oxide.

A systematic investigation of irradiated formaldehyde-nitrogen oxide, acetaldehyde-nitrogen oxide and propionaldehyde-nitrogen oxide systems has been conducted (Kopczynski et al., 1974). The maximum oxidant concentrations decreased almost in proportion to nitrogen oxide concentration below 1.5 ppm for the propionaldehyde-nitrogen oxide system and below 0.6 ppm for the acetaldehyde-nitrogen oxide system. In contrast, the maximum oxidant concentrations increased with decreasing nitrogen oxide concentration down to 0.3 ppm for the formaldehyde-nitrogen oxide system. The corresponding oxidant dosages curves were similar. Both maximum oxidant concentrations and oxidant dosages decreased almost in proportion to aldehyde concentrations at aldehyde concentration levels below 3 ppm for all three systems. The oxidant concentrations or oxidant yields decreased much less than proportionately with a three-fold decrease in the concentration of both the aldehyde and nitrogen oxides.

The results from the various single component, multiple component, and vehicular exhaust irradiations are consistent and lead to several conclusions: (a) control of nitrogen oxides at constant hydrocarbon concentration (increasing ratios of reactants) caused increases in oxidant over a range of nitrogen oxide concentrations, particularly for the slow-reacting organic compounds. Only extremely high levels of nitrogen oxide control appear to provide substantial benefits in terms of oxidant reduction: (b) concurrent control of both organics and nitrogen oxides at constant reactant ratio conditions (particularly at lower reactant ratios) result in modest decreases and sometimes increases in oxidant yields. The inhibition by nitrogen oxides of oxidant formation causes an opposite effect to the reduction in absolute concentration of the reactants (c) control of the concentration of organics is effective in reducing oxidant particularly when the control of organics is large enough to substantially reduce the ratio of organics to nitrogen oxides.

The compositional effects on oxidant formation are very sensitive to the ratio of reactants used. At the most commonly used ratio, 2 to 1, paraffinic hydrocarbons, slower-reacting aromatics, and some aldehydes appear far less effective in participating in oxidant formation than the more rapidly reacting organic species. However, as is shown in Table 4, the systems containing the "less reactive" organics produce more oxidant with increasing ratios of reactants, while the systems containing the "more reactive" organics produce less oxidant at higher ratios of reactants. As is illustrated by the propylene-nitrogen oxide compared to the *n*-butane-nitrogen oxide system, a ratio can be attained where two systems of widely different reactivity produce the same oxidant dosage. In considerable part, the effect of the upward shift in ratio is to reduce the time of inhibition of oxidant formation for the less reactive system by reducing the amount of nitric oxide to be converted. The same effect can be obtained by increasing the nitrogen dioxide to nitric oxide initial ratio, by adding small amounts of more reactive species to less reactive systems or by increasing the time of irradiation of the systems. Therefore, less reactive systems are most effective in oxidant formation when (a) the ambient air systems are at higher ratios of organic species to nitrogen oxide, (b) under conditions when most of the nitrogen oxide is the form of nitrogen dioxide, (c) small amounts of more reactive organic species are initially present, and (d) long irradiation times—regional transport.

In evaluating laboratory results, it is important to utilize realistic atmospheric compositional results. Considerable measurements are available in high density vehicular traffic areas (Kopczynski et al., 1972). Even in such areas, olefins made up only 10 percent of the carbon content and internally double bonded olefins made up only 2 or 3 percent of the carbon content. Paraffinic hydrocarbons constitute over 50 percent of the carbon content, and aromatics constitute over

TABLE 4. Oxidant Dosages from Photooxidation of Organic Substance—Nitrogen Oxide Systems

	Oxidant Dosage (ppm x hr) at Various Organic Substance to Nitrogen Oxide Molar Ratios			
Organic Substance[a]	2 to 1	4 to 1	6 to 1	9 to 1
n-Butane	0.0	0.6	1.3	2.1
Propylene[b]	4.0	3.5	3.0	2.0
Toluene	1.4	1.6	1.5	1.4
m-Xylene	3.6	2.6	2.1	1.7
Formaldehyde	1.1	2.9	3.3	3.6
Acetaldehyde	1.4	2.6	2.7	2.4
Propionaldehyde	3.6	3.7	3.4	2.4

[a]Initial concentration organic substance (except propylene), 3 ppm (molar).

[b]Initial concentration of propylene, 2 ppm (molar).

30 percent of the carbon content. Therefore, measures of oxidant dosage of maximum oxidant parts per million of compound (Kopczynski et al., 1975), combined with representative atmospheric concentrations (Kopczynski et al., 1972) lead to the conclusion that by class aromatics are more important than paraffins, which in turn are more important than olefins.

2. Field Measurements of Ozone in Urban and Rural Areas. The elevated oxidant concentrations in large urban centers has been appreciated for some years beginning with the Los Angeles Basin experience after World War II. Subsequent to installation of oxidant monitoring in Chicago, Cincinnati, Denver, Philadelphia, St. Louis, and Washington, D. C. in the early 1960s, it was appreciated that elevated oxidant concentrations were occurring in other large cities. During the 1960s, confusion about the extent of the oxidant problem existed because of the use of inadequate oxidant instruments sometimes not properly protected against the negative interference of sulfur dioxide. With the advent of the widespread use of chemiluminescent ozone instruments, it became evident that there was a widespread ozone problem.

Monitoring results in the early 1970s began to show a downward trend in ozone concentration levels (Altshuller, 1975; Kinosian and Duckworth, 1973). This downward trend became evident in a number of cities including Los Angeles, Cincinnati, Denver, Philadelphia, St. Louis, and Washington, D. C. Since the monitoring sites were in the central areas of these cities with surrounding high densities of vehicular traffic, the historical evidence for such downward trends is limited to such areas. The reduction in hydrocarbons from vehicular emission control had been modest whereas vehicular emissions of nitrogen oxides had been increasing. One explanation advanced (Korth et al., 1964) based on smog chamber results, attributes the downward trend to decreases in the hydrocarbon to nitrogen oxide ratio at these monitoring sites.

The downward trend in oxidants in the southern California air basin was only evident in the more heavily populated western portion of the air basin (Kinosian and Duckworth, 1973). Increases in oxidant concentration levels were observed through the 1960s at monitoring sites in the eastern portion of the basin. This effect has been attributed both to ozone transport from the western to the eastern portion of the basin and to population growth in the eastern portion of the basin.

In other parts of the country, even limited ozone monitoring in sites downward of the central areas of the large cities did not begin until the 1970s (Altshuller, 1975). In the Washington D.C. area, during the last few years, ozone monitoring just over the district line in Maryland indicated ozone concentrations as high and as frequently elevated as ever experienced at the monitoring site in downtown Washington. Again, transport as well as extensive population growth must be considered as factors. Although transport of ozone and its precursors immediately downwind of the urban centers is a reasonable possibility, direct experimental evidence has become available only in the last 2 years. Several experimental studies supported by EPA have provided much of this evidence.

Washington State University investigators conducted aircraft and ground-level measurements of ozone, nitrogen oxides, hydrocarbons, and carbon monoxide late in September 1973 in the Phoenix area (Westberg and Rasmussen, 1973) and in October 1973 in the Houston area (Westberg and Rasmussen, 1973). In the Phoenix experiments ozone levels in the 0.08 to 0.11 ppm level were measured on three afternoons 70 km downwind of Phoenix than 70 km downwind. The ozone flux on September 25, 1973, 70 km downwind through the 1000- to 5000-ft levels were slightly higher than over Phoenix itself, suggesting both transport of ozone along with additional production from hydrocarbons and nitrogen oxides. Flights made on October 17, 18, and 19, 1973, over Houston (Westberg and Rasmussen, 1973b) and out to 80 km downwind of Houston during an ozone episode demonstrated the presence of ozone concentrations 80 km downwind of 0.16 to 0.18 ppm on October 17, of 0.20 ppm on October 18, and of 0.18 ppm on October 19. On these days, ozone concentrations upwind of Houston were elevated but well below the ozone levels downwind. Also on October 18 and 19, the peak ozone concentrations were measured 30 to 50 km downwind where they attained 0.24 to 0.25 ppm and 0.30 to 0.33 ppm on October 18 and 19. Thus, evidence of a decline from peak values at intermediate distances out to downwind distances of 80 km with ozone concentration still well above background was provided.

The ozone concentration levels measured in the Houston area (Westberg and Rasmussen, 1973b) reached much higher levels during the October 17 to 19 period than the ozone levels measured in the Phoenix area (Westberg and Rasmussen, 1973a). The hydrocarbon concentration levels downwind of Houston averaged about three times higher than those measured at the same distances downwind of Phoenix. The corresponding nitrogen oxide concentrations downwind were about twice as high in concentration downwind of Houston compared to downwind of Phoenix. Therefore, there were higher ozone precursor concentrations available downwind of Houston to participate in ozone production compared to Phoenix at least during the September-October period investigated.

In a midwestern summer study of ozone in rural areas centered in Ohio during June, July, and August 1974 the distribution of ozone was investigated (Research Triangle Institute, 1975; Westberg and Rasmussen, 1974). During this period, maximum ozone concentration levels as well as the frequency that ozone exceeded 0.08 ppm was measured at six urban and five rural sites in Ohio, Pennsylvania, and Maryland (Research Triangle Institute, 1975). The ozone concentrations at the rural sites exceeded both in terms of concentration and frequency ozone levels at the urban sites. At the rural sites, the maximum hour of ozone during the entire period was in the 320- to 400-$\mu g/m^3$ (0.16- to 0.20-ppm) range. The ozone levels during 10 percent of the hours were above 170 to 210 $\mu g/m^3$ (0.08 to 0.10 ppm) at the rural sites (Research Triangle Institute, 1975). A significant distinction between rural and urban sites was in the three- to five-fold higher nonmethane hydrocarbon to nitrogen oxide concentration ratios measured at rural sites. The nitrogen oxide concentrations were below 20 $\mu g/m^3$ 75 to 95 percent of the time at the five rural sites (Research Triangle Institute, 1975). Therefore, depletion of nitrogen oxide compared to hydrocarbons appears to be indicated although the nitrogen oxide concentrations were often at or near the detection limit of 10 $\mu g/m^3$ for the analyzers.

During the midwestern ozone study several flights provided evidence of local downwind perturbations by urban plumes (Research Triangle Institute, 1975). These subregional scale effects were superimposed onto region-wide systems of high ozone concentration 400 or 500 km in diameter. During these flights, the experimental results for ozone and hydrocarbon levels downwind of Columbus and Pittsburgh indicated a specific urban contribution extending 50 to 80 km downwind.

A detailed analysis has been made of results for two days of observations aloft in the southern California air basin (Blumenthal et al., 1974). The movement of air trajectories between the heavily populated western area of this basin and the more lightly populated eastern area can be followed readily. It was concluded that ozone and its precursors were transported from the western area of the basin approximately 100 km downwind into the eastern portion of the basin. Most of the ozone measured at downwind locations was associated with precursors emitted earlier the same day or the previous evening in the western portion of the basin. On the outermost portions of the basin in desert or mountains ozone concentrations of 0.3 ppm were observed despite no significant local sources with the closest sources at least 100 km upwind (Blumenthal et al., 1974).

Bell Laboratories investigators have applied a statistical technique to oxidant results from four sites 27 to 49 km from Philadelphia-Camden (Cleveland and Kleiner, 1975). Surface wind measurements were used along with surface ozone concentrations. On days favorable for photochemical processes and with winds blowing from the urban complex to one of another of the downwind sites, the highest ozone concentrations were measured at these downwind sites. Again, as in the midwest study, ozone concentrations within the urban complex at Camden were slightly lower than at the downwind nonurban sites (Cleveland and Kleiner, 1975).

The analysis of these results from Los Angeles, Phoenix, Houston, Columbus, Pittsburgh, and Philadelphia urban areas lead to the conclusion that downwind transport of ozone and its precursors can occur at least 30 to 100 km downwind of these urban areas. Therefore, a substantial portion of the elevated ozone concentration levels at rural sites can be attributed to either direct transport of ozone and to a greater extent, transport of precursors from the urban areas with photochemical reactions producing additional ozone during transport.

The rural sites in the midwestern study showed elevated ozone concentrations for prolonged periods of time (up to 5 days) despite no consistent evidence of air flow with any specific urban sources downwind (Research Triangle Institute, 1975). These high ozone concentrations existed when the sites were within the center of a synoptic high pressure system. Such a condition results in the region experiencing relative clear skies and weak, poorly organized wind flow near the surface. Despite the inability to associate the elevated ozone concentrations in the rural areas uniformly with specific urban sources or large point sources, the concentrations measured of carbon monoxide, acetylene, and other hydrocarbons from man-made emissions indicated the presence of anthropogenic sources of oxidant precursors. Vertical profiles of ozone, hydrocarbons and carbon monoxide did not support the hypothesis of a significant stratospheric ozone contribution (Research Triangle Institute, 1975; Westberg and Rasmussen, 1974). In fact, the hydrocarbon profiles constructed from the measurements were largely associated with vehicular exhaust and evaporative losses of gasoline. No direct evidence for terpenes or their reaction products was obtained. However, small contributions from natural hydrocarbons or nitrogen oxide sources cannot be excluded. The most likely source of precursors over the rural sites is a complex combination of contributions from adjacent urban areas, small towns, farms, and rural vehicular traffic. Thus far, it is not possible to account for these relative contributions of individual types of anthropogenic sources on a quantitative basis.

Several aspects of the studies already discussed require comment. Why are ozone concentrations so frequently higher downwind of urban areas in the afternoon hours? Particularly in western areas, the explanation can be that the air entering the city from upwind has fewer precursors than the air downwind of the city. Another explanation is that the high nitric oxide production rates within the city consume ozone transported from other upwind sources. High nitric oxide concentrations produced by afternoon traffic in cities after the ozone production rates has decreased (lower solar intensities) also can cause rapid decreases in ozone concentrations. In adjacent rural areas, nitric oxide concentrations continue to be small compared to the preexisting ozone levels. Therefore, ozone formation proceeds from hydrocarbon-nitrogen oxide reactions at a reduced rate late in the afternoon in rural areas, but this rate still may exceed the rate of ozone destruction processes. These conditions qualitatively can account for the later and higher peak ozone concentrations at rural compared to urban areas.

The nitric oxide effect discussed previously can account in part for the observations that Sunday ozone concentration levels are as high or higher than weekday levels in many urban areas (Altshuller, 1975; Cleveland et al., 1974). On Sunday mornings all the precursors are decreased compared to weekday mornings, but nitric oxide is especially low in concentration on Sunday mornings. Such low nitric oxide concentrations result in reduced inhibition of ozone formation with a potential for substantial ozone formation in the morning despite reduced absolute concentrations of precursors. Such a result would be consistent with smog chamber studies indicating that ozone formation will not decrease at all in proportion to absolute decreases in precursor concentration.

As already discussed, ozone near the ground typically is reduced rapidly by reaction with nitric oxide in the late afternoon hours in urban areas. By night the ozone is reduced down to the 10- to 30-$\mu g/m^3$ range within midwestern cities. At rural sites, ozone decreases later and more slowly down to the 40- to 60-$\mu g/m^3$ range at most rural sites in the study of the midwest. Because ozone decreases less overnight, more ozone must still exist in midmorning the next day in rural sites. The slow decrease in ozone observed at rural sites at night must be caused by removal of ozone by soil and by vegetation.

Vertical profile measurements over Los Angeles and midwestern sites show that ozone will survive essentially undepleted at night in layers separated from the surface by a stable surface layer caused by the establishment of a radiation inversion or in coastal areas uncutting by cooler maritime air. Such layers containing ozone and residue precursors aloft can begin reaction the next morning. After the sun rises and surface heating becomes effective, the surface layer rises and eventually loses its identity with increased turbulence later in the morning and during the afternoon mixing ozone from aloft with precursors from the surface. The downward transport of ozone from such stable immediately aloft could contribute significantly to surface ozone concentrations in rural areas. The downward transport of ozone from stable layers over cities or immediately upwind may contribute to surface ozone concentrations in cities also. Such contributions from ozone aloft formed the day before may also contribute to the high Sunday ozone concentration levels within cities.

B. Formation and Composition of Atmospheric Sulfate Aerosols

The results of community health studies of air pollutants (CHESS) conducted at several study sites in the eastern and western United States have been reported (U. S. Environmental Protection Agency, 1974). It was concluded that several morbidity parameters increased significantly in general population groups with increasing atmospheric concentrations of water soluble sulfates. Analysis of aerometric measurements for sulfates strongly indicates that water soluble sulfates are widely dispersed over large areas of the United States (Altshuller, 1973). Such widespread dispersion of sulfates are fully consistent with the modes of formation of sulfates and with the chemical and physical properties of sulfates. This review will emphasize available scientific results on the photochemistry and chemical kinetics of formation of sulfates, the properties of sulfates along with available results on the variations of atmospheric concentrations and distribution of sulfates in the United States.

Although maritime sulfates and direct emissions of sulfates from power production and from industrial sources can contribute, atmospheric sulfates predominantly are formed by atmospheric conversions of sulfur dioxide to sulfates. Because the adverse biological effects of sulfates appear to occur at significantly lower concentrations than sulfur dioxide, the reductions needed in sulfate concentration levels in the atmosphere may require significantly more control of sulfur dioxide than would be necessary to control atmospheric sulfur dioxide concentration levels only.

C. Sources of Sulfate

The sources of ambient air sulfates include natural contributions, directly emitted sulfates from industrial sources, and sulfate formed from sulfur dioxide by means of atmospheric reactions. Natural contributions include sulfates formed by atmospheric oxidation of hydrogen sulfide and organic sulfide to sulfur dioxide and the subsequent gas to aerosol conversion to sulfate particles. Directly emitted sulfates include sulfuric acid from sulfuric acid production plants and calcium or sodium sulfate from industrial operations such as Kraft paper mills. A new direct source of sulfuric acid near roadways can be expected from presently designed catalytically equipped automobiles. However, the predominant source of sulfates in or near cities is the chemical conversion of manmade sulfur dioxide from stationary emission sources to sulfate aerosols. The problems of control of atmospheric sulfates relate to improving source to atmospheric sulfur dioxide relationships and quantitative determination of the contributions of various mechanism of conversion of sulfur dioxide to sulfate. Excluding very localized situations involving large direct industrial sources of sulfate, control of source emissions of sulfates does not appear likely to contribute substantially to the reduction of atmospheric sulfates.

A background of sulfates exists throughout large areas of the eastern United States outside of cities (Altshuller, 1973). This background is in the 5- to 10-$\mu g/m^3$ annual average range at most nonurban monitoring sites. Most of this background does not appear to be of natural origin. This background decreases substantially in areas of the United States in which usage of higher sulfur content fuels is small. For example, background sulfates at monitoring sites in nonurban rural areas in farming regions in the western United States are at least 5 $\mu g/m^3$ lower than corresponding regions in the eastern United States (Altshuller, 1973). Background levels at some nonurban western United States sites are in the 1- to 2-$\mu g/m^3$ range suggested as continental background for sulfate (Altshuller, 1973).

The higher nonurban sulfate concentrations in the eastern United States appear consistent with the source being downwind transport from large urban areas of sulfur dioxide. The conversion rates of sulfur dioxide to sulfate and the estimated lifetimes for sulfate aerosols appear consistent with transport distances of at least several hundred miles. Such transport distances would be sufficient to account for high sulfate concentrations in most nonurban sites in the eastern United States.

A possible exception to the previous discussion are the unusually high background levels of sulfate aerosols reported at Cape Hatteras well off the coast of North Carolina (Altshuller, 1973). Sulfate levels measured at this site have been in the 10- to 12-$\mu g/m^3$ annual average range. Several nonurban coastal sites, both north and south of Cape Hatteras, have sulfate concentration levels about half of those at Cape Hatteras. The North Carolina coast in the area adjacent to Cape Hatteras is swampy and shallow coastal waters exist in this area. It seems reasonable that substantial amounts of natural sulfides may be generated in this area, which is converted to sulfate while being transported to the Cape Hatteras area. This example is discussed in some detail because it is felt that there may be some limited areas elsewhere in the United States impacted by substantial natural contributions of sulfates. However, this assumption needs experimental verification. This possible natural source is unlikely to be of significance in most regions of the United States in accounting for high sulfate concentration levels.

D. Chemical Kinetic Mechanisms for Conversion of Sulfur Dioxide to Sulfate

Several different mechanisms have been proposed and they have been subject to limited experimental verification under laboratory conditions. Some of these mechanisms involve homogeneous, photochemical, gas phase conversions to particles. Other mechanisms involve heterogeneous reactions in liquid droplets or in liquid films on the surfaces of solid particles. Physical adsorption or chemisorption may result in small losses of sulfur dioxide into various solid aerosol particles. It is entirely possible that more than one of these mechanisms may contribute concurrently to conversion of sulfur dioxide to sulfate in the atmosphere. However, there also is evidence that some mechanisms are more likely to be dominant under specific atmospheric conditions. Photochemical processes will be important on warm sunny periods of low humidity. Heterogeneous reactions in droplets or in films on particles will be important on cool periods having very high humidities particularly with fogs or mists present. The various mechanisms also may contribute disproportionately to the formation of different species, that is sulfuric acid, ammonium hydrogen sulfate, or ammonium sulfate.

The simplest mechanism probably is the direct photooxidation of sulfur dioxide in air to sulfate by solar radiation. This reaction has been studied by numerous investigators with considerable differences reported in the quantum yield and oxidation rate for sulfur dioxide (Sidebottom et al., 1972; Smith and Urone, 1974; Cox and Penkett, 1970). The chemical reactive species is believed to be sulfur dioxide in the triplet state, $^3SO_2$ (Sidebottom et al., 1972). The quenching rate constants for reaction of $^3SO_2$ with a wide variety of molecules have been determined. These quenching rate constants vary greatly. For example, the quenching rate constants for olefinic hydrocarbons with $^3SO_2$ are 10 to 20 times greater than with oxygen or water vapor. Based on the assumption that all $^3SO_2$ quenched by oxygen is oxidized, a maximum rate of oxidation of sulfur dioxide in sunlight has been calculated at 2 percent/hour

(Sidebottom et al., 1972). However, most of the experimental rates cluster between 0.1 and 0.6 percent/hour (Smith and Urone, 1974; Cox and Penkett, 1970), with the lowest rates most likely to be correct. Such experimental results suggest that most $^3SO_2$ molecules do not undergo oxidation in quenching reactions with oxygen or other molecules present in the atmosphere. One mechanism suggested involved formation of SO_4 from $^3SO_2$ and O_2 and ozone from reaction of SO_4 with oxygen to form sulfur trioxide. Since neither SO_4 or ozone have ever been observed, the details of this mechanism for producing sulfur trioxide are still unclear (Sidebottom et al., 1972).

No dark reaction has been reported between sulfur dioxide and nitrogen dioxide in the parts-per-million range in air (Smith and Urone, 1974). At some concentration ratios of nitrogen dioxide to sulfur dioxide in air, irradiation of the dry reaction mixture causes sulfur dioxide to react at up to twice the rate of sulfur dioxide in air alone (Smith and Urone, 1974). In similar mixtures at 50 percent relative humidity, the rate of reaction of sulfur dioxide has been reported to increase tenfold over that in the dry mixture (Smith and Urone, 1974). However, the experiments were carried in small flasks. In larger reactors, little, if any, increase in reaction rate has been observed (Miller et al., 1972). These differences in results can be associated with a surface reaction in the liquid film on the sides of the small reaction flasks used with very large surface to volume ratio. The rates of reaction of alkanes and alkenes directly with sulfur dioxide is very slow in the ppm range (Kopczynski and Altshuller, 1962).

The atmospheric photochemical reactions of hydrocarbons and nitrogen oxides in the presence of ultraviolet solar radiation produce intermediates and radicals capable of oxidizing sulfur dioxide to sulfuric acid (Schuck and Doyle, 1959; Endow et al., 1963; Prager et al., 1960; Groblicki and Nobel, 1971; Harkins and Nicksic, 1968). The relative effectiveness of hydrocarbon species in promoting the oxidation of sulfur dioxide is in the order: olefins > alkylbenzenes > alkanes or alkynes. Alkanes, alkynes, and aldehydes have much less capability than olefins or alkylbenzenes to participate in the oxidation of sulfur dioxide in these photochemical reactions. The dark reaction between olefin and ozone is believed to produce a reactive intermediate capable of rapidly oxidizing sulfur dioxide, which may be of particular significance in explaining aerosol formation in irradiated olefin-sulfur dioxide-nitrogen dioxide systems (Cox and Penkett, 1972; McNelis, 1974).

There is experimental evidence from microchemical analysis, infrared spectroscopy, and electron microscopy (Cox and Penkett,

1970; Groblicki and Nobel, 1971; Harkins and Nicksic, 1968) that sulfuric acid or ammonium sulfate produced when sulfur dioxide, lower molecular weight olefins, and nitrogen oxides are irradiated with simulated ultraviolet solar radiation. Higher molecular weight olefins, particularly cyclo olefins or branched-chain olefins and alkylbenzenes, when irradiated with nitrogen oxides in the absence of sulfur dioxide will produce organic aerosols (Prager et al., 1960; Groblicki and Nobel, 1971; Harkins and Nicksic, 1968). As sulfur dioxide is added to such mixtures, the light scattering increases rapidly for the reaction mixtures containing olefins, but much less so for mixtures containing alkylbenzenes. These results suggest that sulfuric acid is a less abundant fraction of the total aerosol loading from alkylbenzene than olefin-containing mixtures. Therefore, atmospheric photochemical reactions can produce as atmospheric aerosols complex organic aerosols as well as sulfuric acid aerosols. Experiments relating reaction time to light scattering or aerosol products indicate that the organic aerosols are produced from the beginning of the reaction time while sulfuric acid aerosol does not form until nitrogen dioxide peaks and ozone forms at significant concentrations. The sulfate, nitrate, and organic aerosols formed both from dark reactions and irradiation are submicron in size. Newly formed particles are in the 0.01- to 0.1-μm range but sizes between 0.1 and 1 mu are formed after aging (Cox and Penkett, 1970; McNelis, 1974; Stevenson et al., 1965; Lee et al., 1971).

In experiments with irradiated propylene-sulfur dioxide-nitrogen oxide mixtures, the rate of sulfur dioxide disappearance has been shown to have the same general relationships as does oxidant to propylene and nitrogen oxide initial concentrations (Groblicki and Nobel, 1971). The rate of sulfur dioxide disappearance increases with increasing propylene concentrations, but decreases with increasing nitrogen dioxide concentration above a very low initial nitrogen dioxide concentration.

A large number of elementary reactions potentially can contribute to the oxidation of sulfur dioxide in the overall hydrocarbon-sulfur dioxide-nitrogen oxide photooxidation (Table 5). These include reactions with hydroxy radicals, perhydroxyl radicals, and various oxygenated radicals and intermediates. The rate of reaction of sulfur dioxide with atomic oxygen from the photolysis of nitrogen dioxide and with ozone is much too slow to account for any significant fraction of overall rate of sulfur dioxide reaction.

The dark reaction between olefin and ozone to produce a reactive intermediate (a biradical or zwitter ion) that oxidizes sulfur dioxide appears to be of particular significance

TABLE 5. Rates of Some Homogeneous Reactions Involving Oxidation of Sulfur Dioxide in Air

	Reaction	Rate
(1)	$SO_2 + O + M \rightarrow SO_3 + M$	Very slow
(2)	$SO_2 + O_3 \rightarrow SO_3 + O_2$	Very slow
(3)	$SO_2\ h\nu \rightarrow SO_2$ (triplet)	Very slow
	$2SO_2$ (triplet) $+ O_2 \rightarrow 2SO_3$	
(4)	$SO_2 + OH \rightarrow HSO_3$	Approximately 0.5%/hr
(5)	$SO_2 + HO_2 \rightarrow SO_3 + OH$	0.5 to 1%/hr
(6)	Olefin + $O_3 \rightarrow$ Biradical + Aldehyde	0.5 to 3%/hr
	Biradical + $SO_2 \rightarrow SO_3$ + Aldehyde	
(7)	Unsaturated HC + $NO_x + SO_2 \rightarrow SO_3$ + Products	1 to 10%/hr

in explaining the sulfur dioxide conversion to aerosols in irradiated olefin-sulfur dioxide-nitrogen oxide systems (Cox and Penkett, 1972; McNelis, 1974). Since this type of dark reaction would not be significant until ozone started forming in the photochemical photoxoidation, it is consistent with the time-concentration of profiles observed in these systems. Although OH radicals are likely products of reaction of ozone with olefin, the rate of sulfur dioxide oxidation by OH is not considered sufficient to account for the observed rates (McNelis, 1974). The rate of aerosol formation is determined by the rate of olefin-ozone reaction. Aerosol formation is a linear function of the product of olefin and ozone indicating second-order kinetics (Cox and Penkett, 1972). The critical intermediate is short-lived with a half life of less than 15 minutes. Aerosol formation rate did decrease with increasing relative humidity (Cox and Penkett, 1972). Sulfuric acid aerosol is formed in the 0.01-μm particle size range (McNelis, 1974). Coagulation shifts the particle size distribution upward at a rate such that most of the particle volume is above 0.1 μm within 1 hour. As a result during this time period the computed visual range decreases markedly (McNelis, 1974). The rate of conversion of sulfur dioxide in the presence of propylene has received the most attention with good agreement on a rate near 0.4 percent/hour (Cox and Penkett, 1972; McNelis, 1974). These reactions would appear to be of significance for urban pollution conditions.

The reaction of sulfur dioxide with HO, RO, HO_2, and RO_2 radicals produced from hydrocarbon-nitrogen oxide-sulfur dioxide-irradiated systems can contribute significantly to the conversion of sulfur dioxide to sulfate. Calvert and McQuigg computed rates of 1 to 1.5 percent/hour for the sum of these reactions of radicals with sulfur dioxide. Since the presence of these radical species, particularly HO and HO_2, do not depend on the presence of substantial concentrations of the more reactive olefins, they are likely to be the key reactive species downwind in urban plumes, plumes from some combustion sources and generally in the troposphere.

The oxidation of sulfur dioxide in aqueous droplets can occur at an appreciable rate if a sufficient concentration of soluble cation with catalytic capabilities is present such as ferric or manganese ions (Johnstone and Coughanowr, 1958; Junge and Ryan, 1958). The oxidation of sulfur dioxide stops when the pH decreases below 2 in the droplets probably due to the low solubility of sulfur dioxide in strongly acid solutions (Junge and Ryan, 1958). If sufficient ammonia is present in air, the ammonia will continue to diffuse into the droplet, neutralize the acid, so the oxidation of sulfur dioxide will continue.

A mathematical relationship has been derived relating the rate of sulfur dioxide oxidation in droplets with time to sulfur dioxide, ammonia, ferric ion, and relative humidity (Friberg, 1974). It was found that the increase rate due to an increase in RH will increase the diffusion time. More water will condense and the size of the droplets increase (Friberg, 1974). Such dilution increases the amount of SO_2 available for oxidation, increases the pH, and dilutes the Fe^{+3}. The net result because of compensation results in $d[SO_2]/dt$ is proportional to $1/[H^+]^3$ or $1/[1 - RH]^3$ since $[H^+]$ is proportional to $[1 - RH]$. As a result, when the RH increases from 80 to 95 percent, the rate of oxidation increases 64-fold (Friberg, 1974). Computation of the overall dependence on temperature from the various reaction rate, dissociation and solubility constants, shows a substantial decrease in rate of sulfur dioxide oxidation in the aqueous droplets with increasing temperature. The relationship predicts a 40- to 80-fold decrease in rate per 10°C increase in temperature between 5 and 30°C (Friberg, 1974).

Another mechanism would be that involving oxidation of sulfur dioxide in liquid films around large insoluble particles of metal oxides (Foster, 1969). As an acidic film is formed at high relative humidities, some soluble catalytic metal ions such as ferric or manganese ions are considered capable of dissolving into the liquid form. This type of mechanism is suggested to explain some of the experimental results obtained at very high relative humidities in power plant plumes with high particle loadings (Gartrell et al., 1963). Experimental rates of conversion for sulfur dioxide were as high as 2 percent/minute. The computed rates obtained were 0.2 to 1.5 percent/minute (Foster, 1969). A noncatalyzed oxidation of sulfur dioxide dissolved in water droplets has been proposed (Scott and Hobbs, 1967; McKay, 1971). This oxidation of sulfite also would be expected to decrease with increased acidity unless ammonia was sufficiently available to reduce the acidity.

Sulfite or sulfate have been detected by ESCA on flyash, coal smoke, and manganese oxide, iron oxide, magnesium oxide, and calcium oxide particles exposed to sulfur dioxide (Hulett et al., 1972). Once adsorbed on atmospheric particles, sulfur dioxide cannot be desorbed, supporting a change in valence state on the solid surface of various particles. However, the amount of sulfur dioxide actually removed on such surfaces (monolayer or greater) is not clear.

As mentioned previously, early investigators of power plant plume conversions of sulfur dioxide to products reported very high rates of conversion. More recent investigators have not been able to duplicate these earlier rates. Rates of sulfur dioxide to sulfate of no more than a few percent per hour was measured in plumes 10- to 20-miles downwind of power plant stacks. The large differences with earlier work has been attributed to the large reduction in particulate loading in power plants with the advent of high efficiency electrostatic precipitators. Recent work with techniques allowing flux measurements on sulfur compounds have been used on a large coal-fired power plant plume in St. Louis during the summer of 1974. The results again show very little conversion during the first 5 to 10 miles of downwind movement of the plume. However, the rate of conversion of sulfur dioxide and of fine particulate formation increases to several percent per hour in the plume after 10 to 20 miles of travel downwind. The results suggest that the rates of conversion may be continuing to increase after the first 20 miles of plume travel downwind.

Concurrent work in the St. Louis area following the urban plumes of St. Louis during the daylight hours during the summer resulted in a computed rate of dry deposition equivalent to 3 to 6 cm/sec (Vaughan et al., 1975). This rate of deposition is much higher than reported over land in England or over water. With this large rate of deposition removal becomes very important compared to conversion of sulfur dioxide for the urban plume. Additional experiments must be made during different seasons of the year and at night as well as during the daytime hours.

E. Atmospheric Concentrations and Distribution of Sulfates

Annual average concentration of water-soluble sulfates in urban sampling in the United States have ranged from as low as 2 to 3 $\mu g/m^3$ up to almost 50 $\mu g/m^3$. Maximum 24-hour average sulfate concentrations at sites in New York City, Philadelphia, and Baltimore have reached to 60- to 80-$\mu g/m^3$ range. At some sites in Los Angeles 24-hour average sulfate concentrations have reached 50 to 60 $\mu g/m^3$. The United States annual average concentration has been near 10 $\mu g/m^3$. The sulfate concentrations are higher on the average at urban sites in the eastern and midwestern United States than at urban sites in the western or southern United States (Altshuller, 1973). This distribution is consistent with the use of higher sulfur content fuels east of the Mississippi River. Industrial activity and power plant capacity also has been greater in the northern than southern part of the United States.

Figure 3 is a plot of United States urban sulfates as a function of sulfur dioxide (Altshuller, 1973). Above 100 $\mu g/m^3$ of sulfur dioxide, no increase in atmospheric sulfates occurred. Below 100 $\mu g/m^3$, the sulfates decreased linearly with decreasing

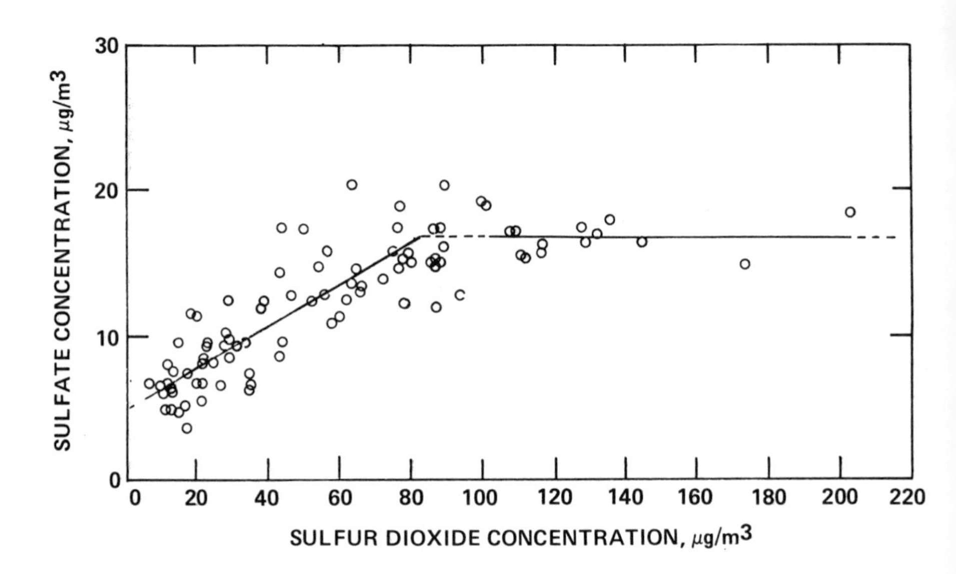

FIGURE 3. Relationships between annual average sulfur dioxide and sulfur dioxide and sulfate concentrations for 18 United States cities (after Altshuller, 1973; reprinted with permission from Environmental Science and Technology 7, 8, 1973. Copyright by the American Chemical Society.)

sulfur dioxide. At zero local sulfur dioxide concentration, a residue of about 5 μg/m^3 of sulfate remained (Altshuller, 1973).

The particle size distribution of sulfate aerosols has been measured at a number of urban sites both in the eastern and western United States (Ludwig and Robinson, 1965; Wagman et al., 1967; Lee and Patterson, 1969). The mass median diameters (MMD) computed for equivalent spheres of unit density from measurements in Chicago, Philadelphia, Cincinnati, Los Angeles, and San Francisco areas usually have been in the 0.2- to 0.4-μm range with a few MMD values in the 0.5- to 1-μm range. All of these measurements are consistent in indicating that 80 percent or more of the urban atmospheric sulfate is associated with aerosol particles below 2 μm. This particle size range for sulfates is entirely consistent with the transport of sulfates on a large regional scale with very little loss of such finely divided particles by dry deposition.

A trend toward increasing particle size of sulfate aerosols with increasing relative humidity rather than absolute humidity has been computed from measurements in Chicago, Cincinnati, and Philadelphia (Ludwig and Robinson, 1965). Therefore, the particle size of the sulfate aerosols appears sensitive to the degree of saturation of condensibility of available water vapor. The sulfate concentration tended to increase with increase in absolute humidity rather than relative humidity in samples from Chicago, Cincinnati, and Philadelphia (Wagman et al., 1967). Thus, the sulfate concentration shows a dependence on the vapor pressure of the water, not its degree of saturation in the air.

The cation concurrently present in sulfate-containing aerosols is of considerable interest. Available concentration and particle size measurements indicate that the abundant metal cations including iron (Lee et al., 1968; Gladney et al., 1974; Lundgren, 1972; Dzubay and Stevens, to be published), aluminum (Gladney et al., 1974; Hidy et al., 1973), sodium (Gladney et al., 1974), magnesium (Lee et al., 1968), and calcium (Dzubay and Stevens, to be published), are found in the coarse size range in impactor samples collected at a number of eastern, midwestern, and western sites. Therefore, the atmospheric sulfate aerosols must consist predominantly of ammonium sulfate, ammonium hydrogen sulfates, or sulfuric acid. The particle size distribution of ammonium ion in measurements in Chicago, Philadelphia, and Cincinnati were very close to those of the sulfate.

A molar ratio close to that of $(NH_4)_2SO_4$ was computed from Chicago aerosol measurements (Wagman et al., 1967). An excess of ammonium ion was available from the Philadelphia and Cincinnati measurements (Wagman et al., 1967). In part, the excess ammonium could have been associated with the nitrate ion which was in the same particle size range also. Similarly, aerosol chemistry measurements at several sites in the southern California air basin provided sufficient ammonium to neutralize the sulfate and nitrate (Hidy et al., 1973). In Swedish measurements, the composition often corresponds in $(NH_4)_2SO_4$, but summertime samples show strong acid content and correspond to the presence of a mixture of NH_4HSO_4 or $(NH_4)_3H(SO_4)_2$ with $(NH_4)_2SO_4$ (Brosset, 1975). At some midwest sites, insufficient ammonium ion has been measured to account for sulfate as $(NH_4)_2SO_4$ (Spicer et al., 1975).

In transport of sulfur dioxide through cities and rural terrain, rates of removal of sulfur dioxide can be significant compared to rates of conversion to sulfate. Removal processes will be minimized for emissions from tall stacks or for air masses isolated from the ground by radiation inversions. Concurrent measurements of conversion and removal of sulfur dioxide from urban plumes over different terrains need continued emphasis.

In the atmosphere, sulfur dioxide, in addition to being converted to sulfate, can be removed by adsorption on vegetation or water or by precipitation or dry deposition onto land or oceans. On a global basis precipitation on land and losses to the oceans constitute the more important processes. On the time scale of regional movements of pollutants, the role of removal processes are less well defined. In many regions in some seasons of the year, precipitation is unimportant. Removal into the oceans is not relevant to pollutant transport within inland regions. Therefore, the processes available for removal of sulfur dioxide during periods of minimum precipitation in an inland region are largely limited to dry deposition. On the other hand, in coastal regions with abundant vegetation and rainfall, multiple removal processes are available. Therefore, the understanding of the processes for transformation and removal of sulfur dioxide on a regional scale must be tailored to the geographical, climatological, and vegetal characteristics of each region.

The lifetime of sulfur dioxide (as other pollutants) on a day by day basis in an urbanized region depends on source characteristics and meteorological conditions. The sulfur dioxide from an elevated plume in dry weather is likely to be removed more slowly than the sulfur dioxide in the lower level urban plume. Sulfur dioxide and other pollutants to the extent they are isolated aloft between a subsidence and radiation inversion at night will be subject to minimal removal rates. On a local scale, manmade surfaces in a large city may be of importance for removal of sulfur dioxide emitted near ground level.

It is evident that much remains to be learned about the processes of transformation and removal on a day-by-day scale of events as air masses containing sulfur dioxide moved out of a city across rural terrain and on to another urban area, 50, 100, or several hundred kilometers away. For example, on this scale just what is the relative contribution of sulfur dioxide from plumes to ground level sulfate concentrations compared to sulfur dioxide from low-level emission sources within the city and in the surrounding region during periods of elevated sulfate concentrations levels? Much additional effort must be focused on such questions if we are to learn how to optimize our control efforts to obtain the greatest benefits for the least expenditure of resources.

III. GENERAL DISCUSSION

Recent investigations emphasize the regional character of both ozone and sulfate distribution patterns. Although ozone and sulfate are much different chemically, they do have several important similarities. Both ozone and sulfates during periods of warm sunny weather will be produced concurrently from photochemical reaction processes in the atmosphere. As more diurnal sulfate measurements become available concurrently with ozone measurements, it will be possible to adequately test such an association statistically. Sulfates once formed will survive for days because of small particle sizes and associated low rates of deposition. Ozone will decay more rapidly near the surface than sulfate. However, both ozone and sulfates when isolated aloft will survive to be carried along downwind and mixed down to the surface the next day. Both ozone and sulfates are formed by intermediate reactions depending often on the participation of the same radical species. However, the sensitivity of sulfates compared to ozone to control of precursor species is not yet well known.

In the winter months, high concentrations of sulfates are known to exist although perhaps with differences in chemical composition (Brosset, 1975). Ozone production is very low in the winter because of low ultraviolet radiation intensities and low temperatures. It is evident that additional mechanisms involving heterogeneous processes discussed earlier are more likely to play a significant part in sulfate production in cold weather.

The regional nature of ozone formation processes already have raised questions as to the impact of such findings on control strategies. Are the existing strategies based on considering the oxidant problem of each city as an independent entity effective approaches? This is an especially significant problem when considering elevated ozone concentrations associated with slowly moving, high pressure fronts. Within high pressure cells, the poorly organized winds slowly stir ozone precursors throughout the region. Under these circumstances, control of hydrocarbons on a large region basis would seem appropriate.

The widespread atmospheric dispersion of submicron particles of sulfur-containing species in the atmosphere along with the substantial lifetimes of such aerosols can complicate engineering approaches to control of sulfur dioxide. It cannot be assumed that such measures as use of tall stacks or use of fuel switching during periods of stagnation will have the same beneficial effects for reduction of atmospheric sulfate concentrations as for sulfur dioxide. Locating new power plants in rural areas in the United States was justified at least in part for the purpose of reducing sulfur dioxide exposures of urban populations. Since sulfate pollution appears to be a regional rather than a local air pollution problem, the siting of power plants in rural areas needed to be examined in terms of their impact on regional sulfate concentration levels. Concurrently, the rates of conversion of near-surface urban sulfur dioxide emissions within and immediately downwind of cities needs further evaluation. The rates of dry deposition of sulfur dioxide must be obtained under a variety of conditions.

It is regretable that the sulfur dioxide control problem, which is already complicated by the impact of fuel shortages, should be further complicated by the sulfate problem. It is crucial that we clearly delineate the relationships between emissions of sulfur dioxide and the atmospheric behavior of sulfates. If appropriate control measures are selected now for control of sulfur dioxide emissions, we may have to live with the adverse consequences of these decisions in the decade to come. The resources involved in these decisions are enormous. Therefore, it is essential that definitive scientific results be available on sulfates to assure that appropriate decisions are made in the United States and other advanced industrial societies.

REFERENCES

1. Allen, E. R., McQuigg, R. D., and Cadle, R. D. 1972. The Photooxidation of Gaseous Sulfur Dioxide in Air, Chemosphere 1, 25-32.

2. Altshuller, A. P. 1975. Evaluation of Oxidant Results at CAMP Sites in the United States, J. Air Pollut. Control Assoc. 25, 18-24.

3. Altshuller, A. P. 1973. Atmospheric Sulfur Dioxide and Sulfate. Distribution of Concentration at Urban and Nonurban Sites in the United States, Environ. Sci. Technol. 7, 709-712.

4. Altshuller, A. P. and Cohen, I. R. 1963. Photooxidation of Acrolein-Nitrogen Oxide Mixtures in Air, Int. J. Air Water Pollut. 7, 1043-1049.

5. Altshuller, A. P. and Cohen, I. R. 1964. Atmospheric Photooxidation of the Ethylene-Nitric Oxide System, Int. J. Air Water Pollut. 8, 611-631.

6. Altshuller, A. P., Cohen, I. R., and Purcell, T. C. 1967. Photooxidation of Hydrocarbons in the Presence of Aliphatic Aldehydes, Science 156, 937-939.

7. Altshuller, A. P., Kopczynski, S. L., Lonneman, W. A., Becker, T. L., and Slater, R. 1967. Chemical Aspects of the Photooxidation of the Propylene-Nitrogen Oxide System, Environ. Sci. Technol. 1, 899-914.

8. Altshuller, A. P., Kopczynski, S. L., Lonneman, W. A., and Sutterfield, F. D. 1970a. A Technique for Measuring Photochemical Reactions in Atmospheric Samples, Environ. Sci. Technol. 4, 503-506.

9. Altshuller, A. P., Kopczynski, S. L., Lonneman, W. A., Sutterfield, F. D., and Wilson, D. L. 1970b. Photochemical Reactivities of Aromatic Hydrocarbon-Nitrogen Oxides and Related Systems, Environ. Sci. Technol. 4, 44-49.

10. Altshuller, A. P., Kopczynski, S. L., Wilson, D., Lonneman, W., and Sutterfield, F. D. 1968. Photochemical Reactivities of Paraffinic Hydrocarbon-Nitrogen Oxide Mixtures upon Addition of Propylene or Toluene, J. Air Pollut. Control Assoc. 19, 791-794.

11. Altshuller, A. P., Kopczynski, S. L., Wilson, D., Lonneman, W., and Sutterfield, F. D. 1969. Photochemical Reactivities of n-Butane and Other Paraffinic Hydrocarbons, J. Air Pollut. Control Assoc. 19, 787-790.

12. Blumenthal, D. L., White, W. H., Peace, R. L., and Smith, T. B. 1974. Determination of the Feasibility of the Long-Range Transport of Ozone or Ozone Precursors, November EPA Contract No. 68-02-1462.

13. Brosset, C. 1975. Acid Particulate Air Pollutants in Sweden, Swedish Water and Air Pollution Research Laboratory, February.

14. Bufalini, J. J. and Altshuller, A. P. 1963. The Effect of Temperature on Photochemical Smog Reactions, Int. J. Air Water Pollut. 7, 769-771.

15. Bufalini, J. J. and Altshuller, A. P. 1967. Synergistic Effects in the Photooxidation of Mixed Hydrocarbons, Environ. Sci. Technol. 1, 135-138.

16. Bufalini, J. J. and Altshuller, A. P. 1969. Oxidation of Nitric Oxide in the Presence of Ultraviolet Light and Hydrocarbons, Environ. Sci. Technol. 3, 469-472.

17. Calvert, J. G. and McQuigg, R. D. 1975. The Computer Simulation of the Rates and Mechanisms of Photochemical Smog Formation, Symposium Proceedings, Int. J. Chem. Kinet., Symposium No. I, 113-154.

18. Cleveland, W. S., Graedel, T. E., Kleiner, B., and Warner, J. L. 1974. Sunday and Workday Variations in Photochemical Air Pollutants in New Jersey and New York, Science 186, 1037-1038.

19. Cleveland, W. S. and Kleiner, B. 1975. The Transport of Photochemical Air Pollution from Camden-Philadelphia Urban Complex, Environ. Sci. Technol. 9, 869-872.

20. Cox, R. A. and Penkett, S. A. 1970. Atmos. Environ. 4, 425.

21. Cox, R. A. and Penkett, S. A. 1972. J. Chem. Soc., Faraday Trans. I, Part 9, 69, 1735.

22. Doyle, G. J., Lloyd, A. C., Darnell, K. R., Winer, A. M., and Pitts, J. N., Jr. 1975. Gas Phase Kinetic Study of Relative Rates of Reaction of Selected Aromatic Compounds with Hydroxyl Radicals in an Environmental Chamber, Environ. Sci. Technol. 9, 237-241.

23. Dzubay, T. G. and Stevens, R. K. 1975. Ambient Air Analysis with Dichotomous Sampler and X-ray Fluorescence Spectrometer, Environ. Sci. Technol., to be published.

24. Endow, N., Doyle, G. J. and Jones, J. L. 1963. J. Air Pollut. Control Assoc. 13, 141.

25. Foster, P. M. 1969. Atm. Environ. 3, 157-175.

26. Friberg, J. 1974. Environ. Sci. Technol. 8, 731-734.

27. Gartrell, F. E., Thomas, F. W., and Carpenter, S. B. 1963. Amer. Ind. Hygiene Assoc. J. 24, 113-120.

28. Gay, B. W., Jr., Hanst, P. L., Bufalini, J. J., Noonan, R. C. 1976. Atmospheric Oxidation of Chlorinated Ethylenes, Environ. Sci. Technol. 10, 58-67.

29. Gladney, E. S., Zoller, W. H., Jones, A. G., and Gordon, G. E. 1974. Environ. Sci. Technol. 8, 551-557.

30. Groblicki, P. J. and Nobel, G. J. 1971. The Photochemical Formation of Aerosols in Urban Atmospheres, Chemical Reactions in Urban Atmospheres, C. S. Tuesday, ed., Elsevier, New York, pp. 241-267.

31. Harkins, J. and Nicksic, S. W. 1968. J. Air Pollut. Control Assoc. 15, 218-221.

32. Hidy, G. M. and colleagues. 1973. Characterization of Aerosols in California Interim Report for Phase I—October 25, 1971 to April 1, 1973. Submitted to Air Resources Board, State of California, December 15, ARB Contract No. 358.

33. Hulett, L. D., Carlson, T. A., Fish, B. R., and Durham, J. L. 1972. Studies of Sulfur Compounds Adsorbed on Smoke Particles and Other Solids by Photoelectric Spectroscopy in Determination of Air Quality, Proceedings of the ACS Symposium on Determination of Air Quality, Los Angeles, California, April 1-2, 1971 (G. Mamantor and W. D. Shults, eds.), Plenum, New York, pp. 179-187.

34. Johnstone, H. F. and Coughanowr, D. R. 1958. Ind. Eng. Chem. 50, 1169-1172.

35. Junge, C. E. and Ryan, T. G. 1958. Quat. J. Ray Meteor. Soc. 84, 46-55.

36. Kinosian, J. R. and Duckworth, S. 1973. Oxidant Trends in the South Coast Air Basin, California Air Resources Board Report, April.

37. Kopczynski, S. L. and Altshuller, A. P. 1962. Int. J. Air Water Pollut. 6, 133-135.

38. Kopczynski, S. L., Altshuller, A. P., and Sutterfield, F. D. 1974. Photochemical Reactivities of Aldehyde-Nitrogen Oxide Systems, Environ. Sci. Technol. 8, 909-918.

39. Kopczynski, S. L., Kuntz, R. L., and Bufalini, J. J. 1975. Reactivities of Complex Hydrocarbon Mixtures, Environ. Sci. Technol. 9, 648-653.

40. Kopczynski, S. L., Lonneman, W. A., Sutterfield, F. D., and Darley, P. E. 1972. Photochemistry of Atmospheric Samples in Los Angeles, Environ. Sci. Technol. 6, 342.

41. Korth, M. W., Stahman, R. C., and Rose, A. H., Jr. 1964. Effects of HC/NO_x Ratios on Irradiated Auto Exhaust—Part I, J. Air Pollut. Control Assoc. 14, 168-175.

42. Leach, P. W., Lang, L. J., Bellar, T. A., Sigsby, J. E., Jr., and Altshuller, A. P. 1964. Effects of HC/NO_x Ratios on Irradiated Auto Exhaust—Part II, J. Air Pollut. Control Assoc. 14, 176-183.

43. Lee, R. E., Jr. and Patterson, R. K. 1969. Atmos. Environ. 3, 249-255.

44. Lee, R. E., Jr., Patterson, R. K., Crider, W. L., and Wagman, J. 1971. Atmos. Environ. 5, 225.

45. Lee, R. E., Jr., Patterson, R. K., and Wagman, J. 1968. Environ. Sci. Technol. 2, 288.

46. Lee, R. E., Jr., Goranson, S. S., Envione, R. E., and Morgan, G. B. 1972. Environ. Sci. Technol. 6, 1025-1030.

47. Ludwig, F. L. and Robinson, E. 1965. J. Colloid Sci. 20, 571-584.

48. Lundgren, D. A. 1970. J. Air Pollut. Control Assoc. 20, 603.

49. McKay, H. A. C. 1971. Atmos. Environ. Sci. 5, 7-14.

50. McNelis, D. N. 1974. Aerosol Formation from Gas-Phase Reactions of Ozone and Olefin in the Presence of Sulfur Dioxide, EPA Publication 650/4-74-034, Office of Research and Development, Research Triangle Park, North Carolina, August.

51. Miller, D. F., Levy, A., and Wilson, W. E., Jr. 1972. A Study of Motor-Fuel Composition Effects on Aerosol Formation. Part II. Aerosol Reactivity Study of Hydrocarbons, Report to American Petroleum Institute (API Project EF-2) from Battelle Memorial Institute, Columbus, Ohio, February.

52. Prager, M. J., Stephens, E. R., and Scott, W. E. 1960. Ind. Eng. Chem. 52, 521.

53. Research Triangle Institute, RTP, North Carolina. 1975. Investigation of Rural Oxidant Levels as Related to Urban Hydrocarbon Control Strategie, (BOA Contract No. 68-02-1386, Task 4)., January.

54. Schuck, E. A. and Doyle, C. J. 1959. Photooxidation of Hydrocarbons in Mixtures Containing Oxides of Nitrogen and Sulfur Dioxide, Air Pollution Board (Los Angeles), Report 29, October.

55. Scott, W. D. and Hobbs, P. V. 1967. J. Atmos. Sci. 24, 54-57.

56. Sidebottom, H. W., Badcock, C. C., Jackson, G. E., and Calvert, J. G. 1972. Environ. Sci. Technol. 6, 72-79.

57. Smith, J. P. and Urone, P. 1974. Environ. Sci. Technol. 8, 742-746.

58. Spicer, C. W., Gemma, J. L., Joseph, D. W., Stickel, P. R., and Ward, G. F. 1975. Report on the Transport of Oxidant Beyond Urban Areas, EPA Contract No. 68-02-1714, Battelle Columbus Laboratories, Columbus, Ohio, pp. 60-65, May.

59. Stevenson, H. J. R., Sanderson, D. E., and Altshuller, A. P. 1965. Int. J. Air Water Pollut. 9, 367-375.

60. U. S. Environmental Protection Agency. 1974. Health Consequences of Sulfur Oxides, A Report from CHESS 1970-1971, EPA-650/1-74-004, May.

61. Vaughan, W. M., Sperling, R. B., Gillan, N. V., and Husar, R. B. 1975. Horizontal SO_2 Mass Flow Rate Measurements in Plumes, presented at Annual Air Pollution Control Association Meeting, Boston, Massachusetts, June.

62. Wagman, J., Lee, R. E., Jr., and Axt, C. J. 1967. Atmos. Environ. 1, 479-489.

63. Westberg, H. H. and Rasmussen, R. A. 1973a. Measurement of Light Hydrocarbons in the Field and Study of Transport of Oxidant Beyond an Urban Area, Monthly Technical Report, EPA Contract No. 68-02-1232, October.

64. Westberg, H. H. and Rasmussen, R. A. 1973b. Measurement of Light Hydrocarbons in the Field and Studies of Transport of Oxidant Beyond an Urban Area, Monthly Technical Report, EPA Contract No. 68-02-1232, November.

65. Westberg, H. H. and Rasmussen, R. A. 1974. Measurement of Light Hydrocarbons in the Field and Studies of Transport of Oxidant Beyond an Urban Area, Monthly Technical Report, EPA Contract No. 68-02-1232, September to December.

Large-Scale Dispersion of Atmospheric Pollutants with Particular Consideration of Anthropogenic Emission of Sulfur and the Global Sulfur Cycle

BERT BOLIN
Department of Meteorology
Arrhenius Laboratory
University of Stockholm, Sweden*

I. INTRODUCTION

The fate of atmospheric pollutants is dependent on the combined effects of source distribution, horizontal and vertical transfer by the air motions and relevant sink processes, either in situ such as chemical reaction or rainout, or occurring at the lower boundary, that is, dry deposition. The way the problem is to be dealt with much depends on the efficiency of the sink processes. If these are acting slowly in comparison with the time required for global mixing, that is, the resulting residence time of the pollutant in the atmosphere is several months or years, the distribution in the atmosphere (troposphere) will be rather uniform and the discussion presented by Junge (1974a) is most relevant. We need not concern ourselves with the characteristics of the dispersion patterns that result from the distribution of sources, but should basically look for what the overall balance of sources and sinks imply.

*Contribution No. 304.

For pollutants with shorter residence times, the problem becomes increasingly complex because the distribution in the atmosphere becomes more variable, and we need to concern ourselves with the dispersion through the atmosphere, that is the transfer of the pollutant from the sources to the sinks. As long as the amounts of pollutants in the atmosphere were small, we could neglect the more detailed considerations of the sink processes in attempts to deduce concentration patterns in the atmosphere and merely concern ourselves with the dispersion implicitly assuming that the concentration in the atmosphere in general would not increase significantly because of the manmade emission. The ultimate fate of the pollutant was considered to be of little interest.

We know today that man is emitting into the atmosphere a number of constituents in amounts comparable to or larger than what is there naturally and many of these have lifetimes of the order of days, weeks, or months. Because of the variable concentrations in the atmosphere of pollutants with such comparatively short lifetimes our knowledge about the atmospheric inventory and the relative importance of various possible sink mechanisms usually is very unsatisfactory. In addition, the natural sources are also often poorly known and accordingly the relative importance of natural and anthropogenic sources for the overall cycles are difficult to assess. Sulfur and the nitrogen oxides (except for dinitrogen oxide) are atmospheric pollutants of this kind and considerable research efforts are required to clarify the short-term geochemical cycles of these compounds and to assess the role of man in this context.

In such efforts, it is important to be able to describe the atmospheric transport processes quantitatively to link various sources and sinks to each other and to assess residence times more precisely. We need to consider both horizontal and vertical transfer and also possibly the fact that they may be dynamically coupled. On the regional and global scales, with which we are concerned, the weather systems and their horizontal and vertical flow fields constitute the essential turbulent elements that bring about this transfer. The more detailed horizontal dispersion of individual smoke plumes, on the other hand, are of secondary importance and may often be neglected. It may well be necessary to use advanced general circulation models to derive more accurately the role of the weather systems in this regard and introduce into such models our best quantitative estimates of the sources, sinks, and distribution of the compounds under consideration. At this exploratory stage, however, when sources and sinks still are poorly known, there is a need for simpler models with the aid of which the relative importance of various possible source and sink mechanisms can be studied using available information on atmospheric concentrations in different parts of the world. The present article will summarize a method employed in the development of a simple statistical model for assessing the magnitude of regional dispersion and the application of this model to the problem of sulfur pollution over Europe (Bolin and Persson, 1975). We shall also be able to make some inferences regarding the sulfur cycle by making use of recent field studies of the efficiency of rainout as a sink for sulfur emitted from oil-fired power plants (Granat, 1975).

II. BASIC THEORETICAL CONSIDERATIONS

The continuity equation for any atmospheric compound may be written in a more complete form than presented in Chapter 1.

$$\frac{\partial \overline{m}}{\partial t} = - V_H \cdot \nabla_H \overline{m} - \nabla_H \overline{V_H' m'} - \frac{1}{\rho}\frac{\partial}{\partial z}(\overline{\rho w' m'}) + \frac{1}{\rho}\overline{E} - \frac{1}{\rho}\overline{S} \qquad (1)$$

where m is the mixing ratio of pollutants, V_H is the horizontal wind velocity, w is the vertical velocity, ρ is the air density, E is the source, S is the sinks, $\overline{(\;)}$ denotes long-term average, and $()'$ denotes departure from long-term average. The five terms on the right-hand side denote changes of $\overline{m}$ due to

1. Advection by the mean wind, which is assumed to be horizontal;

2. Turbulent transfer by horizontal winds (essentially due to synoptic disturbances);

3. Vertical turbulent transfer due to

   a. Mechanically induced turbulence primarily of importance in the surface boundary layer,

   b. Convection,

   c. Large-scale vertical motions associated with synoptic motions;

4. Sources;

5. Sinks.

In addition to the sources and sinks described explicitly by E and S in Eq. 1 the lower boundary (the land or ocean surfaces) may serve as source (evaporation) or sink (condensation or dry deposition). For a more detailed discussion reference is made to Bolin and Persson (1975).

Rather than dealing with the continuity equation 1 explicitly, we shall use a statistical approach. We assume that the transfer processes, as well as the sources and sinks, are statistically independent of each other, and we may then treat the processes described by the five terms in Eq. 1 separately. In reality, this is an oversimplification of the problem. We know, for example, that vertical transfer in the surface boundary layer depends on the vertical wind shear and thus on the horizontal wind. The sink process such as rainout and washout are associated with particular features of the snyoptic disturbances and thus also related to the wind field. In the westerlies, rain is more common when the winds blow from the south than from north and topographical features of the earth surface may also imply more frequent rains for particular wind directions. Even though this assumption of statistical independence thus is not generally valid, it represents a useful first approach. Differences between computed, and observed distributions must, however, be considered in the light of this assumption.

III. SUMMARY OF COMPUTATIONAL RESULTS

With the aid of available wind data over western Europe, the horizontal transfer characteristics have been evaluated by computing a large number of trajectories at different levels in the lower troposphere. The dispersion pattern to a first approximation turns out to be Gaussian around the point that is displaced with the mean wind. The principally elliptical dispersion pattern deviates rather little from being isotropic, that is, circular. Figures 1 and 2 show some of the results that were obtained. For further details reference is made to Bolin and Persson (1975). It should be noted that the curves in Figure 1 rather closely correspond to Fickian dispersion with an eddy diffusivity of $D \approx 1.2 \times 10^{11}$ cm^2 sec^{-1}. This value is considerably larger than normally is used for large-scale diffusion processes in the atmosphere (compare, for example, Bolin and Keeling (1963)). This value is, however, only applicable in middle latitudes, while the lower values usually refer to north-

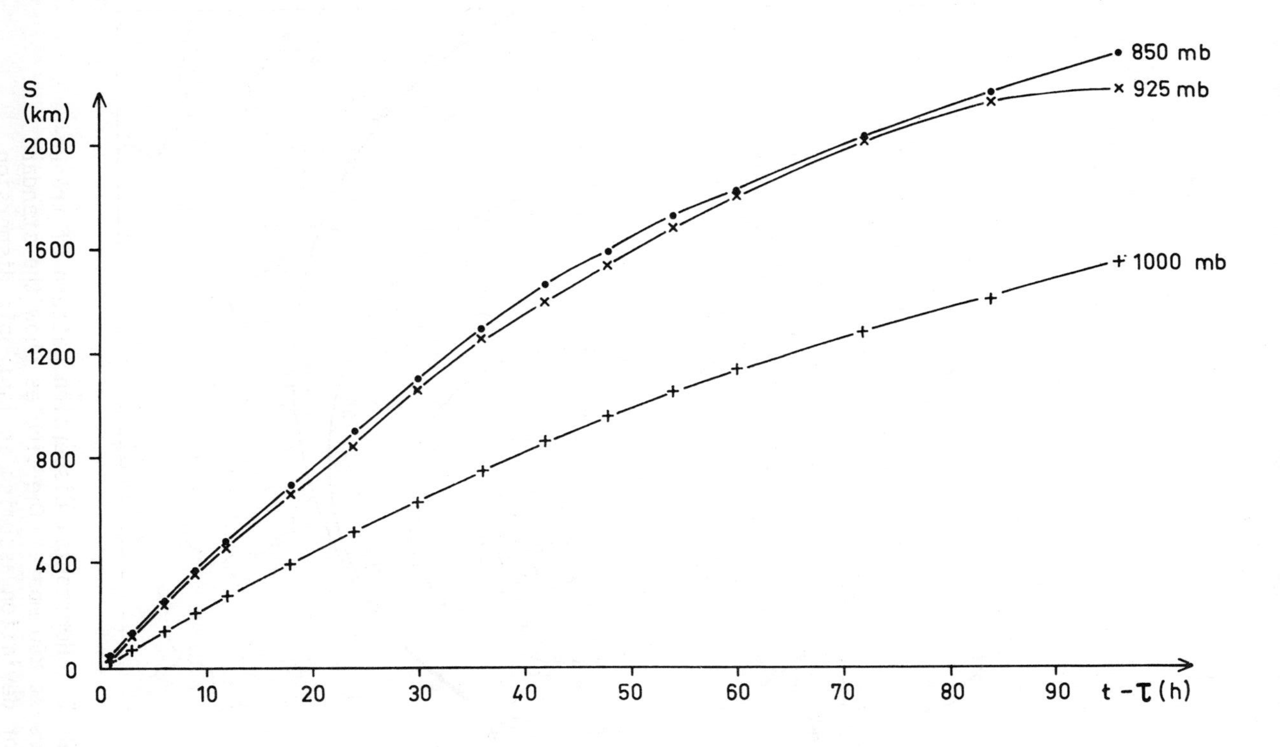

FIGURE 1. Standard deviation of distribution of end points of trajectories released from point in central England as indicated in Figure 2. Trajectories computed every third day October 1, 1972 through March 31, 1973.

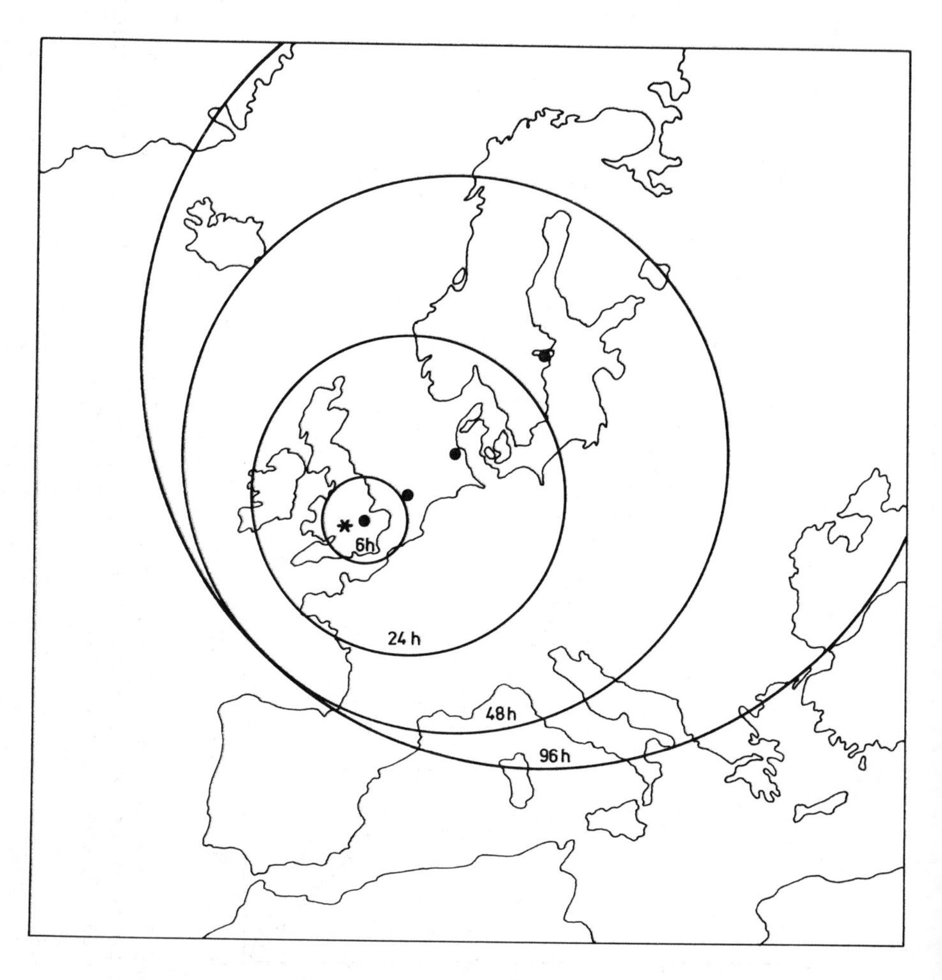

FIGURE 2. Horizontal dispersion pattern of trajectories at 850 mbar. The circles show the standard vector deviation assuming an isotropic dispersion around the mean wind.

south transfer from pole to equator or an appreciable part of the hemisphere. There are good reasons to believe that north-south transfer is considerably slower in polar regions and particularly in subtropical regions. The vertical dispersion was evaluated assuming a logarithmic wind profile in the surface layer of the atmosphere implying a linearly increasing eddy diffusivity up to a level about 100 m, above which the diffusivity was assumed to be constant (equal to 10 $m^2 sec^{-1}$). Emissions were assumed to take place at 85-m elevation. At the earth's surface, dry deposition was permitted to occur with a transfer (or deposition) velocity v_s which in the example shown here was assumed to be 1 cm/sec. In the free atmosphere removal by precipitation was assumed with rates of elimination β = 0.04, 0.02, and 0.01/hr. Figure 3 shows a sample computation (β = 0.02/hr) in which we see the gradual vertical dispersion due to the turbulent processes and the decrease of the total amount present in the atmosphere due to the sink processes.

The anthropogenic sulfur sources over western Europe are reasonably well known (see Strömsöa, 1973; Figure 4) and attempts were next made to compute the steady-state sulfur concentration in the air and the associated deposition patterns due to rainout and washout on one hand, and dry deposition on the other (for details see Bolin and Persson, 1975) using the emission values given in Figure 4. Figures 5 and 6 show wet and dry deposition if, assuming emission height to be 85 m everywhere, the deposition velocity v_s = 1 cm sec^{-1} and the rainout rate β = 0.02/hr. Figure 7 shows the observed pattern of wet deposition of sulfur over Europe after deduction of the part that originates from sea salt. The data have been obtained from the European rain chemistry network. Figure 8 compares the computed vertical concentration profiles with measurements over Germany (Jost, 1974) and south Sweden (Rodhe and Grandell, 1972).

IV. DISCUSSION

The maps and graphs shown here only represent a few samples of the different combinations of rates for wet and dry deposition that were tried. The general observations and conclusions that can be drawn from these experiments are the following:

1. The best agreement between calculated and observed wet deposition is obtained with a characteristic average residence time of sulfur in the atmosphere before rainout of 2 to 3 days. A longer, say 4-day, residence time or more would yield too small a wet

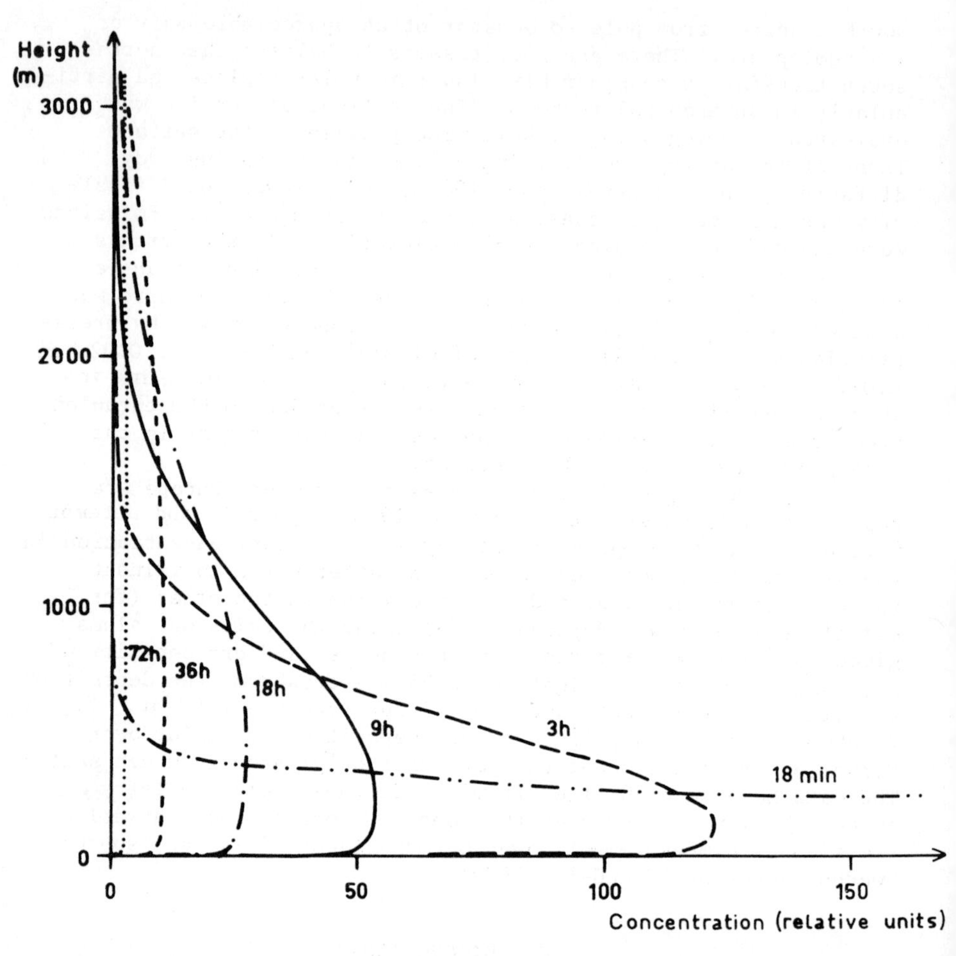

FIGURE 3. Vertical concentration profile as a function of time. Emission is assumed to take place at a height of 85 m, rainout rate β = 0.02/hr, vertical dispersion in the lowest 100 m is characterized by a logarithmic wind profile with the friction velocity u_x = 0.30 m/sec, and a constant eddy diffusivity of 10 m^2 sec^{-1} above 100 m. Dry deposition velocity at the earth surface has been assumed V_s = 1 cm/sec.

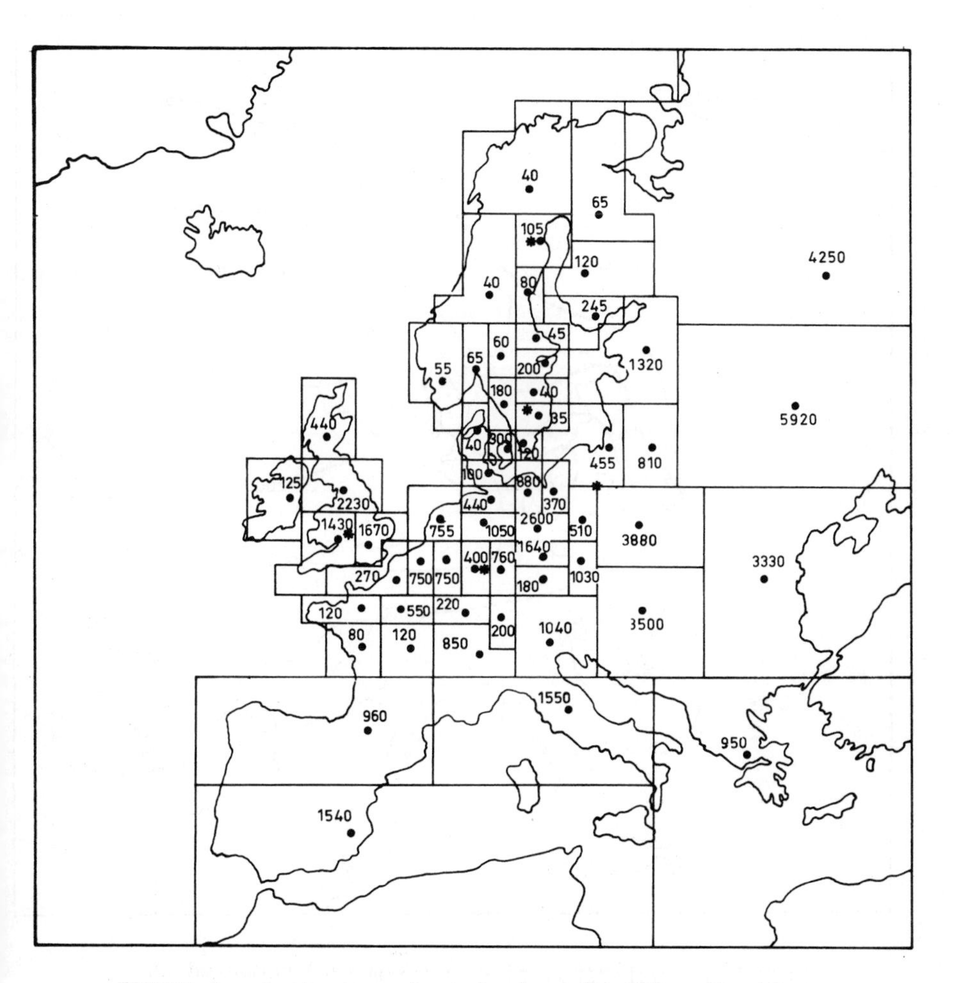

FIGURE 4. Anthropogenic emission of sulfur dioxide over Europe in units of 10^3 tons SO_2/yr.

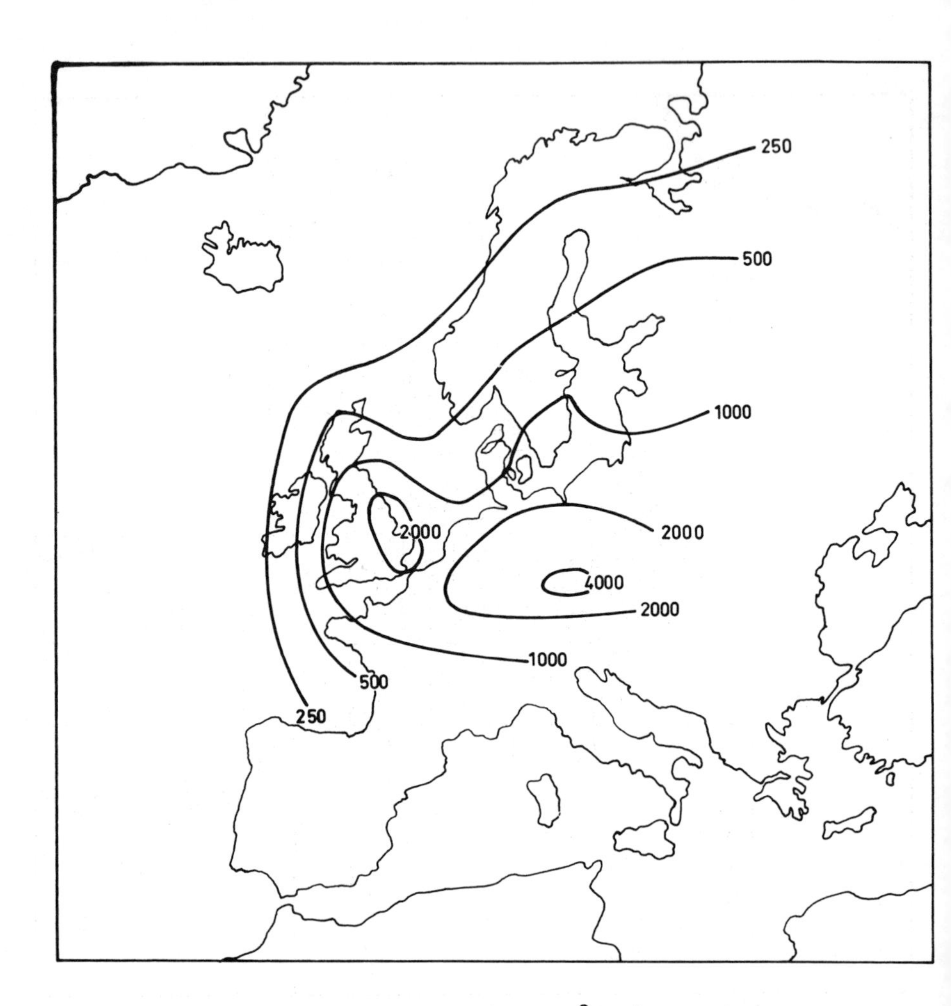

FIGURE 5. Wet deposition (mg $S/m^2/yr$) computed as resulting from anthropogenic sources given in Figure 4 assuming an emission height of 85 m, rate of removal by rainout or washout $\beta = 0.02/hr$, deposition velocity V_s = 1 cm/sec, and vertical diffusion otherwise as given in Figure 3.

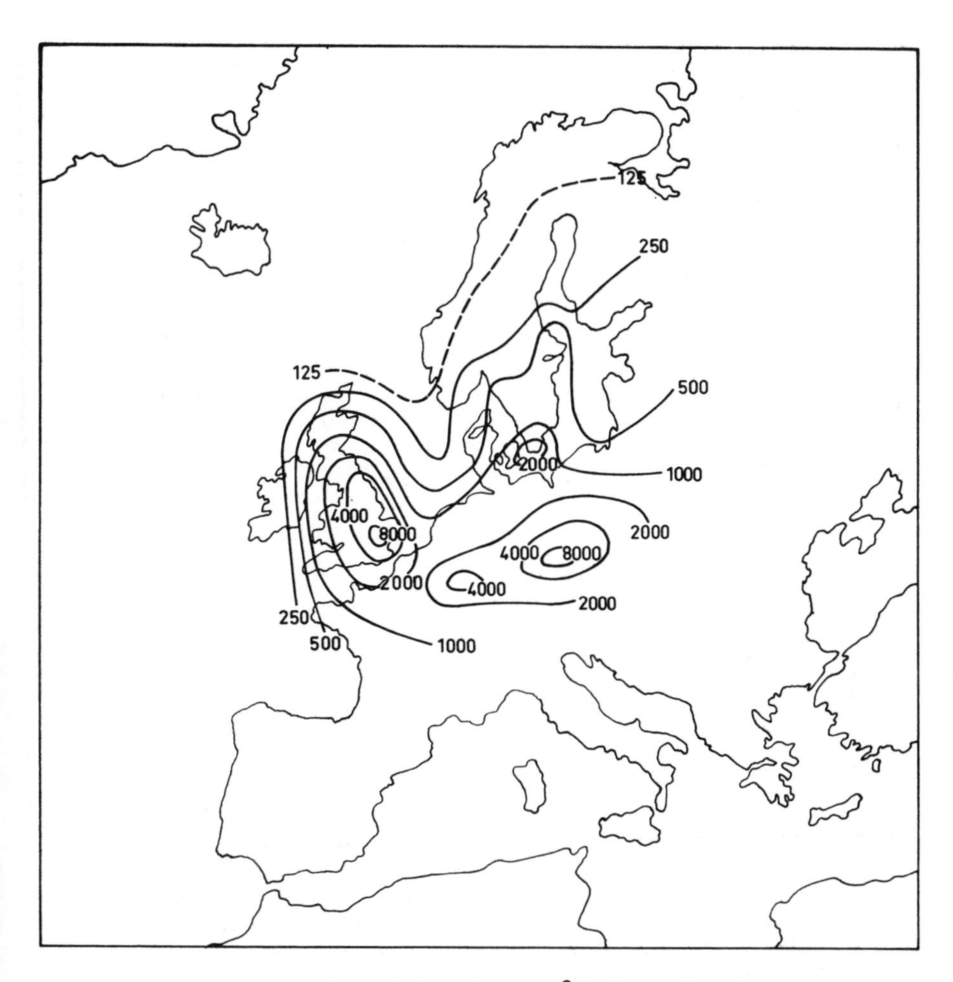

FIGURE 6. Dry deposition (mg $S/m^2/yr$) computed as resulting from anthropogenic sources as given in Figure 4 assuming otherwise sink mechanisms as given in Figure 3.

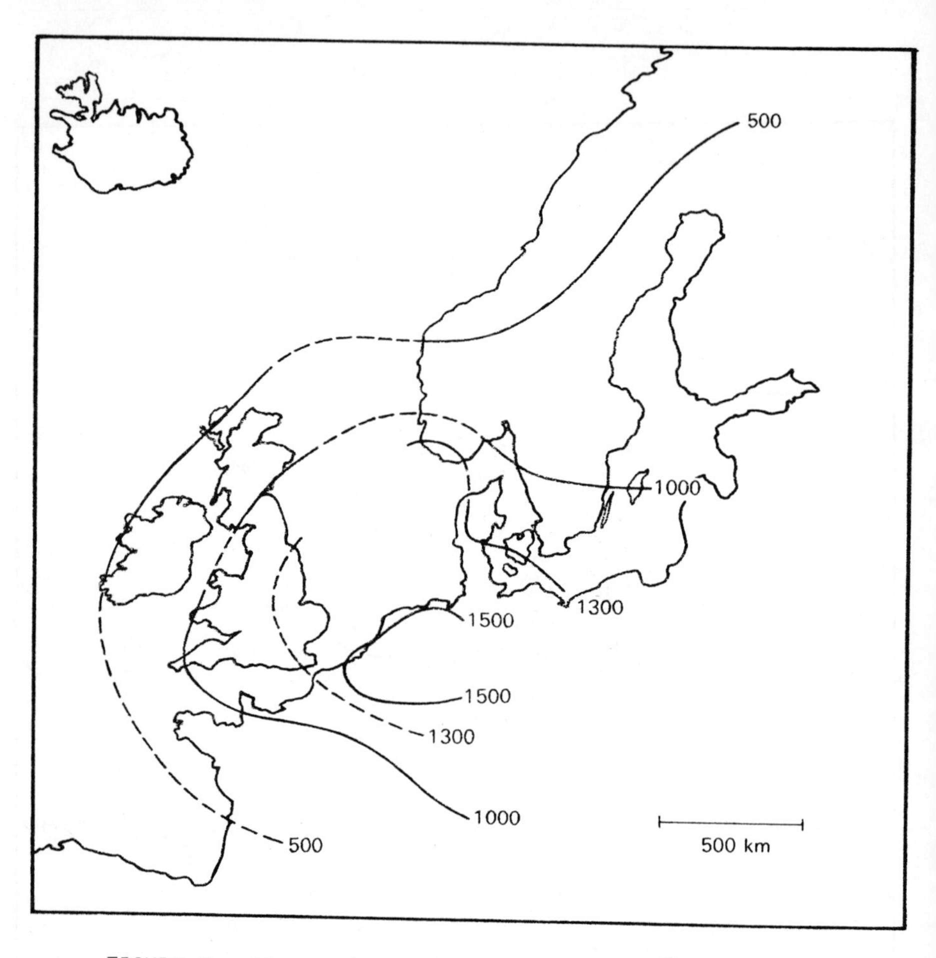

FIGURE 7. Observed wet deposition (mg $S/m^2/yr$) of excess sulfur estimated by deducting sulfur of oceanic origin (using observed Na-concentrations) from observed deposition by rain.

deposition over the central parts of the area, and too slow a decline of the wet deposition as a function of distance from the region of maximum emission. Also, this would imply a transport of a major part of the sulfur emitted far away from the source area, several thousand kilometers, which seems not likely to be the case.

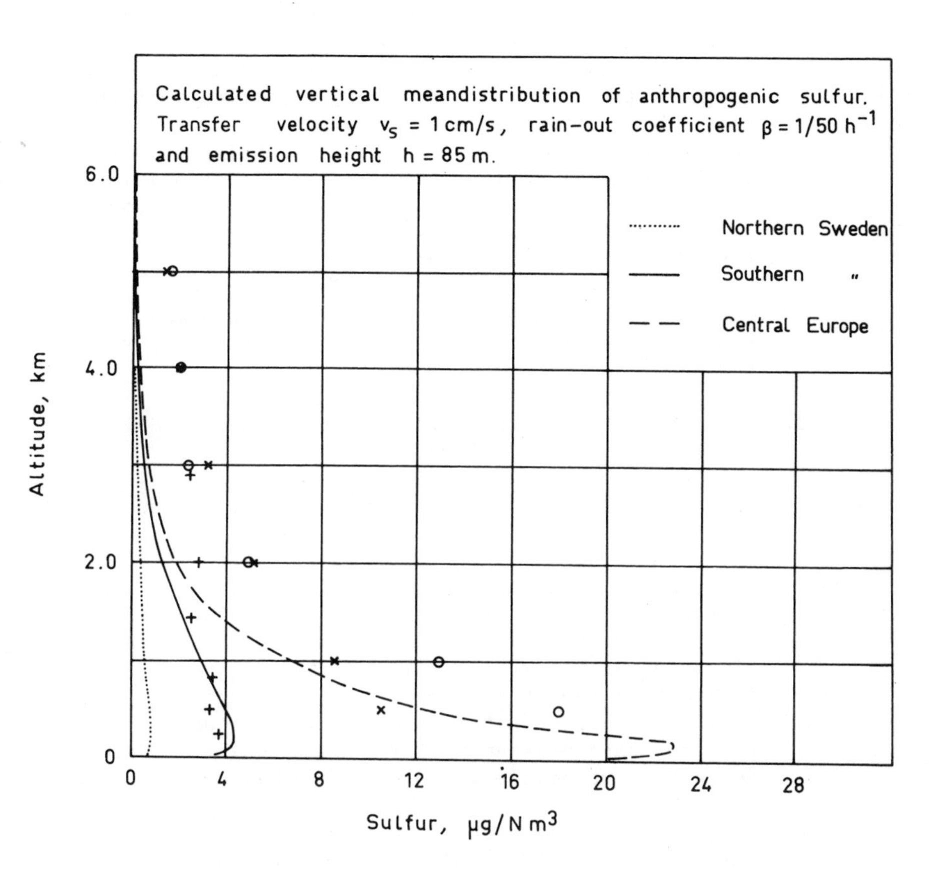

FIGURE 8. Observed mean distribution of sulfur is indicated, o winter and x summer over central Europe (after Jost, 1974). + yearly mean over southern Sweden (after Rodhe, 1972a).

2. Some dry deposition most likely takes place, otherwise wet deposition should be larger than observed. Our knowledge about the characteristic deposition

velocity to be used to describe the dry deposition process is still not satisfactory. The value of 1 cm sec^{-1} could both be one half or several times larger without changing the wet deposition pattern to an extent that would be incompatible with observed wet depositions. It is interesting to note in this context that an increase of the deposition velocity beyond a few centimeters per second does not further increase the dry deposition much (except in the vicinity of the sources if these are at low level), since the limiting process then is the vertical transfer within the atmosphere rather than the final transfer to the earth's surface.

3. It is by now well known that most of the sulfate in the atmosphere is found in particles with a radius of about 0.1/μm. The collection of efficiency of objects at the earth's surface of such particles is small, which is also the main reason why such particles do not further increase by coagulation. Therefore, most of the dry deposition is probably due to transfer of sulfur dioxide to the earth's surface and its absorption on vegetation and water surfaces. It might be of some interest to repeat the computations described previously but then treat sulfur dioxide and sulfate separately assuming for example quite different deposition velocities. So far this has not been done.

4. It is to be noted that 10 to 20 percent of an emitted amount of sulfur still is airborne after 4 days accepting the most likely values for v_s and β as arrived at previously. As is obvious from Figure 2, the spatial probability distribution function of the emission in central Europe after this lapse of time extends far outside Europe, which implies that what remains after this time cannot be distinguished from background amounts of sulfur due to other, distant anthropogenic sources or natural sources.

5. Figure 3 shows a computed decrease of sulfur concentration with elevation that is more rapid than the observed one. Admittedly, the observations at levels above a few kilometers often yield values close to the detection limit of the observational

method used. Nevertheless, the difference between observed and computed values may well be real. If so, it probably implies either that the vertical eddy diffusivity in the atmosphere above the surface boundary layer is considerably larger than was assumed in these computations or that the rainout and washout rates have been assumed too large. The latter can, however, hardly be much less since a considerably smaller value for β would by necessity yield much more horizontal dispersion before rainout and thus a less satisfactory agreement between computed and observed patterns of wet deposition (Figures 5 and 7). It seems hardly likely that $\beta < 0.01$/hr (cf. Bolin and Persson, 1975).

6. Extensive measurements of sulfur deposition by rain around an oil-fired power plant have recently been completed in Sweden (Granat, 1975). The result of these observations indicate that rainout and washout are not very efficient sink mechanisms if the emissions from the stack contain few particles. Within a circle of 35 km only 2 to 5 percent of the emission during rain were recovered in the rain that fell in the environments and not more than twice these values were found out to a distance of 60 km from the source. With prevailing winds, this corresponds to a removal of merely a few percent during the first hour or two, that is, a rate <u>during rain</u> of only about 0.02 to 0.05/hr. In view of the fact that rain occurs only during about 10 percent of the time, the removal efficiency of "clean" SO_2 seems to be quite small. It should be remarked that the pH of the rainwater in the seven cases of rain studied varied between 4.0 and 4.6 and that rainout and washout may be more efficient mechanisms when the rain water is less acid. Statistics compiled for southern Sweden shows a medium pH value of about 4.4 and the cases investigated may be considered as rather close to normal conditions in this part of Europe.

Högström (1973) on the other hand, has maintained that rainout and washout are quite efficient removal mechanisms for sulfur dioxide and arrived at a rate of about 0.2/hr. It should be emphasized that he studied conditions around the city of Uppsala and that this particulate matter might have been playing a more important role in the transformation of sulfur dioxide into sulfate in the raindrops. Also rain in the area of Uppsala usually is somewhat less acid, but no determinations of pH are available from the cases that were studied. Högström's interpretation is, however, also critically dependent on the background concentrations of sulfate in the rain that were assumed to prevail. As is clear from the work by Granat (1975), Högström's results probably should be considered as an upper limit for the efficiency of removal by rain.

In the light of the foregoing discussion, noting again the difference between computed and observed concentrations in air as revealed by Figure 8, and recalling that the average duration of rain (in south Sweden) is 5 to 10 hours (Rodhe and Grandell, 1972), one cannot escape the conclusion that 10 to 20 percent of the anthropogenic emissions, perhaps even more, may escape removal by rainout or dry deposition during the 3 to 4 first days after emission and that most of this sulfur then may be found well above the boundary layer. In view of the fact that rain clouds to a considerable degree are supplied by air through horizontal convergence in the lowest layers of the atmosphere, the likelihood of wet removal thereafter may have diminished considerably.

There is one other circumstance that also seems to indicate that the lifetime of sulfur, having excaped the removal in the lowest layers of the atmosphere may be quite long. Junge (1974b) has considered the physical-chemical processes that create a sulfate layer in the stratosphere and finds that a sulfur dioxide mixing ratio at the tropopause of about 10^{-10} is required and that the flux of sulfur dioxide into the stratosphere due to turbulent motions needs to be about 25,000 ton/yr. These values must, of course, be compatible with concentrations and vertical fluxes in the troposphere. According to Friend (1973), the annual input of sulfur dioxide into the atmosphere due to natural processes (including the oxidation of hydrogen sulfide) and combustion of oil and coal is about 170 Mton/yr. This figure is quite uncertain, but the most likely value is between 125 and 225 Mton/yr. A very rough idea of what internal consistency implies in this regard may be obtained by considering the one-dimensional diffusion as employed by Junge (1974b).

$$\frac{\partial}{\partial z}\left(\rho D \frac{\partial m}{\partial z}\right) - \beta \rho m = 0 \qquad (2)$$

where ρ is the air density, D is the vertical eddy diffusivity coefficient, and β is the rate of removal, and assume that

$$\rho = \rho_o e^{-z/H} \tag{3}$$

where H is the scale height. The solution of this equation is

$$m = m_1 e^{-z/H_1} + m_2 e^{-z/H_2} \tag{4}$$

in which expression m_1 and m_2 are determined by satisfying two boundary conditions. We may specify the mixing ratio m or the flux $\rho D\, \partial m/\partial z$ at the lower boundary ($z = 0$) or at the tropopause ($z = z_T$). More detailed computations of this kind will be described elsewhere. It suffices to state here that if $D \leq 20$ $m^2 sec^{-1}$, we must have $\beta^{-1} > 6$ to 10 days in order to have a ratio of $m(z_T)/m(0) \geq 0.05$, which seems a rather likely value according to Junge (1974b). It should be remarked, however, that we have implicitly assumed that the flux of sulfur from the earth's surface to the tropopause is in the form of SO_2, which need not necessarily be the case. Thus Crutzen has advanced the hypothesis (personal communication) that sulfur may exist in a much more inert form (for example CS_2) and that the sulfur flux into the stratosphere, which is necessary to maintain the sulfur layer, takes place in the form of such a gas flux.

It is clear from the preceding discussion that even though the computations of regional dispersion as described give us a further insight into the sulfur cycle, considerable uncertainties still exist. Accurate and representative measurements of SO_2 (and sulfate) in the atmosphere are needed in order to eliminate those uncertainties. It seems particularly important to obtain values from those parts of the atmosphere, where the industrial influence still is quite small, that is, the Antartica and the southern hemisphere in general, the tropics, and the upper troposphere all over the globe. Already a limited number of analyses would be of great significance.

REFERENCES

1. Bolin, B. and Keeling, C. D. 1963. Large Scale Atmospheric Mixing as Deduced from Seasonal and Meridional Variations of Carbon Dioxide, J. Geophys. Res. 68, 3899-3920.

2. Bolin, B. and Persson, Ch. 1975. Regional Dispersion and Deposition of Atmospheric Pollutants with Particular Application to Sulfur Pollution over Western Europe, Tellus 27(3).

3. Friend, J. P. 1973. The Global Sulfur Cycle. In Chemistry of the Lower Atmosphere (I. Rasco, ed.), pp. 177-201, Plenum, New York.

4. Granat, L. 1975. Atmospheric Deposition Due to Long and Short Distance Sources—With Special Reference to Wet and Dry Deposition of Sulfur Compounds Around an Oil-Fired Power Plant, Met. Inst. Univ. Stockholm, Report AC-32.

5. Högström, U. 1973. Residence Time of Sulfurous Air Pollutions from a Local Source During Precipitation, Ambio 2, 37-41.

6. Junge, C. E. 1974a. Residence Time and Variability of Tropospheric Trace Gases, Tellus 26, 477-488.

7. Junge, C. E. 1974b. Sulfur Budget of the Stratospheric Aerosol Layer. Proc. Int. Conf. on Structure, Composition, and General Circulation of the Upper and Lower Atmospheres and Possible Anthropogenic Perturbations, IUGG. Atmospheric Environmental Service, Ottawa, Canada.

8. Jost, D. 1974. Aerological Studies of the Atmospheric Sulfur Budget, Tellus 26, 206-212.

9. Rodhe, H. 1972. Measurements of Sulfur in the Free Atmosphere over Sweden 1969-1970, J. Geophys. Res. 77, 4494-4499.

10. Rodhe, H. and Grandell, J. 1972. On the Removal Time of Aerosol Particles from the Atmosphere by Precipitation, Tellus 24, 442-454.

11. Strömsöe, S. 1973. The LRTAP Emission Survey. Cooperative Technical Programme to Measure the Long-Range Transport of Air Pollutants, Norwegian Air Research Institute, Oslo, pp. 250-260.

The Formation of Organic Aerosols in a Fast Flow Reactor

MARTIN LIPELES, DWIGHT A. LANDIS,* and GEORGE M. HIDY*
Rockwell International Science Center
Thousand Oaks, California

I. INTRODUCTION

The presence of haze continues to be one of the most important manifestations of air pollution in urban areas. Haze in urban areas has been subject to increasing study in the past several years. Some of the work has included observational programs in the field (Hidy et al., 1974), while in other work laboratory experiments have been undertaken to study the formation mechanism of these aerosols (Wilson et al., 1972; Groblicki and Nebel, 1971; Kocmond et al., 1973). Although considerable progress has been made, the details of mechanisms of formation and growth of airborne particles in polluted atmospheres remains poorly understood. This chapter presents a novel approach to

*Present address: Environmental Research and Technology, Inc., Western Technical Center, 741 Lakefield Road, Westlake Village, California, 91361.

studying mechanisms of aerosol formation in the atmosphere. Several important results from this method are presented.

The flow reactor employed maintains laminar flow in an irradiated section with a residence time in the neighborhood of 10 seconds. The object of using a laminar flow reactor is to investigate aerosol-forming reactions under controlled, steady flow while minimizing the interaction of the walls with the reacting gases. The philosophy of this method has been described previously in the contract report for the first part of this effort (Lipeles et al., 1973).

With the flow reactor, an experimental program to study condensation nuclei formation was carried out with eight different hydrocarbons, including 1-hexene, cyclohexene, toluene, m-xylene, 1-dodecene, 1,3-hexadiene, 1,5-hexadiene, and α-pinene. With the exception of toluene, condensation nuclei were found for all hydrocarbons, their concentration depending upon which hydrocarbon is being used and its concentration. The effect of each hydrocarbon on the ozone production in the reactor was measured as a function of hydrocarbon concentration. In addition to these experiments, several systems were studied with various combinations of SO_2, NH_3, and water vapor added to the reactant mixture.

The most striking feature of the data was that each system produced different results, although some consistent patterns could be discerned. In each case, there was a minimum hydrocarbon concentration at which production of nuclei increased dramatically. In addition, there is a plateau region where the nuclei concentration changes little for increasing hydrocarbon concentration. Over the range of hydrocarbon concentrations studied, however, the ozone production varied considerably. In some cases, the ozone production increased over the entire span of hydrocarbon variation. In other cases, ozone reached a peak at high hydrocarbon concentration or decreased over most of the hydrocarbon concentrations examined. In all cases where SO_2 was added, the condensation nuclei production increases; however, the concentrations levels at which this was achieved varied from as low as 6 ppb to as high as 100 ppb. The introduction of ammonia at 5 to 10 ppm concentration increased the condensation nuclei number in each case it was tried. On the other hand, for the two olefins with which it was added, water vapor at 50% relative humidity decreased or eliminated aerosol formation.

For many of the cases studied, aerosol samples were collected on filters. Typically, 30 to 125 μg of condensed material were collected on these filters. For three of these cases—aerosols produced with α-pinene, aerosols produced with 1-dodecene, and aerosols produced with 1-hexene and ammonia—infrared spectra were obtained from samples extracted from these filters. These spectra show the presence of carbonyl compounds and possible C—O linkages as well as possible organic nitrates.

II. APPROACH OF THE FLOW REACTOR EXPERIMENT

The development of a technique to improve the study of the formation and composition of photochemical aerosols has been initiated (Lipeles et al., 1973). The most important feature of this technique is the use of a flow reactor to eliminate the walls from affecting the reacting gases. In this way we also gain reproducibility without preconditioning the reactor. The other important feature is to employ every practicable method to achieve a total material balance for the aerosol particles and for the gas phase. We carried out this approach for simple cases with, for example, only one hydrocarbon present.

In order for the reactor to be truly wall free, it must satisfy certain criteria. It must be true that no material produced at the walls is collected or detected as a reaction product. It is also necessary that no material, either initial or produced in the reaction, is lost to the walls of the reactor, including both products and intermediates. This last point is important in that the loss or destruction of an intermediate at the walls may change the ultimate products. This leads to the last criterion, which is that no material produced at the walls react in the reactor to remove any key component of the system.

The satisfying of these criteria for a flowing system, however, requires a very short residence time compared with the expected overall reaction times. For example, for the tube used in this study only 10 seconds of residence time are allowed for meeting these criteria. The potential solution to these problems in studying the aerosol formation and growth involves two types of what we shall refer to as time scaling. These are now defined.

1. Differential time scaling is the stepping to a later time in a complex reaction by setting all reactant intermediate and product concentrations to those that will pertain at the selected time.

2. Total time scaling is the adjusting of the initial conditions (concentrations, light intensity, temperature, tec.) of a complex reaction so that the overall reactions will take place at a faster

rate without changing the final products, though possibly changing their relative final concentrations.

Our study employs both of these methods simultaneously and their achievements are now discussed. The differential scaling involves no intrinsic error provided that all of the concentrations of reactants, intermediates, and products can be reproduced and achieved; once this has been achieved the reaction proceeds just as it would have if the induction period that was skipped had actually taken place. There are two problems with this method, however. The rate of reaction for the remaining portion of the reaction scheme may still be too slow for the time available. If this deviates only by a small factor, then a study of the period of time available will allow one to then proceed differentially to another point in time. If the complete reaction takes place over only a few such steps, the reaction mechanism may be studied and understood over this period. However, if this is not possible, one may further resort to total time scaling superimposed on differential scaling. The other problem of differential scaling is that it may be difficult to determine the exact conditions at a later time in a complex reaction. Toward this end, we are helped by the fact that most intermediates come into equilibrium rapidly and, therefore, do not need to be added so only the principal reactants and products need be considered.

Total time scaling is more complex and does not hold in general. It should be immediately clear, however, that any complex set of reactions that have a rate-limiting step of nonzero order in any reactant may be scaled by increasing the concentration of that reactant. If the rate limiting step depends on several reactants, then higher-order scaling is possible up to the order of the rate-limiting step. For example, if the rate-limiting step is third order and each reactant in this step is scaled by a factor of m, then the rate scales by a factor of m^3. Of course, the reactants in the rate-limiting step are not necessarily the initial reactants, but if the rate-limiting reactants depend in some simple way upon the initial reactants, then the scaling can be related to the initial reactants.

Since no complete model exists for photochemical aerosol production, it is difficult to work out possible scaling relations for aerosols. However, it is known that aerosols do not form until after most of the NO has been converted to NO_2 and there is no appreciable O_3 concentration. Thus in our experiments we apply the concepts of differential and total time scaling as a first approach to the problem keeping in mind the limitations of these concepts. The concept of differential time scaling is applied by starting with NO_2-hydrocarbon system rather than a NO-hydrocarbon system. Since O_3 reactions with olefins produce aerosols and O reactions with olefins in the presence of O_2 should have similar chemistry, the concept of total time scaling is attempted by increasing the NO_2 concentration and light intensity (thereby increasing the O atom and O_3 concentration) and by increasing the hydrocarbon concentration from the atmospheric concentration values.

In the appendix is presented the quantitative justification, on the basis of molecular diffusion, for the selection of residence time in the reactor.

III. EXPERIMENTAL PROCEDURES

Figure 1 shows a diagram of the flow reactor. It consists of a main 2.4-m section, a 15-cm probe section, a 1.2-m outlet section, and a gas input system. The main reactor section is composed of two concentric Pyrex tubes. The inner tube is 15-cm I.D. and the outer one is 18.8-m O.D. The outer tube is about 20 cm shorter than the inner. They are held in place by a plastic ring at each end to which they are sealed with RIV Silastic cement. Water is admitted to the space between them through the plastic rings.

The illuminated section is nominally 20-cm long and is located approximately 150 cm from where the gas mixture enters the tube. This illumination is provided by three Hanovia 674-A medium pressure mercury vapor lamps, axially arranged in an aluminum housing. The light intensity in the reactor produces a k_d for NO_2 of 4.8 min^{-1}.

The input end of the reactor is reduced to 12-cm O.D. This occurs beyond the water jacket. The gas mixture is introduced at this end through four 0.6-cm holes in a stainless steel flange. Before reaching this flange, the gas is equilibrated to the reactor temperature by passing through a spiral condenser in whose water jacket the reactor cooling water flows.

For all of the experiments described here, a fixed probe was employed. It was 1 cm in diameter and 6 cm beyond the end of the illuminated section. The proble then passes out of the reactor through the wall of the 15-cm section. This section is followed by the 1.2-m end section, the purpose of which is to keep end disturbances away from the probe. The output of this last section flows into a conical hood.

The gases are mixed in a stainless steel manifold followed by a long mixing tube leading to the condenser used for temperature adjustment. The gases for this mixture come from several

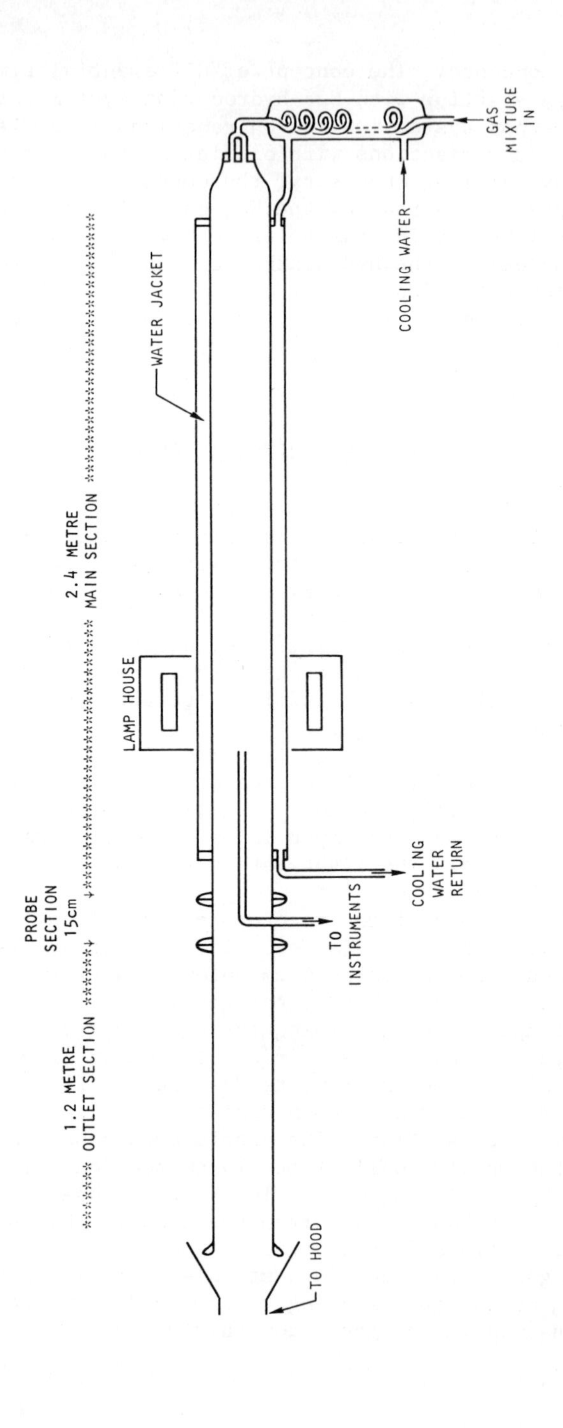

FIGURE 1. Schematic diagram of the flow reactor.

sources. The air is made up from our building supply of dry nitrogen and oxygen. The NO_2, SO_2, NH_3, and 1-hexene are obtained in tanks as dilute mixtures with nitrogen. Except for 1-hexene, all hydrocarbons were added by saturating dry nitrogen at a controlled temperature, either in an ice bath or in a regulated constant temperature bath. Water vapor was added by diverting some of the nitrogen flow through a fritted saturator.

For each hydrocarbon considered, the production of condensation nuclei was investigated over a range of hydrocarbon concentrations beginning with a few ppm. The onset of nuclei formation varied with hydrocarbon, but once it was observed, the concentration of hydrocarbon was then increased gradually. At each concentration, a period of time from 5 to 30 minutes was allowed before measurement to insure a stable valid value. In those cases, as we describe later in which no stable value was obtained in 30 minutes, then an average value over the next 5 minutes was recorded. After each condensation nuclei reading the probe output was switched for an ozone reading

The condensation nuclei were measured with an Environment One condensation nuclei counter, and ozone was measured by a REM chemiluminescent ozone monitor. The probe flow output passed through a three-way valve so that one could switch from one instrument to the other. The nuclei counter required a flow of 2.1 liter/min, while the REM required 1 liter/min. Both of these flows were in large excess over the isokinetic sampling flow, and all of the data was collected with a simple 9-mm diameter probe. This was not expected to be a problem for particles as small as ours; however, we do attribute some instabilities in formation of particles for some cases to the flow instability caused by the probe.

The aerosol samples were collected on 25-mm Gelman Type A glass fiber filters in a standard millipore filter holder. The filters were cleaned and dried by baking them at least overnight at 425°C. A regulated flow of 3 liter/min was sampled from the probe through the filter for 16 hours (overnight). The level of the condensation nuclei was checked before and after the filter collections and found to be constant. The filters were weighed before and after collection on a Mettler Type M microbalance.

IV. RESULTS

One of the most striking aspects of the observations made in this program is the great variation in results from one hydrocarbon to another. This is particularly evident in the ozone production as a function of hydrocarbon concentration.

These data are not directly aerosol measurements; however, we will argue that there is a connection between ozone and aerosol production.

In Figure 2, the data on ozone production, as a function of hydrocarbon concentration for six different hydrocarbons, are shown. We immediately see that photolysis of the initial NO_2

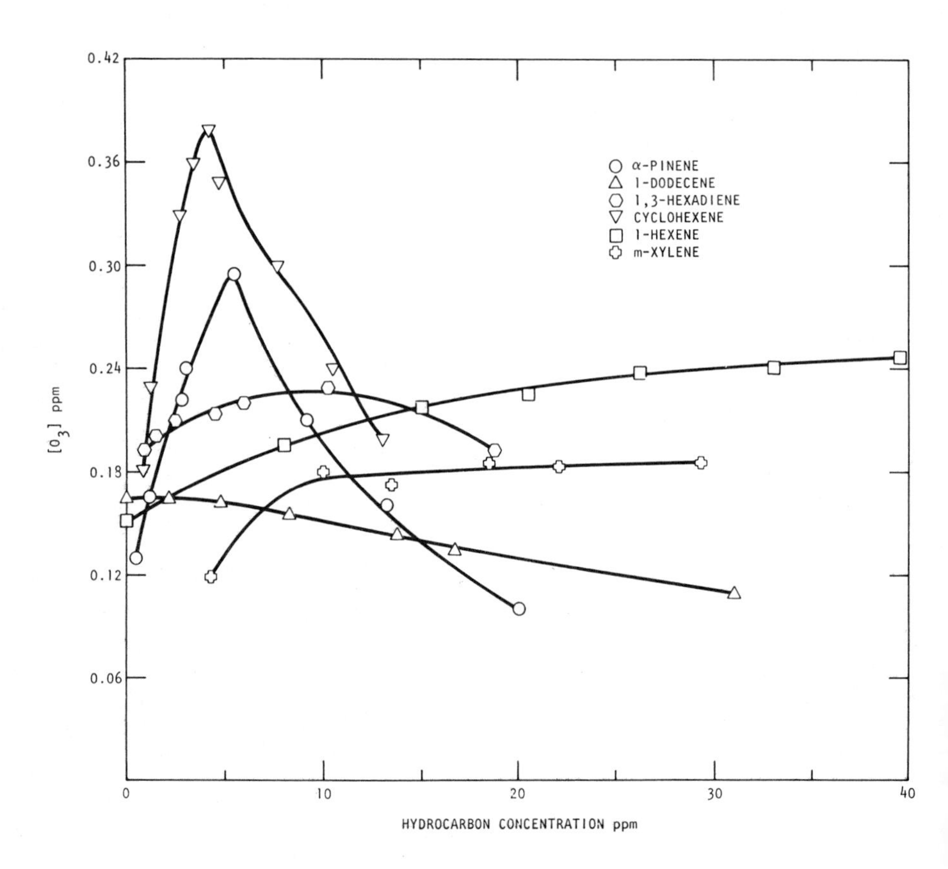

FIGURE 2. Ozone production as a function of hydrocarbon concentration for various hydrocarbons irradiated in dry air containing 0.8 ppm NO_2.

cannot account for all of the ozone production since in several cases the ozone concentration increases with increasing hydrocarbon concentration. Thus in 10 seconds of residence time, sufficient photochemistry is taking place to dramatically increase ozone production in some cases. It is also clear that as hydrocarbon concentration increases the ultimate result is a decrease in ozone production over a fixed residence time for the majority of cases studied.

In Figure 3, the nuclei production data for all runs with a hydrocarbon, NO_2, and dry air are shown. From this figure we immediately see the pattern of the data that, except for m-xylene is a family of parallel curves all exhibiting the same rapid rise with hydrocarbon concentration followed by a plateau region. The pattern for m-xylene is similar, but its curve is not parallel to the other curves.

In the cases of cyclohexene and 1-hexene, the aerosol production was not stable as a function of time. If the reactor was operated with no flow through, the probe and then the probe flow was started through the condensation nuclei counter, a reading was obtained that gradually decreased to zero over a several-minute time span. If at this point the probe flow was turned off and left off and the procedure repeated 0.5 hour later, the aerosol count would return. In the case of 1-hexene, this effect was repeatable and the data shown in Figure 3 represent the peak values obtained in this way. With cyclohexene, the data were not reproducible and no measurement was taken. For several of the other hydrocarbons, a similar behavior was observed below a threshold; but once a critical concentration was reached, the measurements were stable. The initial points in Figure 3 reflect this critical concentration.

Small amounts of SO_2 were added with three different hydrocarbons present, m-xylene, 1-dodecene, and 1,3-hexadiene. In all three cases, an increase in condensation nuclei was observed. In the cases of 1-dodecene and 1,3-hexadiene, only a few parts per billion of SO_2 was required to produce almost an order of magnitude increase. However, with m-xylene, 0.1 ppm was required to produce a small increase in nuclei production.

The addition of NH_3 to three different hydrocarbons also increased the condensation nuclei formed in each case. For m-xylene the increase is only 50 percent, while for 1-hexene it reaches a factor of seven, and for 1-dodecene there was an order of magnitude increase. In each case, this occurs in the range of 3 to 10 ppm of ammonia.

The five cases for which water vapor was added are summarized in Table 1. In all these cases water vapor was added at approximately 50 percent relative humidity, and no measurements were made at other water vapor concentrations.

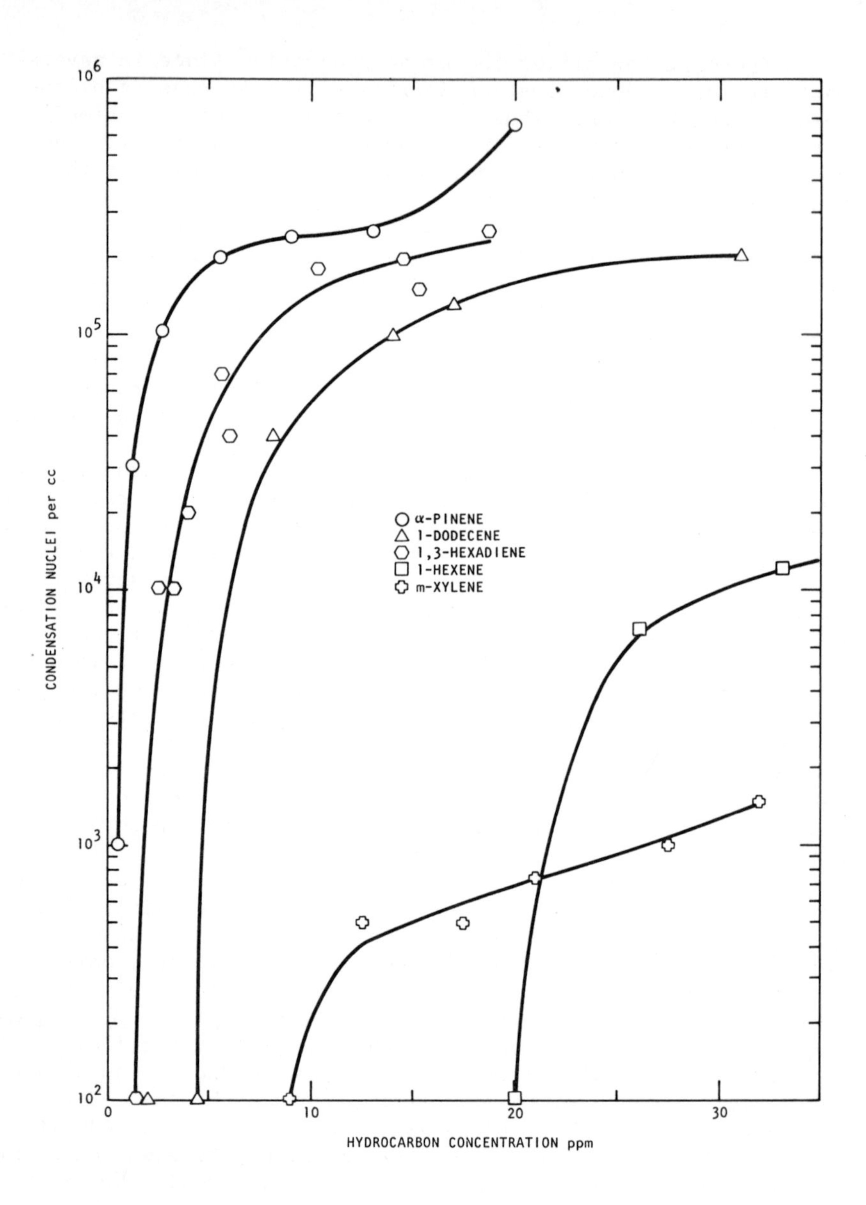

FIGURE 3. Condensation nuclei production as a function of hydrocarbon concentration for various hydrocarbons irradiated in dry air containing 0.8 ppm NO_2.

TABLE 1. Data on the Effect of Water Vapor on Aerosol Formation in Dry Air and in Air at 50 Percent Relative Humidity

	Concentration of Reactants			Products			
				Dry		With Water Vapor	
Hydrocarbon	Hydrocarbon (ppm)	NH_3 (ppm)	SO_2 (ppb)	O_3 (ppm)	Nuclei (per cm^3)	O_3 (ppm)	Nuclei (per cm^3)
1-Dodecene	32	---	---	0.13	2×10^5	0.13	ND[a]
	33.5	2	---	0.13	3×10^5	0.14	2×10^4
	32	---	4	NM[b]	5×10^6	NM	1×10^4
	43.3	2	6	NM	$>10^7$	NM	ND
1-Hexene	18.9	---	---	0.19	2×10^3	0.19	ND
	18.9	1.3	---	0.19	6×10^4	0.19	ND

[a]Not detected.

[b]Not measured.

Also, only a few single experiments were performed for each case. In all cases the presence of water vapor at 50 percent relative humidity decreased or eliminated aerosol production. In those cases in which aerosols were present, there was an instability in the condensation nuclei number.

For five of the hydrocarbons and eight cases, filter collections of aerosols were made to determine the actual mass concentration of material produced. These collections are summarized in Table 2. There are several significant facts to be observed in these results. There is not a direct correlation between the number of nuclei formed and the weight gain observed. This is in fact expected due to differences in growth and nucleation rates in the different systems. However, we found a large variation from sample to sample for each hydrocarbon. The largest variation was found with 1-dodecene where the average weight gain was 81.8 μg with an average deviation of 25 μg. All others were much better than this, but there is clear indication that stability of the reactor system over long periods of time must be explored in future work.

Some significant observations are that adding SO_2 to 1-dodecene increased the nuclei count by almost an order of magnitude but did not increase the weight gain. This is discussed later. Also observed but not understood, with NH_3 and SO_2 present along with 1-dodecene, nuclei counts are extremely high, typically $10^7/cm^3$, but no material was found on the filters. In three cases infrared spectra were obtained for solvent extracts of the filters. These are presented in Figures 4 through 6.

The spectrum in Figure 4 was obtained from the α-pinene aerosol sample and is shown together with a spectrum obtained by Wilson, Schwartz, and Kinzer (1972) at Battelle. The general features of these two spectra are strikingly similar. The absorption in our spectrum, near 3600 cm^{-1} is identified as predominantly due to contamination by water in the extraction process, which used carbon tetrachloride as a solvent.

The characteristic C—H absorptions due to stretch around 2900 cm^{-1} are present, although the peak at 2780 is somewhat low. The bending modes at 1400 and 1335 are also present, but the CH_2 scissoring at 1465 cm^{-1}, if present at all, is not clear. The sharpest peak is the carbonyl stretch at 1725 cm^{-1}, which may have a shoulder at 1745 cm^{-1}. Another important feature is the broad peak centered at 1100 cm^{-1}. This is most likely a C—O or O—O type absorption, but it is too broad to be definitely identified. The sharp peak at 2330 cm^{-1} is also present in the

TABLE 2. Filter Collections of Aerosol Samples[a]

Hydrocarbon	Hydrocarbon concentration (ppm)	CNC (per cm^3)	Weight gain (μg)	Mass Concentration ($\mu g/m^3$)
1-Hexene	50	4×10^4	30	10.4
	50	4×10^4	54	18.8
1-Hexene + NH_3	50	5×10^4	45	16.6
	50	5×10^4	65	22.6
1-Dodecene	15	10^5	63	21.9
	15	10^5	74	25.7
	15	10^5	91	31.6
	15	1.8×10^5	140	48.6
	15	3×10^5	41	14.2
1-Dodecene + SO_2	12	10^6	91	31.6
1-Dodecene + SO_2 + NH_3	12	10^7	2	--
	12	10^7	0	--
1,5-Hexadiene	70	3×10^4	38	13.2
1,3-Hexadiene	15	2×10^5	55	19.1
	15	2×10^5	53	18.4
α-Pinene	20	7×10^5	216	75.0

[a] 16-hour collections at 3 liter/min: for all cases $[NO_2]$ = 0.8 ppm; where present $[SO_2]$ = 5.7 ppb; $[NH_3]$ = 2.0 ppm.

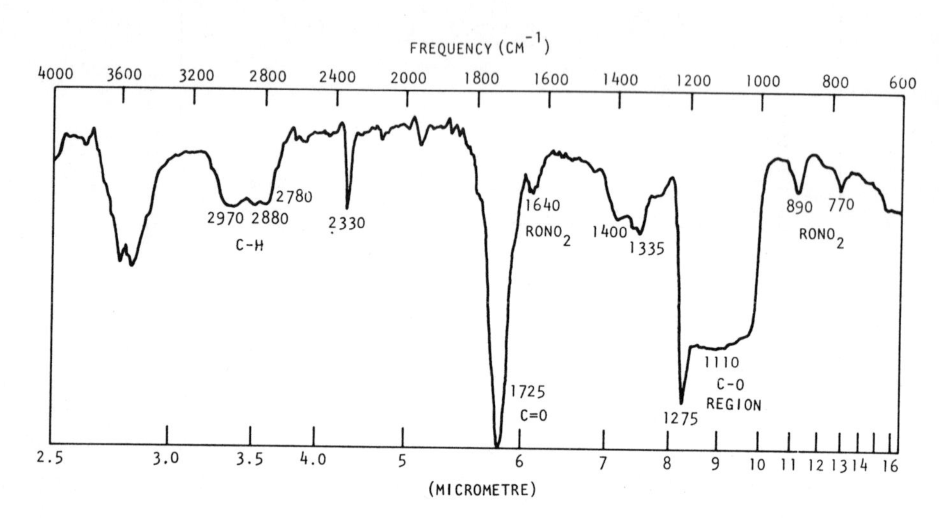

AEROSOL PRODUCED BY IRRADIATING α-PINENE AND NO_2 IN DRY AIR

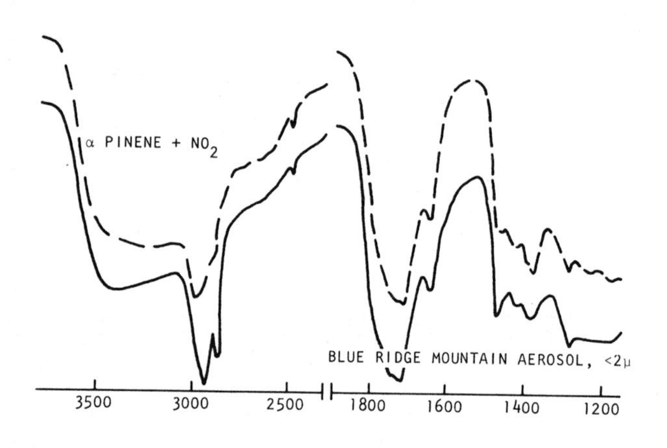

IR SPECTRUM FROM BATTELLE STUDY

FIGURE 4. Infrared spectrum of aerosol produced by irradiating α-pinene and NO_2 in dry air. (Top) frequency (cm^{-1}) compared with similar results from Wilson, Schwartz, and Kinzer (1972) at Battelle.

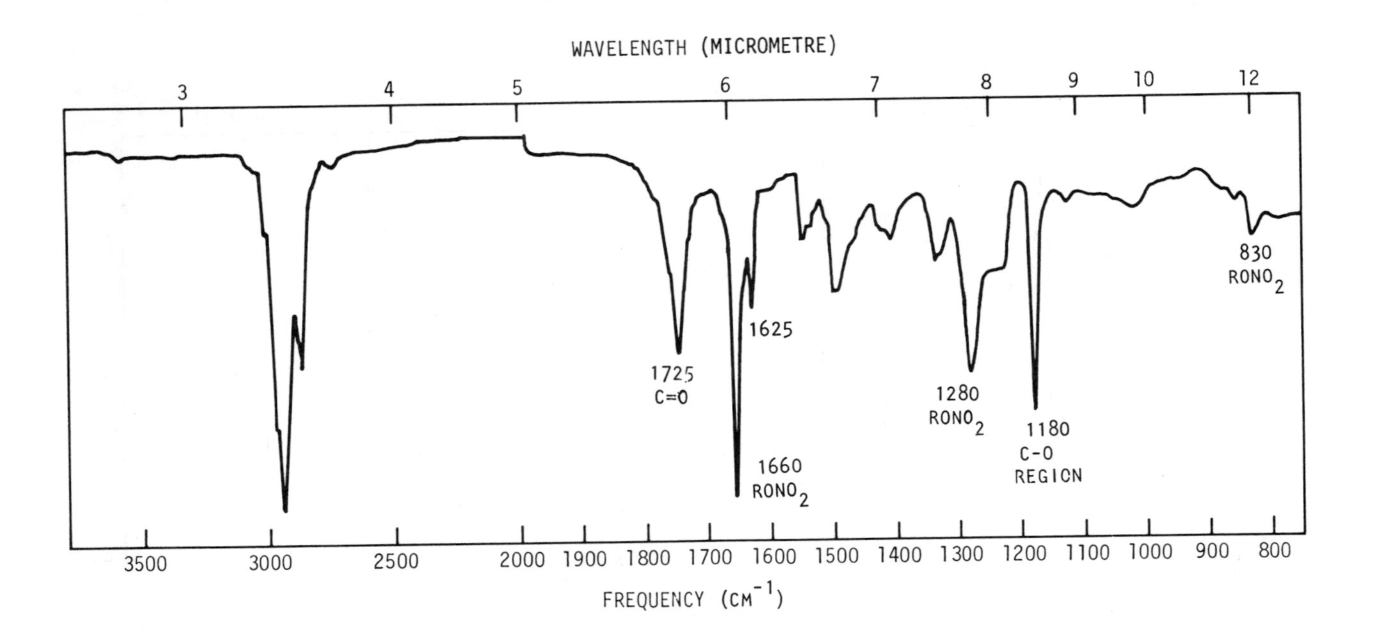

FIGURE 5. Infrared spectrum of aerosol produced by irradiating 15 ppm of 1-dodecene and 0.8 ppm of NO_2 in dry air.

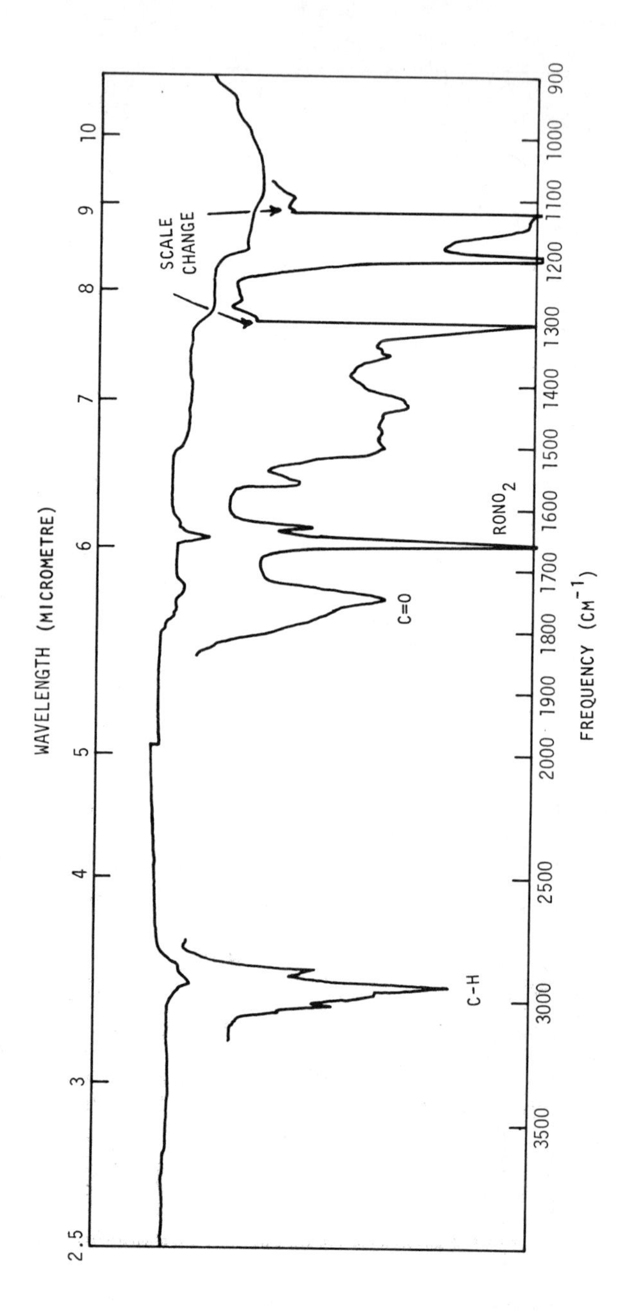

FIGURE 6. Infrared spectrum of aerosol produced by irradiating 50 ppm of 1-hexene, 0.8 ppm of NO_2, and 2 ppm of NH_3 in dry air.

Battelle spectrum but lies in a region with a few absorptions. The Battelle group identified the 1640 band as either a C=C or an organic nitrate. If this is in fact the asymmetric stretch of the NO_2 group, then the symmetric stretch, the NO stretch, and the NO_2 bending should be found. The symmetric stretch may be on the shoulder of 1335 cm^{-1} peak and the 890 cm^{-1} and 770 cm^{-1} could be the N—O stretch and the NO_2 bending, although both are slightly higher in frequency than would be expected. This nitrate identification must, as a result, be regarded as very tentative and if present not a major constituent.

Figure 5 shows the spectrum obtained with aerosol produced from 1-dodecene. There are some major differences between this spectrum and that from the α-pinene. The C—O band at 1725 cm^{-1} is not as strong even as the C—H stretch bands. The strength of this component, along with the feature at 1280 cm^{-1}, suggests the asymmetric and symmetric nitrate groups, and the single band N—O stretch could be the band at 830 cm^{-1}. This is the most likely analysis of this portion of the spectrum. However, this does not include the 1625 cm^{-1} band, which we also see in Figure 6 associated with the 1660 cm^{-1} band. Once again there is a strong absorption in the C—O region, but this band at 1180 cm^{-1} is surprisingly sharp.

The spectrum shown in Figure 6 was obtained with aerosol from 1-hexene and NH_3. It is very similar from 1600 cm^{-1} to 3500 cm^{-1} to the 1-dodecene case. The remainder is marked by a very diffuse absorption with some detail superimposed. It is difficult to say any more about this spectrum than the 1-dodecene case. Both these spectra are marked, however, by a conspicuous absence of an OH absorption. There does not appear to be any group originating from the presence of NH_3, that is, NH_4^+ or an amino group.

V. DISCUSSION

Ultimately, the goal of this research is to determine how gaseous organic material in the atmosphere becomes part of aerosol material. The emphasis has been on building a technique with which to approach this problem. The important questions to be asked are: What is the material in the aerosols; What are the precursor molecules in the atmosphere; What are the pathways from precursors to aerosols; and What are the kinetics of the process? Each of the aspects of the data and observations of this program is now discussed with these questions in mind as well as the general question of the validity of the technique.

It is probable that the ozone production is related to hydrocarbon reactivity, but we do not yet have enough cases to test such a correlation. In the three cases where a drop occurs, we have attempted to fit the decrease in ozone with increasing hydrocarbon concentration to a simple first-order mechanism. The result is that, except for the last three points in the case of α-pinene, the decrease in ozone production does not fit a simple first-order removal. The complete explanation of these effects requires further study. However, the conçlusion that photochemical reactions involving the hydrocarbons are taking place, is important in justifying the validity of the flow reactor concept with short residence times.

It is useful to make a model to know whether one is dealing with homogeneous nucleation of a single component or binary nucleation or some more complex phenomena. However, it is possible to show that the condensation nuclei production data is qualitatively consistent with a simple model. If we assume that we are dealing with the homogeneous nucleation of a single component, we may write an expression for the rate of formation of nuclei (Byers, 1965)

$$J = AS^2 e^{-B/\ln^2 S}$$

where S is the supersaturation ratio for the nucleating component (the ratio of the actual vapor pressure to the equilibrium vapor pressure) and A is a function of the density of the liquid in the drop, its surface tension, its equilibrium vapor pressure for a flat surface, its molecular weight, and the temperature, while B is a function of the surface tension, the density, the molecular weight, and the temperature. If we further assume that the nuclei concentration is linear in the production rate and the supersaturation ratio is linearly proportional to the initial hydrocarbon concentration, then we may use this equation to make qualitative predictions about the condensation nuclei concentration. The rapid rise in nuclei concentration as a function of hydrocarbon concentration occurs when the exponential term dominates. However, as S gets larger the rate of change of the exponential term shows and eventually the nuclei production rate is dominated by the AS^2 factor. Unfortunately, A, B, and S are determined in part by properties of the nucleating substance, which in this case is unidentified, that makes quantitative comparison difficult. We feel that the form of our nuclei data is consistent with homogeneous nucleation. The form of data shown in Figure 3 may be fitted by selected values of A and B and a relation between S and the initial hydrocarbon concentrations, but it remains to be determined whether a set of parameters may be selected that is also consistent with other physical and chemical knowledge of the system. We suspect that the onset of the plateau is too rapid to be accounted for by coagulation. Other explanations will have to include instrument artifacts.

As we saw before, the average weight gain of the filter when sampling 1-dodecene, NO_2, and dry air is 81.8 μg for approximately 10^5 particles/m^3. In 16 hours at 3 liter/min, 2.9×10^{11} particles would be collected and for 81.8 μg, which is 2.8×10^{-16} g/particle. Assuming the density of the condensed material of 1 gm/m^3 we find a particle average radius of 0.04 μm. This is a reasonable value in comparison to atmospheric size ranges. It is interesting that the addition of SO_2, for the case of 1-dodecene, produced a large increase in the condensation nuclei, but as we see from the data of Table 2 there is no proportional increase in weight gain. Possibly in the presence of SO_2 a new condensable material (e.g., H_2SO_4) is produced that has a smaller critical nucleus and hence a higher rate of nuclei production. The bulk of the material should remain similar to that obtained without SO_2; further data are required to establish this and eliminate other conceivable explanations of this observation.

Another effect that must be explained is that of instabilities in aerosol production in the cases of 1-hexene and cyclohexene. This may be related to the stirring effect observed by Wilson (1971). He reported that once formed, aerosols are not appreciably removed by stirring, but their formation is quenched by stirring. The sampling probe in our study was not operating isokinetically during measurements of condensation nuclei, so some flow shear was present at the probe inlet. If this instability is sufficient to quench aerosol production (as the stirring does) and if the nuclei are not formed until they reach the vicinity of the probe, their formation will be quenched by the operation of the probe. Furthermore, the time scale for the onset of the quenching will not be determined by the average flow but rather by the rate at which the flow instability grows after the initial disturbance. These time constants were found to be large in flow visualization measurements taken in the reactor. If, on the other hand, the aerosols are produced before the flow disturbed-region, they are probably not affected significantly by the flow-disturbed region.

A promising aspect of the present work is the ability to collect and analyze samples. The three infrared spectra presented here are only a preliminary view of what can eventually be done. The most significant aspect of our spectra is that there is no strong acid component identified (in the α-pinene case, this interpretation is not as certain); however, C=O is definitely present, and some C—O linkage as well as an organic nitrate is consistent with the results. The only limit on mechanistic models to fit such simple facts is one's imagination; however, starting from a recently proposed gas phase, ozone-olefin reaction mechanism of O'Neal and Blumstein (1973), one can construct a plausible hypothesis. They suggest that after the initial attack on the double bond by ozone, the molozonide formed, instead of undergoing a Criegie split, becomes a biradical:

```
R1       H               R1         R2              R1        R2
  \     /                  \       /                  \      /
   C=C        + O3  ⇌      HC  -  CH          ⇌        HC  -  CH
  /     \                   |      |                   |      |
H        R2                 O      O                   O      O
                             \    /                    •       \
                               O                                O
                                                                •
                          (Molozonide)               (Biradical)
```

One possible path at this point that O'Neal and Blumstein suggest is the abstraction of hydrogen by the two-oxygen chain from the carbon with one oxygen (α-hydrogen abstraction). One possible result of this is

```
      •
      O      R2                          O     R2
       \    /                            ||   /
  R1 — C  -  CH      ———————→     R1 —  C  -  CH      + OH
       |     |                                |
       H     O                                O
            /                                 •
          .O
```

Radicals such as these, when terminated by an NO_2, would produce the features we see in the infrared spectra. A few steps of polymerization with the olefin present are also possible, leading to heavier, more condensable species. Another possible product from such a rough mechanism is a peroxyacylnitrate, which also has all the required functional groups.

The fact that mechanisms can be constructed to fit the infrared data that involve ozone focuses on ozone as a key in forming aerosols with olefins and NO_2 present in air. This has been discussed before, for example by Groblicki and Nebel (1971). By juxtaposing some of our condensation nuclei data with atmospheric data, we obtain some further evidence for the importance of ozone-olefin reactions in aerosol production. Using data in the α-pinene case obtained by varying $[NO_2]$ and plotting condensation nuclei versus ozone produced, we obtain in Figure 7 a strong correlation between condensation nuclei and ozone concen-

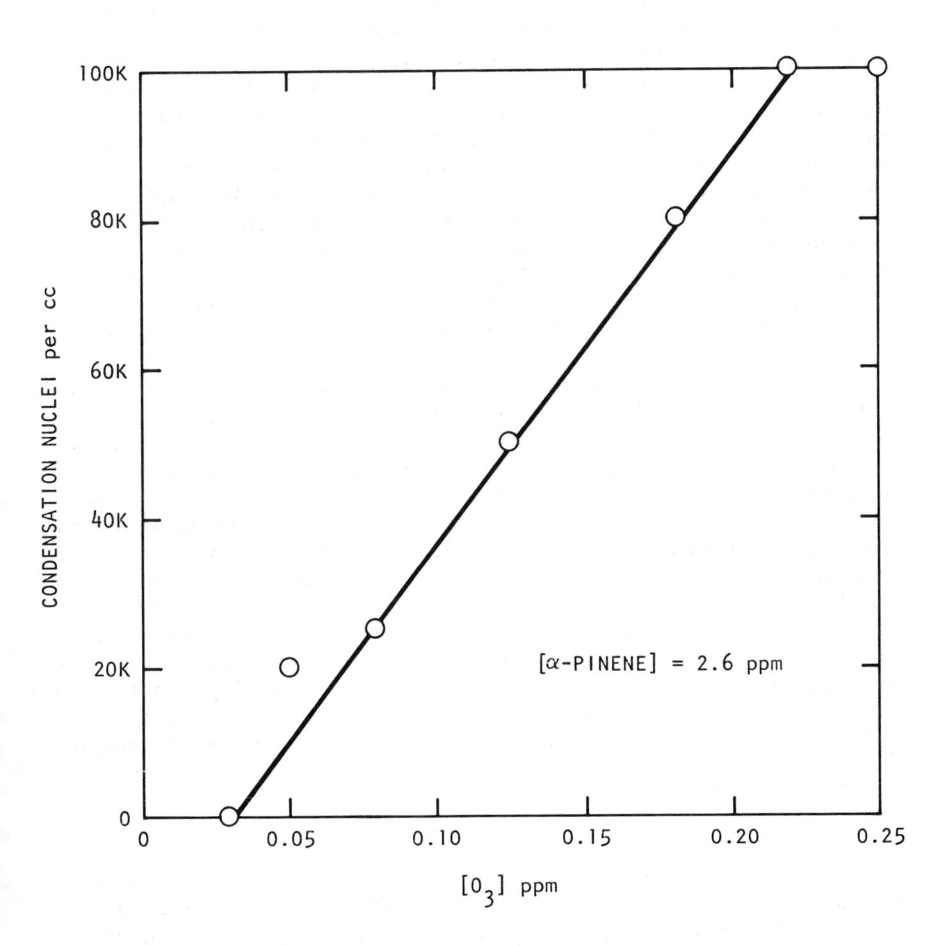

FIGURE 7. Dependence of condensation nuclei on O_3 concentration with fixed α-pinene concentration.

tration for a fixed hydrocarbon concentration. Although α-pinene is not a typical urban hydrocarbon, it was the only one for which we had sufficient data to make the comparison. Figure 8 shows the correlation between peak ozone and b_{scat} (the extinction coefficient for scattering of light by aerosols) from the California Aerosol Characterization Experiment for the summers of 1972 and 1973 (Hidy et al., 1974). We feel that these correlations are not coincidental and that, in particular, our data is pointing to a strong contribution of ozone-olefin reactions to aerosol production in urban atmospheres.

The conditions in the atmosphere are much more complex than those in our reactor. In particular, there are many more gas phase components present and there are both natural and anthropogenic primary aerosols. As far as complexity of the gas phase is concerned, we have done no experiments that test the synergistic effects of multiple hydrocarbons present, however, we have made preliminary experiments with trace inorganic components. The influences of water vapor, sulfur dioxide, and ammonia suggest the significance of complications in aerosol production in the presence of a variety of trace constituents found in smog.

In order to relate our condensation nuclei production results to the atmosphere, one must realize that we only see results of homogeneous nucleation. These products, if they nucleate homogeneously, will condense even more readily on already existing nuclei in the atmosphere. Thus any material we see is to be definitely expected in the atmosphere, while the absence of an effect for our experiments does not preclude condensation under atmospheric conditions. With this in mind, we observe that molecular size plays an important role in aerosol production with large molecules more active, but structure is important also. 1,3-hexadiene is more active than 1-dodecene, and 1,5-hexadiene is relatively inactive. It is important to establish the concentrations of larger, more complex olefinic molecules in the atmosphere, which has not been done at present.

ACKNOWLEDGMENTS

We wish to thank the Coordinating Research Council for support of this work under Contracts CAPA-8-71(1-72) and CAPA-8-71(1-73) and the Environmental Protection Agency for support under Contract 68-02-0562 and 68-02-0771.

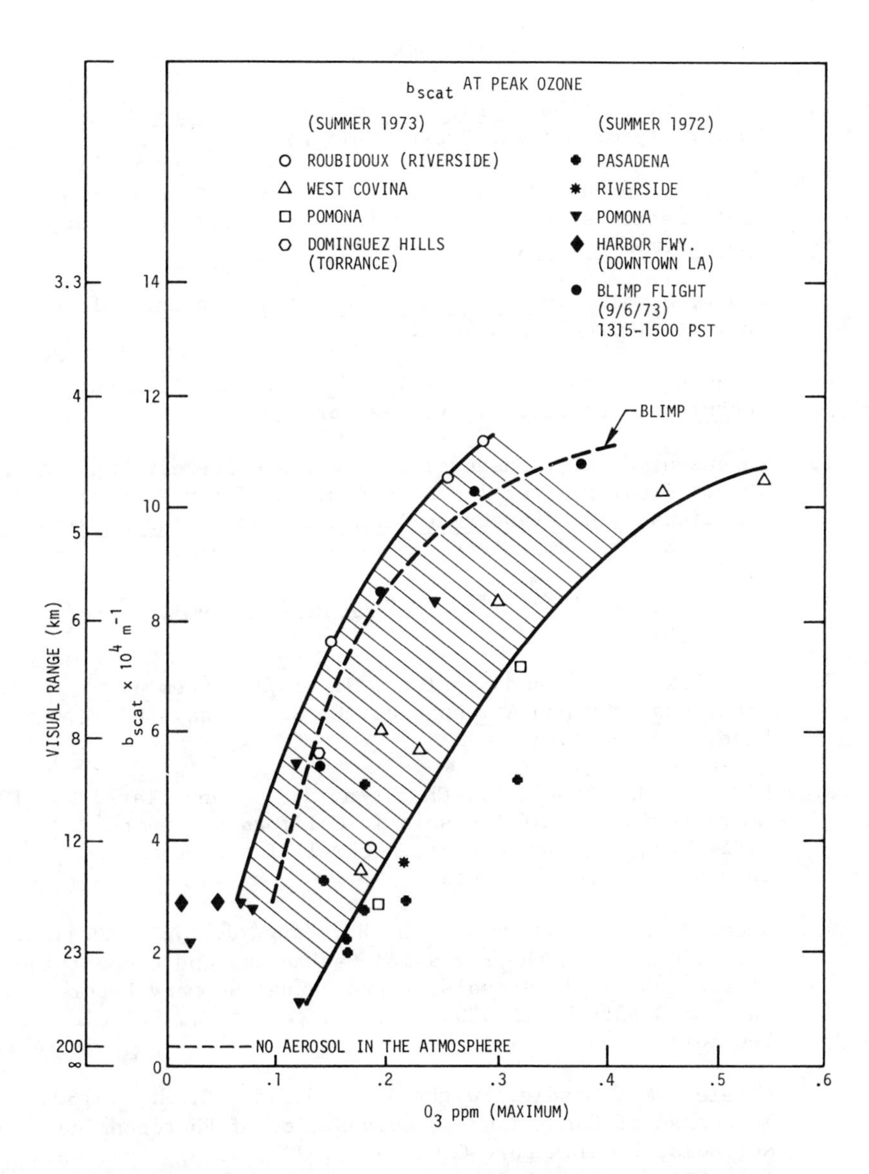

FIGURE 8. Dependence of aerosol light-scattering coefficient on peak ozone concentration as observed at the same time in various places in California during the California Aerosol Characterization Experiment.

REFERENCES

1. Byers, H. R. 1965. Elements of Cloud Physics, University of Chicago Press, Chicago.

2. Cher, M. and Hollingsworth, C. S. 1969. Chemiluminescent Reactions of Excited Helium with Nitrogen and Oxygen, Adv. Chem. Ser. 80, 118.

3. Crank, J. 1957. The Mathematics of Diffusion, Oxford University Press, New York.

4. Dushman, S. 1962. Scientific Foundations of Vacuum Technique, 2nd ed., Wiley, New York.

5. Ferguson, E. E., Fehsenfeld, F. C., and Schmeltekopf, A. L. 1969. Flowing Afterglow Measurements of Ion-Neutral Reactions. In Advances in Atomic and Molecular Physics (D. R. Bates, Ed.), Academic Press, New York.

6. Fuchs, N. A. 1964. The Mechanics of Aerosols, Pergamon, New York.

7. Groblicki, P. J. and Nebel, G. J. 1971. Chemical Reactions in Urban Atmospheres (C. S. Tuesday, Ed.), Elsevier, New York.

8. Hidy, G. M., Appel, B., Charlson, R. J., and Clark, W. 1974. Characterization of Aerosols in California, Report SC524.25FR, Science Center, Rockwell International, Thousand Oaks, California.

9. Kocmond, W. C., Kittelson, D. B., Yang, J. Y., Demerjian, K. Determination of the Formation Mechanisms and Composition of Photochemical Aerosols, First Annual Summary Report, Calspan Report No. NA5365-M-1, Calspan Corp., Buffalo, New York.

10. Lipeles, M., Landis, Dwight A., and Hidy, G. M. 1973. Mechanism of Formation and Composition of Photochemical Aerosols, Final Report CAPA-8-71(1-72), July. Available from Coordinating Research Council, New York.

11. National Research Council. 1929. Int. Critical Tables 5, 62.

12. O'Neal, H. E. and Blumstein, C. 1973. Int. J. Chem. Kinet. 5, 397.

13. Wilson, W. E., Schwartz, W. E., and Kinzer, G. W. 1972. Haze Formation—Its Nature and Origin, Final Report EPA-CPA70-Neg. 172, January.

14. Wilson, W. E., Jr. 1971. Discussion section after Groblicki and Nebel (1971).

APPENDIX

In order to understand diffusional problems in a chemical flow reactor, we must consider both loss of material and appearance of interfering substances. Each of these cases has subcases. The loss of an initial reactant will lead to an apparent increase in its reaction rate, while the loss of a product will lead to an apparently lower formation rate. In the latter case, gas phase products may be lost to the walls by molecular diffusion as in the case of loss of reactants, but aerosols may be lost by the analogous Brownian diffusion and also by gravitational settling. For the interference case, the appearance of contaminants from the wall may produce new products not otherwise obtained from the experimental reacting mixture, or the contaminant may react with and remove an initial reactant, a product, or a key intermediate. In the subcase of removal, the result may be an apparently faster or slower reaction rate, but it may also be the insidious removal of an intermediate by converting it to an otherwise present species. This kind of an interference would be almost impossible to detect but could dramatically change the experimental results. These various cases are summarized in Table A.1.

All of these cases, except for gravitational settling of aerosols, are described by the following partial differential equation for diffusion in a steady flow through a tube (Ferguson et al., 1969, generalization of Eq. 6):

$$v(r) \frac{\partial C}{\partial z} = \frac{D}{r} \frac{\partial}{\partial r} \left(r \frac{\partial C}{\partial r}\right) + R(C,r,z) \qquad (1)$$

where $C = C(r,z)$ is the concentration of the species in question, D is its diffusion coefficient in air, r is the radial position, z is the axial position, $v(r)$ is the mean velocity (in z direction only for laminar flow), and R is the chemical rate of production or removal of species in question.

In general, this equation cannot be solved in closed form due to the nonlinearities introduced by the term $R(C,r,z)$. In fact for the typical case in question in this program, this nonlinear term couples the equation to a large set with one equation for each species. However, if $R = 0$ or is linearly dependent upon C (i.e., $R = \pm kC$), then Eq. 1 separates into an equation for the z dependence and one for the r dependence. In both cases, the z equation has a simple solution. The r equation depends upon the factor $v(r)$. If it is constant, $v(r) = v_o$,

TABLE A.1. Wall Interference in the Flow Reactor

I. Loss to the Walls[a]

A. Diffusional loss of initial reactant leading to apparent higher reaction rate.

B. Loss of product leading to apparent lower formation rate.

1. Diffusional loss of gas phase molecules.
2. Brownian diffusional loss of aerosols.
3. Gravitational sedimentation of aerosols.

II. Diffusion Contamination from the Walls[a]

A. Appearance of material which produces contaminant products or increases the concentration of some intermediate or product otherwise present.

B. Appearance of material which reacts to remove key reactants and intermediates.

[a]Note: A reaction catalyzed at the walls falls simultaneously in both I and II. The catalyzed reactants are lost and the products formed represent a contaminant source.

then the resulting r equation is Bessel's equation of order zero (Ferguson et al., 1969). If $v(r)$ represents laminar flow, that is, $v(r) = v_o(1 - r^2/a^2)$ where a is the radius of the tube, then the r equation may be reduced to the confluent hypergeometric equation (Cher and Hollingsworth, 1969). It has been shown that for a flow reactor assuming that $v(r) = v_o$ only results in a small change in the solution for the case of $R = 0$ (Ferguson et al., 1969).

In order to simplify the analysis of the wall problems, we make some extreme simplifications of the problem to keep $R = 0$. These represent "worst cases" and are not physically realizable. However, we are then assured that the true problem is considerably less important than the calculated one. Since for many of the entries in Table A.1 the wall effects will be unimportant, these overstatements of the problem are totally satisfactory.

For the cases in which the wall effects produce limitations, the stringent nature of the worst case assumptions must be kept in mind.

For case 1(A) we assume that the reactant must be present during the entire residence time, but that it is trapped at the walls. In other words, it will be diffusing away for the entire residence time but its presence as a reactant is most important at the end of the period. On the other hand for 1(B) cases, we assume that the product is formed immediately and has the remainder of the time to decay. Of course, if such a product is other than an intermediate posing as a case 1(A) reactant, then the residence time is just made shorter. For subcase 2 we assume that the aerosol consists of a very small nuclei that only have an effect near the end of the time, while for subcase 3 we assume the aerosol has grown to be very large initially. For both cases 2(A and B) we assume a source at the wall such that the concentration at the wall is comparable to the other important intermediates or aerosol precursors but that it does not have an effect until it reaches the end of the residence period. Now with these assumptions, we may quantitatively explore the limitations of the flow reactor. These are extreme cases so the solutions are ideal bound on the true behavior in the flow reactor. In some cases, the change to more realistic assumptions will be easy; in many, however, this represents a difficult problem.

Once R has been taken as zero and $v(r) = v_o$, Eq. 1 may be separated and solved exactly. We obtain (Crank, 1957)

$$C = \sum_{n=1}^{\infty} A_n J_o(\alpha_n r) e^{-D\alpha_n^2 t} \tag{2}$$

where $J_o(x)$ is the zero-order Bessel function, $a\alpha_n$ is the zero of J_o (i.e., $J_o(a\alpha_n) = 0$) and the A_n's are to be determined by the boundary conditions, and the variable z has been replaced by t such that $z = v_o t$. For case 1 we have as boundary conditions: $C = 0$ for $r = a$ and $t \geq 0$ and $C = f(r)$ for $t = 0$ and $0 < r < a$. Now the solution is (Crank, 1957)

$$C = \frac{2}{a^2} \sum_{n=1}^{\infty} e^{-D\alpha_n^2 t} \frac{J_o(r\alpha_n)}{J_1^2(a_n)} \int_o^a r\, f(r) J_o(r\alpha_n) dr \tag{3}$$

The solution consists of a sum of modes determined by the zeros of J_o. Since α_n increases rapidly from one zero to another and we see that $D\alpha_1^2t$ is already appreciable, then in a reasonable time all the higher modes drop out and only the α_1 mode remains. Thus for case 1(A), we assume that when the flowing mixture enters the irradiated portion of the reactor, its initial distribution is given by $f(r) = C_oJ_o(r\alpha_1)$. Equation 3 now reduces to

$$C = C_oJ_o(r\alpha_1)\ e^{-D\alpha_1^2t} \tag{4}$$

The first root of J_o is at $a\alpha_1 = 2.405$, and a is taken to be 2.5 cm for now. D ranges from 0.05 cm^2/sec for a typical small organic molecule (National Research Council, 1929) to 0.282 for H_2O (Dushman, 1962) all in air at STP. On the axis ($r = 0$) $J_o = 1$ so $C/C_o = e^{-D\alpha_1^2t}$. Thus, for a 1-second residence time, the loss ranges from 4.5 to 23 percent for the range of D specified. The loss is sharply dependent upon the residence time growing to 21 percent for D = 0.05- and 5-cm^2/sec residence. However, for increasing diameter, the improvement is dramatic since α_n is inversely proportional to a^2. Thus for D = 0.282, t = 1 second, and a = 5 cm, the loss drops to 6.3 percent. A series of these results are summarized in Table A.2. Once again it must be noted that the assumption of total absorption at the walls is stringent for our experiments. The solution of the equations becomes more complex for all other cases, except that of no absorption, in which case diffusion has little effect.

Case 1(B.1) is somewhat more complicated than 1(A). If the product is formed in a reaction dominated by a reactant that is absorbed at the wall, then it is produced with an initial distribution proportional to $J_o(r\alpha_1)$ and the solution is the same as that discussed previously. On the other hand, if the product is formed by a group of reactants uniformly distributed (i.e., no absorption at the wall for reactants) then higher modes ($A_n \neq o$, $n > 1$) must be included. Since the α_n's increase rapidly with n, the higher modes damp out quickly. For example, if the first mode has a 10 percent loss for a given case the next mode will have a 43 percent loss and the third mode a 74 percent loss. For the initially uniform case the first mode will represent only 67 percent of the sum and the remainder of the terms will oscillate in sign making approximate solution difficult. Thus we conclude that for a product initially uniformly distributed and with a moderately large diffusion coefficient the higher modes will all be lost so a loss of 33 percent must be added to

TABLE A.2. Diffusional Loss in a Flow Reactor (see text for symbol definitions)

D^a (cm^2sec)	a (cm)	t (sec)	C/C_o	Loss (%)
0.282	2.5	1	0.7703	23
0.05	2.5	1	0.9548	4.5
0.282	2.5	5	0.2712	73
0.05	2.5	5	0.7935	21
0.282	5	1	0.9368	6.3
0.05	5	1	0.9885	1.2
0.282	5	5	0.7216	28
0.05	5	5	0.9438	5.6

[a]Note D = 0.282 represents H_2O in air at 1 atm and 16°C and D = 0.05 is a typical low molecular weight hydrocarbon in air at STP.

those estimated previously. Fortunately the types of products that accumulate in aerosols are much larger molecules and have small diffusion coefficients, but it is certainly true that such products are collected on the walls.

Although the losses to the walls for extreme cases can be large, this is not a serious defect in a flow reactor as long as the reactions being studied are not too much slower than the rate of loss to the walls. Even for very accurate rate measurements, such losses are regularly accounted for in measurements of rate constants in flow reactors (Ferguson et al., 1969; Cher and Hollingsworth, 1969).

For case 1(B.2), the Brownian diffusion of aerosols, the molecular calculations can be carried over exactly. For our calculations, however, we have gone to the aerosol literature directly for the required solutions. We find that for laminar flow in a tube (Fuchs, 1964)

$$\frac{\bar{n}}{n_o} = 1 - 2.56\mu^{2/3} + 1.2\mu + 0.177\mu^{4/3} \qquad (5)$$

where $\bar{n}$ is the mean number of aerosol particles at the end of a tube and n_o is the number at the entry to the tube,

$$\mu = \frac{DL}{a^2 V} \tag{6}$$

where D is again the diffusion coefficient, L is the length of the tube, a is the radius of the tube, and V is the mean velocity of the flow. This solution is valid for small μ, which is applicable to aerosol behavior.

If we now define the axial residence time t as

$$t \equiv \frac{L}{2V} \tag{7}$$

so that combining Eqs. 6 and 7:

$$t = \frac{a^2 \mu}{2D} \tag{8}$$

If we now take $\bar{n}/n_o = 0.99$ and solve Eq. 5 for μ, we find $\mu \cong 2.5 \times 10^{-4}$. For our reactor a = 2.5 so $t = 1.95 \times 10^{-3}/D$. To find D we note that

$$D = k\tau B \tag{9}$$

where k is the Boltzman constant, τ is the temperature (°K), and B is the mobility. The mobility is given by

$$B = \frac{\left(1 + A\frac{\ell}{r} + Q\frac{\ell}{r}e^{-br/\ell}\right)}{6\pi\eta r} \tag{10}$$

where η is the viscosity of the medium; ℓ is the mean free path of molecules of the medium; r is the radius of the particle; A, Q, and b are constants with values; and A = 1.246, Q = 0.42, b = 0.87. For 0.01-μm radius particles in air at STP, $B = 3.4 \times 10^9$, so $D = 1.41 \times 10^{-4}$ cm^2/sec giving t = 14 seconds, which means in 14 seconds only 1 percent of the 0.01-μ radius particles will be lost.

Particles may also be lost by gravitational settling. Similarly to the above a standard expression is also extracted from Fuchs (1964). Since it is unlikely that this would be a problem we just state the result here for completeness. For a 1 percent loss under the same conditions as in the diffusion cases and for 0.1-μm particles, we have a residence time of 650 sec.

Now we move on to the cases 2(A and B) that we can quantitatively treat together. Once again the solution is Eq. 2, but with new boundary conditions: $C = C_o$ for $r = a$ and $t \geq 0$ and $C = 0$ for $t = 0$ and $0 < r < a$. The solution satisfying these boundary conditions is (Crank, 1957):

$$\frac{C}{C_o} = 1 - \frac{2}{a} \sum_{n=1}^{\infty} e^{-D\alpha_n^2 t} \frac{J_o(r\alpha_n)}{\alpha_n J_1(a\alpha_n)} \tag{11}$$

This solution can be expressed in terms of two dimensionless parameters Dt/a^2 and r/a. Curves of C/C_o versus r/a for various values of Dt/a^2 have been presented by Crank (1957). For $C/C_o = 0.01$ and $v = 0$, $Dt/a^2 \cong 0.05$, so we have

$$t = \frac{0.05a^2}{D}$$

for our range of D from 0.05 to 0.282 cm^2/sec and $a = 2.5$ cm, we find residence times of from 6 down to 1 second. Again, it must be noted that this is a restrictive calculation. Any product molecule large enough to be important in aerosol formation probably has a smaller diffusion coefficient than that of our present range, so less of it will reach the detector. More important is the fact that if the product diffusing from the walls is sufficiently important in producing aerosols, it will have so reacted on its way from the wall to the detector forming an aerosol which has a much lower diffusion rate. Thus the foregoing figures are really for gas phase contaminants and in the case of aerosol production, or for that matter inhibition, they are more limiting than necessary. Nevertheless, for the feasibility study described in this report, these contamination calculations were considered to be the limiting case, for if results can be obtained with these limitations, then relaxing them will only make the experiments easier to conduct.

Aerosol Formation in Simple Photochemical Systems

WARREN C. KOCMOND* and J. Y. YANG
Calspan Corporation
Buffalo, New York

DAVID B. KITTELSON and KENNETH T. WHITBY
Mechanical Engineering Department
University of Minnesota
Minneapolis, Minnesota

KENNETH L. DEMERJIAN
U. S. Environmental Protection Agency
Environmental Sciences Research Laboratory
Research Triangle Park, North Carolina

*Present address: Desert Research Institute, University of Nevada, Reno, Nevada 89507.

I. INTRODUCTION

For the past two decades, scientists have been studying complex problems involving air pollution. The initiation of smog chamber experiments that can simulate sunlight and atmospheric conditions with the introduction of selected pollutants has helped in our basic knowledge of the chemistry and aerosol behavior involved. But there is little known about the complex mechanisms by which gas to particle reactions take place. Frequently there is relatively poor reproducibility between chamber product rate studies run by different laboratories, and the results of one study may be rather inaccurate and misleading. Furthermore, there has been little attention given to the aerosol formation mechanisms involved in photochemical smog.

A new approach to smog chamber research with an emphasis on aerosol measurements was initiated by Clark (1972, 1975) in an investigation of SO_2 photooxidation. Here, through measurements of particle number, surface, and volume concentrations and size distributions, the details of aerosol growth were examined, and it was found to be possible to determine SO_2 photooxidation rates from aerosol measurements.

Subsequently, a joint Calspan-University of Minnesota smog chamber study was initiated. Its primary objective is to elucidate the physical and chemical processes contributing to the production of photochemical aerosols in polluted atmospheres. Emphasis has been placed on obtaining repeatability of test results in chambers of widely different physical dimensions and in systems ranging from SO_2 + clean air to HC + NO_x + SO_2 mixes. At both laboratories, accurate assessment of aerosol behavior in relatively simple systems has been stressed.

The 20,800-ft^3 Calspan chamber represents the largest photochemical reaction vessel currently available in the United States. The chamber is especially well suited to studies of aerosol behavior, since wall effects are minimized and the settling height is large. The 600-ft^3 University of Minnesota chamber is fabricated of FEP Teflon. Its flexible walls eliminate the necessity of dilution during sampling, which must be done in rigid wall chambers of the same size.

Results from some of the key systems investigated in this study are discussed within this chapter. The simplest system investigated, SO_2 + clean (filtered) air, is used as a guide in describing aerosol formation and growth and decay mechanisms in photochemical systems. Aerosol behavior in several HC + NO and HC + NO + SO_2 systems are also discussed. The hydrocarbons chosen for study were toluene, 1-hexene, *m*-xylene, and cyclohexene. The data are treated in that order.

For the most part, good agreement was found in the experimental data from the two chambers. Because of the large body of data generated, only a few examples typical of the systems studied can be provided here. A more detailed account can be found in Kocmond et al. (1975). The important conclusions to be derived from this study, however, are provided at the conclusion of this chapter.

II. EXPERIMENTAL FACILITIES

The smog chambers used at Calspan and the University of Minnesota have been discussed elsewhere (Kocmond et al., 1973; Clark, 1972, 1975) and are not treated in detail here. Briefly, however, the Calspan chamber consists of a cylindrical chamber 30 feet in diameter and 30 feet high, enclosing a volume of 20,800 ft^3 (590 m^3). The chamber walls are coated with a specially formulated fluoroepoxy, which has surface adhesion characteristics very similar to those of FEP Teflon. Illumination within the chamber for this investigation was provided by 28.6 kw of fluorescent daylight and blacklamps installed inside 24 lighting modules and arranged in eight vertical channels attached to the wall of the chamber. The measured light intensity is $k_{d[NO_2]} \sim 0.23\ min^{-1}$. (More recent lighting modifications have since been made to give a k_d for NO_2 of $\sim 0.35\ min^{-1}$.) The lighting modules are covered with 0.25-inch Pyrex glass and are sealed from the chamber working volume.

Air purification is provided by a recirculation system that can continuously filter the air through a series of absolute and activated charcoal filters. Nearly all gaseous contaminants and particulate matter can be removed from the chamber air in about 4 hours of filtration. Filtered air generally contains less than 0.01 ppm NO_x, 0.2 ppm C nonmethane HC, and no measurable SO_2 or ozone.

Instrumentation used to monitor aerosol behavior and reactant concentrations within the chamber includes the Bendix Model 8002 chemiluminescent ozone analyzer, Model 8101-B nitrogen oxides analyzer, Model 8300 sulfur analyzer, and the Model 820 reactive hydrocarbon analyzer; a Hewlett-Packard 5750 gas chromatograph, a Thermo-Systems Model 3030 Electrical Aerosol Analyzer (EAA), an MRI integrating nephelometer, a Gardner Associates' small particle detector, and a GE condensation nucleus counter.

The University of Minnesota smog chamber is a cylindrical vessel fabricated of 0.01-inch DuPont FEP Teflon and encompasses a volume of 625 ft^3. The illumination system consists of 72 GE

F40BL fluorescent lamps mounted in vertical pairs on 36 evenly spaced supports. Aluminum foil has been attached behind the lamps to increase the uniformity and intensity of the light. Light intensity is measured to be $k_{d[NO_2]} \sim 0.20$ min^{-1}.

The air purification system consists of an absolute particle filter, an activated charcoal scrubber, silica gel dryer, humidifier, and final filter. Ambient laboratory air is purified by pumping it through the purification system at about 15 ft^3/min. Air passing through the purification system is exposed to only nonreactive metal, glass, and Teflon duct surfaces in order to minimize sources of contamination.

Gas analysis instrumentation consists of a Meloy Model SA160-2 flame photometric total sulfur analyzer, a Bendix Model 8101-B NO_x analyzer, a REM Model 612B chemiluminescent ozone analyzer, and a Hewlett-Packard Model 5700 gas chromatograph.

Observations of aerosol behavior were made with two versions of a portable electrical aerosol analyzer. The "laboratory prototype" analyzer, used for early joint Calspan-University of Minnesota workshops and the first 45 experiments at the University of Minnesota, has been described by Liu and Pui (1974). A second version, the "commercial prototype" (Thermo-Systems Model 3030), was used for the remainder of the investigation. A description of this instrument is in Liu and Pui (1975).

Both analyzers are based on the "diffusion charging-mobility analysis" principle described by Whitby and Clark (1966). The aerosol-laden air flows through the charger, a region containing unipolar ions which have been produced by a corona discharge. The aerosol particles emerge from the charger carrying a negative charge and are introduced into the mobility analyzer. In this section, a positive voltage on a collection rod causes all particles with electrical mobilities greater than a certain critical value to be precipitated. Those particles with smaller mobilities flow past this section and are collected by an absolute filter. An electrometer, which is connected to the filter, measures the current carried by the charged particles. The mobility spectrum and, therefore, the size distribution can be inferred from the electrometer as a function of collecting rod voltage. A complete set of readings takes about 2.5 minutes. The instrument can measure particle diameters in the size range of approximately 0.004 to 0.75 μm.

III. GENERAL MECHANISMS OF AEROSOL FORMATION, GROWTH, AND DECAY IN THE SO_2-CLEAN AIR SYSTEM

In order to discuss and interpret the results of this study, brief reviews of the mechanisms governing aerosol formation, growth, and decay in the SO_2-clean air system are necessary. The SO_2 system is discussed because it is the simplest photochemical aerosol system studied and the only one which is reasonably well understood.

In most photochemical aerosol studies, the usual test procedure is to first purify the sample air and then introduce known concentrations of pollutants into the chamber. The sample is then irradiated while observations are made of aerosol formation, particle size distributions, and gaseous behavior. Achieving adequate air purity prior to an experiment is essential, since even trace amounts of gaseous contaminants can lead to unwanted chemical reactions and the formation of substantial aerosol.

Three main mechanisms govern aerosol growth in these systems: homogeneous nucleation, condensation, and coagulation. Nucleation refers to the formation of new particles in a supersaturaturated vapor mixture. The rate of homogeneous nucleation increases rapidly with increasing supersaturation of the condensing vapor. In the systems studied here, no particles were present initially; hence all particles observed are formed by homogeneous nucleation.

Once formed, these particles continue to grow by diffusional deposition of supersaturated vapors onto their surfaces. The rate of condensation depends primarily upon the supersaturation of the condensing species, its diffusion coefficient, and the size of the particle upon which the condensation is taking place. Condensation does not influence particle number concentration but leads to increases in particle surface and volume concentrations. The third important process governing particle concentrations is coagulation. It is the process whereby particles undergoing Brownian motion collide and adhere to one another. The rate of coagulation is proportional to the square of the particle number concentration. The process of coagulation leads to a decrease in particle number and surface concentration, but the volume concentration in unaffected.

Under some circumstances, other processes could influence particle concentrations. In the atmosphere, convection and gravitational settling can be important. In the experiments described here, however, neither convection (losses to the chamber walls) or gravitational settling are believed to be important.

A. The SO_2-Clean Air System

The photooxidation of SO_2 in the atmosphere is believed to be first order in SO_2. In the systems studied here, the rate of photooxidation of SO_2 is low, that is, a fraction of a percent per hour. Consequently, the SO_2 concentration remains essentially unchanged during an experiment (up to 12 hours in the large chamber), and the products of SO_2 oxidation are produced at a constant rate. The main product of SO_2 oxidation in clean air is believed to be SO_3; and if moisture is present, sulfuric acid droplets will ultimately be formed.

All of the SO_2 photochemical aerosol systems studied here and previously (Clark, 1972, 1975; Kocmond et al., 1973) exhibit the same general behavior. This is illustrated in Figure 1.

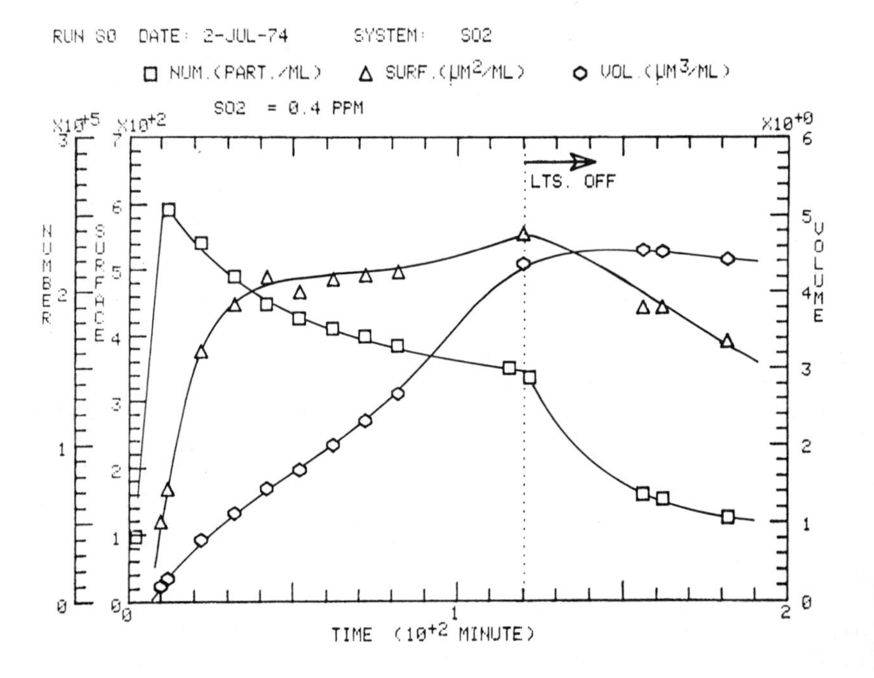

FIGURE 1. Aerosol development in the SO_2-clean air system. Lights are turned on at time zero.

After the reactants are introduced into the chamber, mixed, and the lights turned on, an initiation period follows that ranges in length from less than 1 minute to 20 to 30 minutes, depending on the SO_2 concentration, presence of background contaminants, and humidity conditions in the chamber. During this initiation period, there is no observable production of aerosol; that is, the nuclei count remains at its baseline level of approximately 50 particles/cm^3, and the electrometer current in the EAA is essentially zero.

Friend et al. (1973) have suggested that the initial formation of sulfuric acid nuclei takes place through the following reactions:

$$SO_3 + H_2O \rightarrow H_2SO_4 \tag{1}$$

$$H_2SO_4 + H_2O \rightarrow H_2SO_4 \cdot H_2O \tag{2}$$

$$H_2SO_4(H_2O)_n + H_2O \rightarrow H_2SO_4(H_2O)_{n+1} \tag{3}$$

SO_3 production reactions will start as soon as the lights are switched on. During the initiation period, SO_3 and H_2SO_4 concentrations slowly build up. Nucleation may be thought of as occurring either through Reactions 1 through 3 or by binary vapor nucleation (Reiss, 1950). The rate of nucleation depends very strongly on the degree of supersaturation of the nucleating species (Takahashi et al., 1975).

The nuclei initially formed are probably of the order of 15 Å in diameter, too small to be detected by either the CNC or EAA. However, once formed, these nuclei can grow by condensation. The initiation period then is the time required for particles to nucleate and grow to sufficient size to be detected by the CNC (25 Å).

Once nucleation begins, another path for gas to particle conversion is made available, that is, condensation. This additional path tends to reduce the supersaturation and thus the nucleation rate. At the same time, the increased number concentrations will lead to an increased rate of coagulation. Both of these factors cause the number concentration to increase less rapidly, reach a maximum and finally decrease. This behavior is illustrated in the plot of N against time in Figure 1.

The total aerosol surface area, S (μm^2 cm^{-3}), behaves in a somewhat different manner. As the aerosol grows, new surface is produced by nucleation and by condensation on existing particles.

At the same time, surface is being lost by the coagulation of particles. In most experiments, the rate of production and the rate of loss of surface achieve a balance through a significant part of an experiment and a dynamic equilibrium surface is achieved. This behavior is apparent in Figure 1.

Aerosol volume (um^3/cm^3) is produced by condensation and nucleation but is unchanged by coagulation. Hence, our model predicts a steady growth of volume with time as long as condensation and nucleation processes continue. If the stationary-state approximation is applied to the condensing and nucleation species, the rate of production of aerosol should be directly proportional to the rate of production of these species and thus the rate of photooxidation of SO_2. Since the latter rate should be essentially constant, the rate of aerosol volume formation should also be constant, which is the observed behavior illustrated in Figure 1.

This fairly simple model appears to be adequate to describe the aerosol formation when SO_2 is photooxidized in clean air. The simplicity stems primarily from the fact that the nucleating and condensing species are formed at an essentially constant rate. When the lights are turned off, the reactions forming SO_3 should stop and, hence, nucleation should rapidly cease. Normally the lights are extinguished late in an experiment as shown in Figure 1. The removal of the nuclei production mechanism leads to a marked increase in the rate of decay of number concentration.

When the lights are extinguished, surface concentrations must decrease because nucleation and condensation stop and only coagulation influences surface concentration. Volume growth rate decreases until an essentially constant volume is attained. This happens because the only gas to particle conversion processes taking place are nucleation and condensation. When these processes cease, the total aerosol volume (or mass) must remain constant, as long as the chamber lights are off.

In more complex systems containing NO, NO_2, and hydrocarbons, the aerosol behavior becomes considerably more complex. However, the same three physical processes, nucleation, condensation, and coagulation, will govern behavior in these more complex systems.

B. Calculation of SO_2 Photooxidation Rate

For the SO_2 + clean air photochemical system (Clark, 1972, 1975), it has been found that the rate of volume production approaches a constant value, that is, a plot of volume against time yields a straight line. The reasons for this are as follows: Once an equilibrium surface has been achieved, the concentration of SO_3 in the gas phase will approach a steady-state value. Then the rate of oxidation of SO_2 to SO_3 will be equal to the rate of removal of SO_3 to the condensed phase to form sulfuric acid droplets. The rate of production of sulfuric acid aerosol, corrected for molecular weight change and water concentration, must be equal to the rate of photooxidation of SO_2, which is a constant during the linear growth phase of the experiments. Thus the slope of the straight line volume growth curve may be related directly to the rate of photooxidation of SO_2. The governing equation is

$$- \frac{d[SO_2]}{dt} = \frac{dv}{dt} \times \rho \times \underline{P} \times \frac{MW_1}{MW_2} \qquad (4)$$

where ρ is the density of the sulfuric acid droplet, $\underline{P}$ is the weight fraction of H_2SO_4 in the drop, MW_1 is the molecular weight of SO_2, and MW_2 is the molecular weight of H_2SO_4. The quantities ρ and $\underline{P}$ are determined from data given by Bray (1970) assuming that water vapor in the gas phase is in equilibrium with water in the aerosol droplets.

Equation 4 is true, if one assumes that (a) all SO_3 formed reacts with water to form sulfuric acid; (b) virtually all the sulfuric acid is in the condensed phase; (c) the sulfuric acid aerosol droplets are in equilibrium with the water vapor in the gas phase; and (d) the aerosol droplets are a pure aqueous H_2SO_4 solution. Assumptions (a) and (c) have not been experimentally verified, though reasonable, and assumption (b) has support based on the work of Doyle (1961). The validity of assumption (d) is dependent on the reactants present. For SO_2 in clean air, it should be valid.

Using Eq. 4 and the measured value of dv/dt, the rate of SO_2 photooxidation can be determined. In the large Calspan chamber, SO_2 photooxidation rates of about 0.2 percent/hr^{-1} are typically observed, while rates of about 0.1 percent/hr^{-1} are found at the University of Minnesota. Generally, a somewhat accelerated or "final" rate is observed in both chambers after the first hour or so.

Variations in the SO_2 photooxidation rate in either chamber (within the range of a few tenths of a percent) are not tied to any obvious experimental variable, such as SO_2 concentration, chamber size, or relative humidity. The history of previous experiments does influence the results in that higher rates are usually observed after completing experiments involving reactive hydrocarbons. We have also noted a conditioning effect in both chambers with somewhat lower rates being observed after repeated

SO_2 irradiations. Friend et al. (1973) suggests that Aitken nuclei are formed via the SO_2—O atom oxidation in the presence of water vapor, and that Aitken nuclei are not formed to any extent by the interaction of SO_2-excited states with O_2, H_2O, or O_2 and H_2O. If this is correct, then any calculated SO_2 photooxidation must be due to chamber contamination. According to Friend et al., no observable nuclei should be formed in a contamination-free SO_2-clean air system.

In most smog chambers, the only source of O atom would be from the photolysis of background NO_2. However, in order to achieve the SO_2 photooxidation rates reported here (via O atom attack), NO_2 background levels far exceeding those commonly observed in extensively filtered air would be required. A second contribution, that of SO_2 photooxidation by the HO_2—SO_2 reaction, is not so easily determined. Hydroperoxy radical concentration is usually several orders of magnitude greater than that of O atoms in the atmosphere (and in most chambers) and therefore may contribute significantly to the SO_2 photooxidation rate. In addition, recent results (Davis, 1974) indicate that OH can be an important contributor to the SO_2 oxidation process. The contributions that these reactions may have on SO_2 photooxidation have not been fully determined yet.

IV. AEROSOL FORMATION IN HYDROCARBON-CONTAINING ATMOSPHERES

As part of this investigation, aerosol behavior and chemical conversion data were obtained for various HC + NO and HC + NO + SO_2 systems using realistic concentrations of reactants. The hydrocarbons chosen for study were toluene, 1-hexene, *m*-xylene, and cyclohexene. Nitrogen oxide (NO) was used in place of NO_2 in order to more closely simulate photochemical processes responsible for aerosol formation in an urban environment. The normal concentrations of reactants used in the experiments were $\sim$0.35 ppm HC, 0.15 ppm NO, and 0.05 ppm SO_2. In addition to these tests, a number of SO_2 + clean air experiments were performed as part of normal chamber characterization and contaminant monitoring procedures.

Summaries of the pertinent aerosol and chemical data for the Calspan and duplicate University of Minnesota experiments are provided in Tables 1 through 4. The chemical data summarizes the initial experimental conditions for each test and also the maximum ozone concentration observed and the time to maximum NO_2 concentration, $t[NO_2]_{max}$. The main measure of <u>chemical</u> reactivity of both the hydrocarbon + NO and hydrocarbon + NO +

TABLE 1. Summary of Aerosol Data—Calspan

Run No.	System	RH (%)	N_{max} (10^3 particles cm^{-3})	SE ($\mu m^2 cm^{-3}$)	$\frac{dv}{dt}[SO_2]$ ($\mu m^3 cm^{-3} hr^{-1}$)	$\frac{dv}{dt}[max]$ ($\mu m^3 cm^{-3} hr^{-1}$)	SO_2 Photox. (% hr^{-1})	Comments
6	Toluene + NO	30	31	640	--	2.2	--	No vol. first 4 hr
30	Toluene + NO	20	13	750	--	2.6	--	No vol. first 4 hr
29	Toluene + NO + SO_2	30	160	>750	0.78	1.5	0.32	First 4 hr[a]
7	Toluene + SO_2	20	210	800	1.17	3.2	0.45	First 50 min[a]
5	1-Hexene + NO	40	140	610	--	2.1	--	No vol. first 5 hr
21	1-Hexene + NO	37	120	215	--	0.5	--	No vol. first 6 hr
18	1-Hexene + NO + SO_2	37	140	>1500	0.61	5.8	0.16	First 6 hr[a]
20	1-Hexene + SO_2	35	360	950	0.75	3.2	0.25	First 60 min[a]
15	m-Xylene + NO	38	84	1150	--	14.1	--	No vol. first 60 min
14	m-Xylene + NO + SO_2	29	260	2700	0.92	25.0	0.32	First 60 min[a]
17	m-Xylene + SO_2	35	280	384	0.84	1.6	0.23	First 60 min[a]
10	Cyclohexene + NO	38	36	3500	--	110	--	No vol. first 90 min
12	Cyclohexene + NO	30	42	2450	--	75	--	No vol. first 3 hr
9	Cyclohexene + NO + SO_2	30	170	4200	0.74	105	0.28	First 2 hr
13	Cyclohexene + SO_2	35	270	1300	1.20	10	0.49	First 30 min
1	0.52 ppm SO_2	25	550	>1450	5.61	--	0.21	First 30 min[a]
					10.60		0.39	First 2 hr
4	0.55 ppm SO_2	30	390	4400	4.49	--	0.16	First 30 min[a]
					13.60		0.48	First 2 hr
2	0.05 ppm SO_2	37	230	>575	0.65	--	0.23	First 30 min[a]
					2.35		0.79	First 2 hr
3	0.05 SO_2	40	290	>675	1.04	--	0.36	First 40 min[a]
					2.31		0.79	First 2 hr

[a] Time over which aerosol growth rate was used in computing SO_2 photooxidation.

TABLE 2. Summary of Aerosol Data—University of Minnesota Duplicate Tests

Run No.	System	RH (%)	N_{max} (10^3 particles cm^{-3})	SE (μm^2 cm^{-3})	$\frac{dv}{dt}[SO_2]$ ($\mu m^3 cm^{-3} hr^{-1}$)	$\frac{1}{SO_2}\frac{dv}{dt}$ ($\mu m^3 cm^{-3} hr^{-1} ppm^{-1}$)	$\frac{dv}{dt}[max]$ ($\mu m^3 cm^{-3} hr^{-1}$)	SO_2 Photox. (%/hr)
65	Toluene + NO + SO_2	28	188	330[a]	1.23	11.3	11.3	0.23
76	Toluene + NO	47	4.2	550[a]	--	--	24.5	--
77	Toluene + NO + SO_2	57	170	1850	1.17	29.9	27.4	0.39
87	Toluene + NO	30	10	340	--	--	8.6	--
88	Toluene + NO + SO_2	24	160	1600[a]	1.22	30.5	16.6	0.67
60	1-Hexene + NO + SO_2	28	74	1200[a]	0.40	5.7	21.2	0.12
78	1-Hexene + NO + SO_2	55	230	1530	0.53	13.9	19.1	0.18
92	1-Hexene + NO	33	8.8	31	--	--	0.09	--
93	1-Hexene + NO + SO_2	32	150	1300	0.29	8.5	18.0	0.16
81	m-Xylene + NO	75	23	1600	--	--	73	--
82	m-Xylene + NO + SO_2	54	230	2800	0.49	10.4	67	0.14
89	m-Xylene + NO	26	21	1000[a]	--	--	38	--
91	m-Xylene + NO + SO_2	26	230	1600	0.46	10.0	31	0.21
83	Cyclohexene + NO	51	0.9	320	--	--	50	--
94	Cyclohexene + NO	31	2.7	510	--	--	65	--
95	Cyclohexene + NO	29	1.9	620	--	--	190	--
96	Cyclohexene + NO + SO_2	28	280	5400	0.43	9.7	250	0.20

[a]Equilibrium surface not reached.

TABLE 3. Summary of Chemical Data - University of Minnesota Duplicate Tests.

Run No.	System	RH (%)	HC[b] (ppm)	SO_{2i} (ppm)	NO_i (ppm)	$[NO_2]_{max}$ (ppm)	$t[NO_2]_{max}$ (min)	$[O_3]_{max}$ (ppm)
65	Toluene + NO + SO_2	28	0.35	0.108	0.30	0.145[a]	460	0.2[a]
76	Toluene + NO	47	0.35	--	0.152	0.095	210	0.30
77	Toluene + NO + SO_2	57	0.38	0.039	0.155	0.115	160	0.362
87	Toluene + NO	30	0.35	--	0.155	0.140	130	0.402
88	Toluene + NO + SO_2	24	0.35	0.040	0.17	0.122	155	0.315[a]
60	1-Hexene + NO + SO_2	28	0.35	0.07	0.16	0.123	395	0.162[a]
78	1-Hexene + NO + SO_2	55	0.35	0.038	0.165	0.130	255	0.438
92	1-Hexene + NO	33	0.35	--	0.12	0.104	280	0.290
93	1-Hexene + NO + SO_2	32	0.35	0.034	0.122	0.125	350	0.302[a]
81	*m*-Xylene + NO	75	0.35	--	0.155	0.144	80	0.343
82	*m*-Xylene + NO + SO_2	54	0.35	0.047	0.151	0.130	94	0.361
89	*m*-Xylene + NO	26	0.35	--	0.132	0.142	68	0.379
91	*m*-Xylene + NO + SO_2	26	0.35	0.046	0.117	0.115	70	0.262
83	Cyclohexene + NO	51	0.35	--	0.13	0.101	90	0.32[a]
94	Cyclohexene + NO	31	0.35	--	0.103	0.108	60	0.20
95	Cyclohexene + NO	29	0.35	--	0.124	0.128	103	0.254
96	Cyclohexene + NO + SO_2	28	0.35	0.045	0.133	0.130	85	0.241

[a]Maximum not reached by end of irradiation period.

[b]ppm by volume.

TABLE 4. Summary of Chemistry Data - Calspan.

Run No.	System	RH (%)	HC^b (ppm)	SO_{2i} (ppm)	NO_i (ppm)	$t[NO_2]_{max}$ (min)	$[O_3]_{max}$ (ppm)
6	Toluene + NO	30	0.35	--	0.170	400	0.285
30	Toluene + NO	20	1.17	--	0.530	480	0.380
29	Toluene + NO + SO_2	30	0.35	0.05	0.146	330	>0.225
7	Toluene + SO_2	20	0.35	0.05	b	b	0.047
5	1-Hexene + NO + SO_2	40	0.33	--	0.150	420	>0.200
21	1-Hexene + NO + SO_2	37	0.33	--	0.180	420	0.275
18	1-Hexene + NO	37	0.33	0.07	0.178	430	--
20	1-Hexene + NO + SO_2	35	0.33	0.055	b	b	0.052
15	*m*-Xylene + NO	38	0.34	--	0.150	100	0.222
14	*m*-Xylene + NO + SO_2	29	0.34	0.055	0.150	105	0.305
17	*m*-Xylene + SO_2	35	0.34	0.07	b	b	0.030
10	Cyclohexene + NO	38	0.33	--	0.138	120	0.190
12	Cyclohexene + NO	30	0.33	--	0.140	190	0.192
9	Cyclohexene + NO + SO_2	30	0.33	0.05	0.220	180	0.325
13	Cyclohexene + SO_2	35	0.33	0.06	b	b	0.011

[b] ppm by volume.

SO_2 systems is taken as $t[NO_2]_{max}$. A shorter time implies a more reactive system.

In addition to these conventional measures, HC reactivity can also be described in terms of aerosol behavior. The aerosol data summaries show maximum particle concentration (N_{max}), maximum surface concentration (S_{max}), and two rates of volume production. In the photochemical systems studied, the rate of volume production is directly proportional to the rate of condensation of involatile species formed by photochemical reactions. Volume against time plots for these experiments have a rather complex shape. Consequently, two volumetric production rates are defined and tabulated in Tables 1 and 2, namely $(dv/dt)SO_2$, the slope of the essentially linear volume against time curve which is established early in the HC + NO + SO_2 or HC + SO_2 experiments and $(dv/dt)_{max}$, the maximum rate of aerosol volume production.

A. Aerosol Formation in the HC-NO System

The HC + NO systems behave quite differently from the simple SO_2 + clean air experiments. Each hydrocarbon + NO experiment can be divided into two phases. In the first phase, NO is converted to NO_2, and some oxidation of the hydrocarbon occurs. Ozone (O_3) starts to appear near the end of this phase as NO concentrations become very low. The second phase begins as soon as the initial NO has been oxidized out of the system and NO_2 reaches its maximum concentration. During this phase [NO] remains low, and $[NO_2]$ gradually decreases as NO_2 is converted to higher oxides, acids, peroxyacyl nitrates, and other nitrogen compounds; ozone grows rapidly and approaches a maximum; and aerosol formation takes place.

A typical example of this two-phase behavior is shown in Figure 2 in which aerosol and chemistry data from a 1-hexene + NO experiment is given. It may be seen from the data that the NO disappears and NO_2 maximizes in about 300 minutes. This time is considered as the duration of the first phase. The plot also shows that by the end of this phase some ozone has started to appear and the hydrocarbon concentration has started to decrease rapidly.

Early in the second phase, aerosol growth begins and proceeds rapidly. The same physical mechanisms control aerosol formation in this system as in the SO_2 + clean air system, that is, nucleation, condensation, and coagulation. Note the sharp rise in aerosol concentration at this point and also the corresponding increase in the surface and volume concentration of aerosol. For all of the hydrocarbon + NO systems studied, the

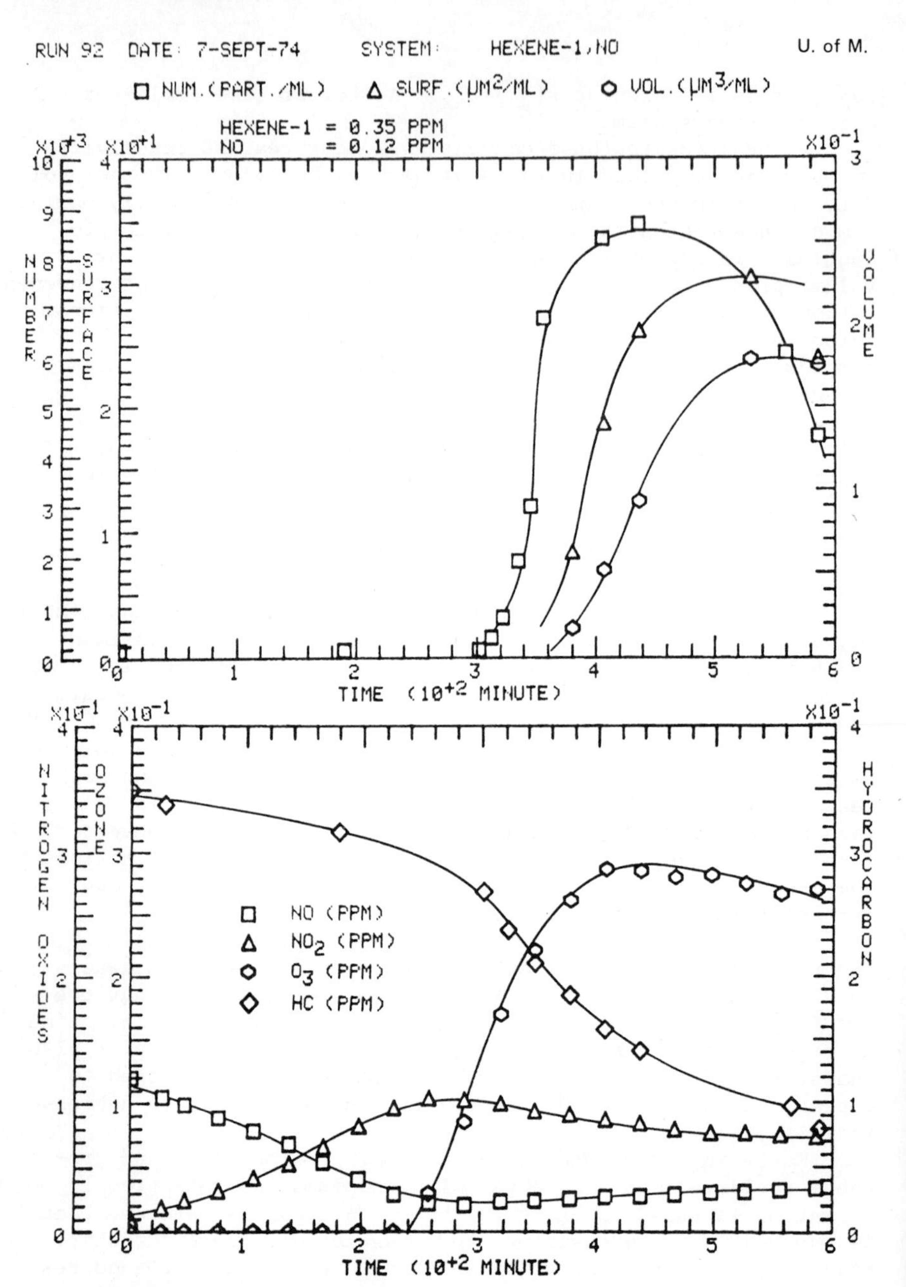

FIGURE 2. Aerosol development and chemistry data for a 1-hexene + NO system.

number of particles formed was less than in the SO_2 + clean air case; but in every case (except for hexene + NO), the volume was always larger, implying the production of fewer but larger particles. Another important difference between the aerosol behavior of the hydrocarbon + NO systems studied and the SO_2 + clean air systems is the shape of the volume against time curves. They are no longer linear. Volume grows rapidly early in the second phase; later the rate decreases and eventually volume becomes essentially constant or even drops. The net rate of aerosol volume production is the difference between the rate of condensation and the combined rates of reevaporation and wall loss. Previous estimates of wall losses (Kocmond et al., 1973) suggest that they should not be important in these chambers. Hence constant volume implies that the chemical species driving condensation have become depleted and the particles are in dynamic equilibrium with the gases in the chamber. Two possible explanations for the decreasing volume are that either continuing reactions in the gas phase have reduced gas phase concentrations of condensed species to such an extent that they reevaporate or reactions take place in or on the particles themselves, which lead to the formation of more volatile species which subsequently evaporate.

The chemical behavior of the system in the second growth phase is also shown in Figure 2. Production of ozone proceeds rapidly and ozone concentration maximizes at about the same time that the aerosol volume curve reaches its plateau; NO_2 decreases continuously while 1-hexene is oxidized out of the system.

The aerosol and chemical behavior shown in Figure 2 is typical of all HC + NO experiments performed. The most pronounced difference between the four hydrocarbons studied was in the rates of NO oxidation and the rates of aerosol formation. In terms of NO oxidation and aerosol formation rates, toluene was found to be the least reactive hydrocarbon in the Calspan chamber, followed closely by 1-hexene. In the Minnesota studies, 1-hexene proved to be less reactive than toluene. This is, perhaps, the most significant difference between the sets of data generated by Calspan and the University of Minnesota.

The *m*-xylene + NO data generated by Calspan and the University of Minnesota compare very favorably. Both aerosol and chemical measurements showed the *m*-xylene + NO systems to be considerably more reactive than either 1-hexene + NO or toluene + NO. Times for NO_{2max} were only 80 and 68 minutes for the University of Minnesota experiments and 100 minutes for the Calspan experiment (see Tables 3 and 4). In each of the experiments, no aerosol was produced until approximately the time of NO disappearance and the rapid formation of ozone. The volume and surface concentration for these tests is substantially higher than for the toluene + NO and 1-hexene + NO cases.

The cyclohexene + NO runs show cyclohexene to have about the same reactivity as m-xylene in terms of NO oxidation rate but much greater reactivity than any of the other hydrocarbons tested in terms of aerosol production. In the cyclohexene cases, the aerosol growth was almost explosive once oxidation of NO was complete. Both the surface and volume concentrations of aerosol were much higher than any other system, even though the number of particles was actually less. This implies the presence of extremely large particles and, indeed, this is the case since substantial visibility losses were noted after only 2.5 hours of irradiation. Aerosol and chemistry plots from a Calspan and University of Minnesota experiment are shown in Figures 3 and 4. The volume production rates $(dv/dt)_{HC}$ of 75 to 110 $\mu m^3/cm^3/hr^{-1}$ for the Calspan chamber and between 50 and 190 $\mu m^3/cm^3/hr^{-1}$ for the University of Minnesota are the highest we have measured and substantially higher than those observed in typical urban polluted atmospheres. As well as oxidizing NO quickly and leading to rapid aerosol formation, the cyclohexene itself was quickly oxidized out of the system, as the data in Figures 3 and 4 show.

From the chemistry data for all the hydrocarbon experiments, a family effect is apparent. Cyclohexene and 1-hexene, both olefins, have curves of basically the same shape: a gradual decrease with time until after the NO is oxidized out of the system, followed by rapid decay as ozone builds and aerosol forms. Toluene and m-xylene, both aromatics, have different shapes. In both cases, hydrocarbon concentration decays gradually and at a more or less constant rate. No change is evident in the decay curve once ozone appears and aerosol formation begins.

The experiments at the University of Minnesota were done over a range of relative humidities and some humidity effects are also evident. With the toluene + NO system and the m-xylene + NO system, pairs of experiments were performed at different humidities but otherwise similar conditions. The data in Table 2 show that for a toluene + NO system performed at 47 percent RH (run 76), the maximum aerosol production rate was 24.5 $\mu m^3/cm^3/hr^{-1}$, whereas for run 87 at 30 percent RH, the maximum aerosol production rate was only 8.6 $\mu m^3/cm^3/hr^{-1}$. Similarly, for m-xylene + NO, run 81 at 75 percent RH, the $(dv/dt)_{max}$ was 73 $\mu m^3/cm^3/hr^{-1}$, whereas run 89 at 26 percent RH produced a $(dv/dt)_{max}$ of only 38 $\mu m^3/cm^3/hr^{-1}$. Thus more rapid aerosol formation was found to take place at high relative humidities. This is probably due to a higher water content in the particles formed under these conditions.

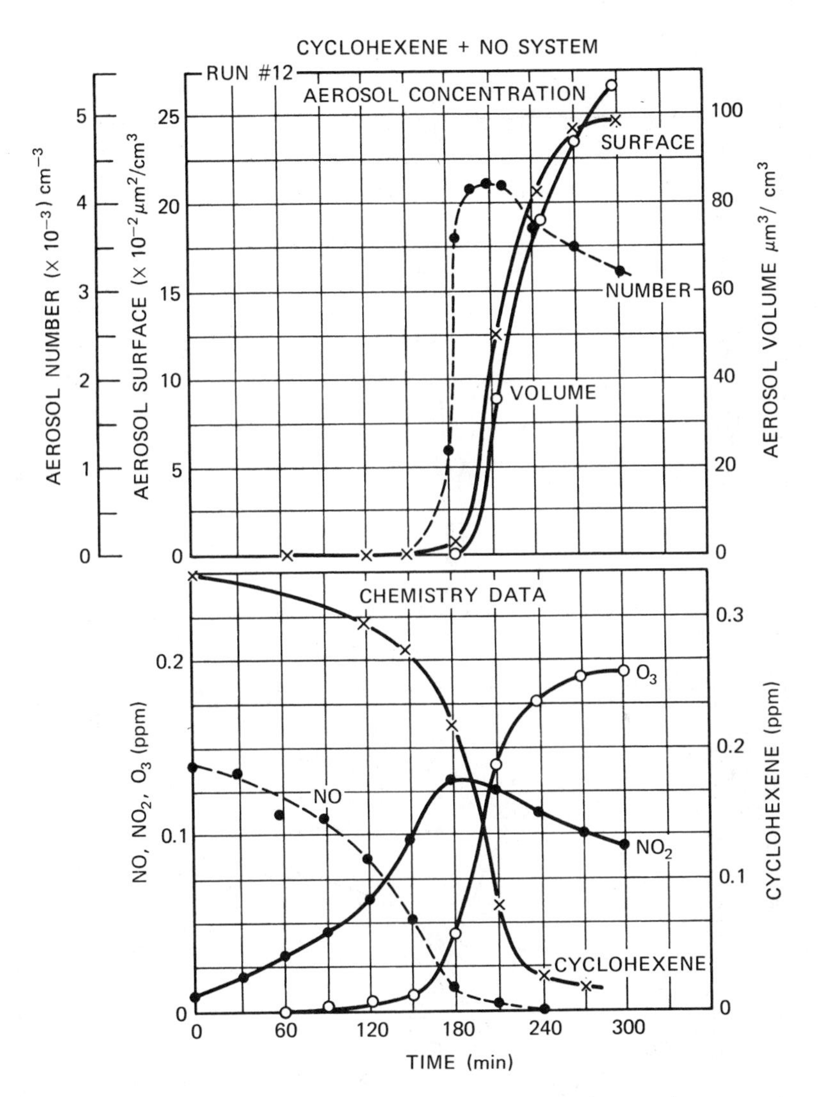

FIGURE 3. Aerosol development and chemistry data for a cyclohexene + NO system

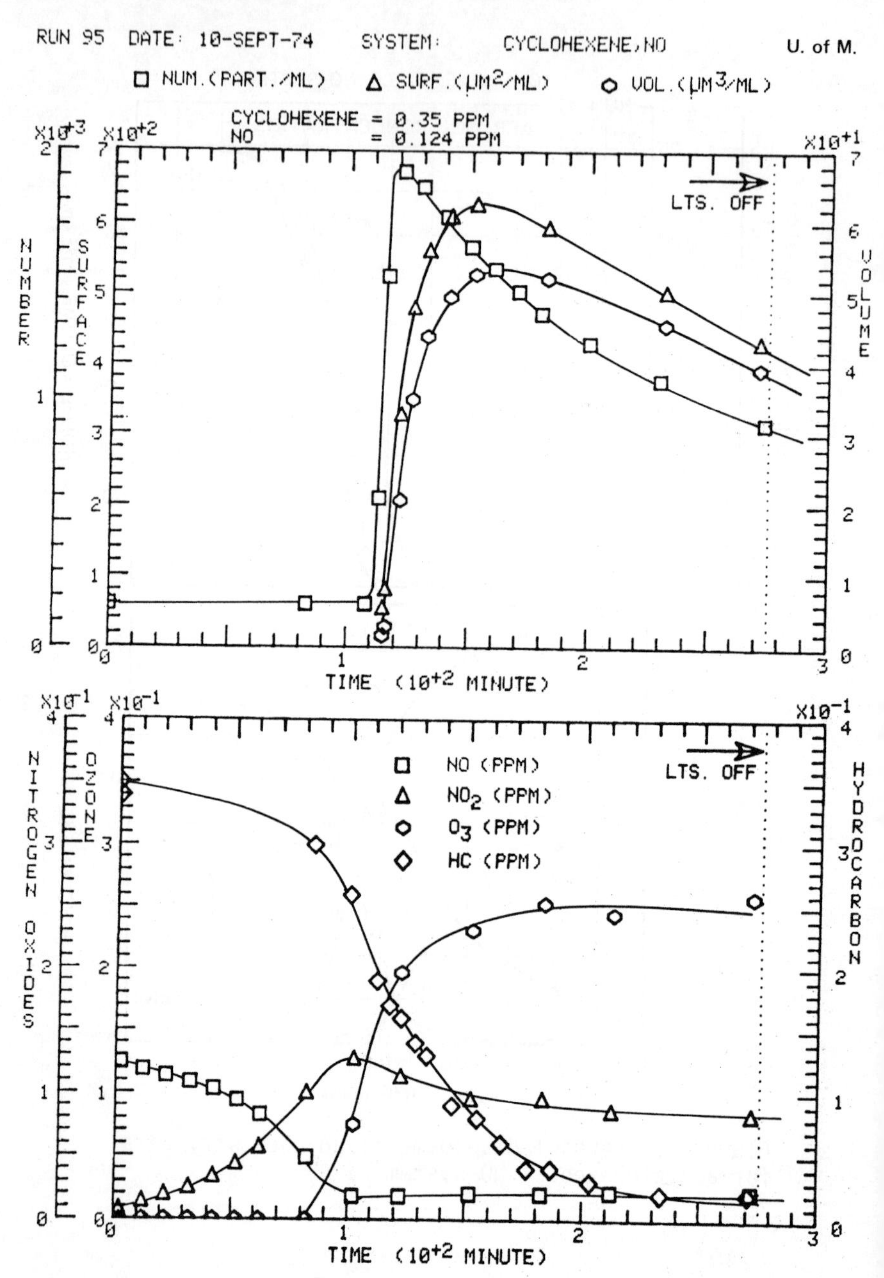

FIGURE 4. Aerosol development and chemistry data for a cyclohexene + NO system.

B. Hydrocarbon + NO + SO_2 Experiments

An additional degree of complexity is introduced when SO_2 is added to the HC + NO system. In terms of the chemical species that we monitor, NO, NO_2, O_3, and hydrocarbon, the experiments with and without SO_2 are virtually identical. The aerosol behavior, however, is much different when SO_2 is added. During the first phase of the experiment, aerosol growth occurs in a manner similar to that of a simple SO_2 + clean air experiment, that is, an essentially linear curve of volume against time. As the second phase of the experiment begins, the rate of volume production increases markedly. During this phase, the growth curves are more like the hydrocarbon + NO experiments, except that the aerosol number concentrations are higher. In terms of qualitative behavior, a hydrocarbon + NO + SO_2 experiment behaves almost as though it resulted from a linear combination of the hydrocarbon + NO system and the SO_2 + clean air system. Quantitatively, however, more aerosol is usually formed in both growth phases than would be predicted by a simple linear combination of the hydrocarbon + NO system and the SO_2 + clean air system. The details of these interactions are discussed briefly below as each hydrocarbon system is treated individually.

For the toluene + NO + SO_2 system, very little effect was observed in the Calspan experiment over that produced by toluene + NO alone. Initial aerosol growth was observed soon after the start of irradiation due to SO_2 oxidation and the formation of H_2SO_4 particles. However, once oxidation of NO was complete, more than 6 hours later only a slight increase in aerosol production was detected.

At the University of Minnesota, the initial phase of aerosol growth in the toluene + NO + SO_2 system was very similar to Calspan's; that is, before the oxidation of NO was complete, aerosol growth was like that of a SO_2 + clean air system, but faster. Apparent SO_2 oxidation rates ranged from 0.23 to 0.67 percent hr^{-1} at the University of Minnesota compared to a rate of 0.32 percent hr^{-1} for the Calspan experiment.

In comparing the toluene + NO + SO_2 experiments with toluene + NO, it was found that the nature of aerosol growth was quite different in the second phase. Especially in the University of Minnesota cases, the experiments reveal that, although aerosol volume production in the second growth phase is similar with and without the addition of SO_2, much larger surface and number concentrations are produced when SO_2 is present. Thus the particle diameters must be much smaller. (This effect is discussed in more detail in Section V.)

In the 1-hexene + NO + SO_2 experiments, the addition of SO_2 was found to appreciably increase aerosol formation but had little effect on the chemical behavior. Aerosol and chemistry data typifying this type of system is shown in Figure 5. During the initial growth phase, while NO is being converted to NO_2, there is a rise in number concentration due to SO_2 oxidation and H_2SO_4 aerosol formation, but the surface and volume production is quite low and proceeds almost as if only SO_2 were present. For this experiment the apparent SO_2 oxidation rate during the initial growth phase was about 0.17 percent hr^{-1}.

During the second phase of growth, at about the same time that ozone begins to appear, the rate of aerosol production again increases. Number concentration begins to grow and the rates of surface and volume production are greatly enhanced. The shapes of the surface and volume concentration curves in this growth phase are quite similar to those obtained in the pure 1-hexene + NO system, but the actual concentrations are higher (see Tables 1 and 2 for comparisons). The increases, especially in the University of Minnesota data, are larger than would be expected from a simple linear combination of the aerosol formed by the SO_2 pure air system (for comparable SO_2 concentrations) and that from the hexene + NO system. Synergistic interactions must take place which lead to enhanced aerosol formation in the combined system. Chemical composition data would be particularly useful in helping to explain these results: however, in the absence of this data, some possible explanations may be:

1. The sulfuric acid aerosol formed during the first growth phase might act to catalyze the formation of aerosol from 1-hexene + NO reaction products.

2. Gas phase products of SO_2 photooxidation might act to accelerate the formation of aerosol from 1-hexene + NO reaction products.

3. Gas phase 1-hexene + NO products might act to greatly increase the rate of SO_2 photooxidation and thus aerosol formation.

Schemes 1 and 2 both depend upon the interaction of SO_2 photooxidation products with the 1-hexene + NO system. The interaction could lead to an acceleration of the rate of formation of whatever hydrocarbon-related species condensed in the second growth phase or to an alteration of the gas phase reaction paths leading to the formation of a nonvolatile, more readily condensible species. A combination of these two effects could also occur. Only a tiny fraction of the 1-hexene that is

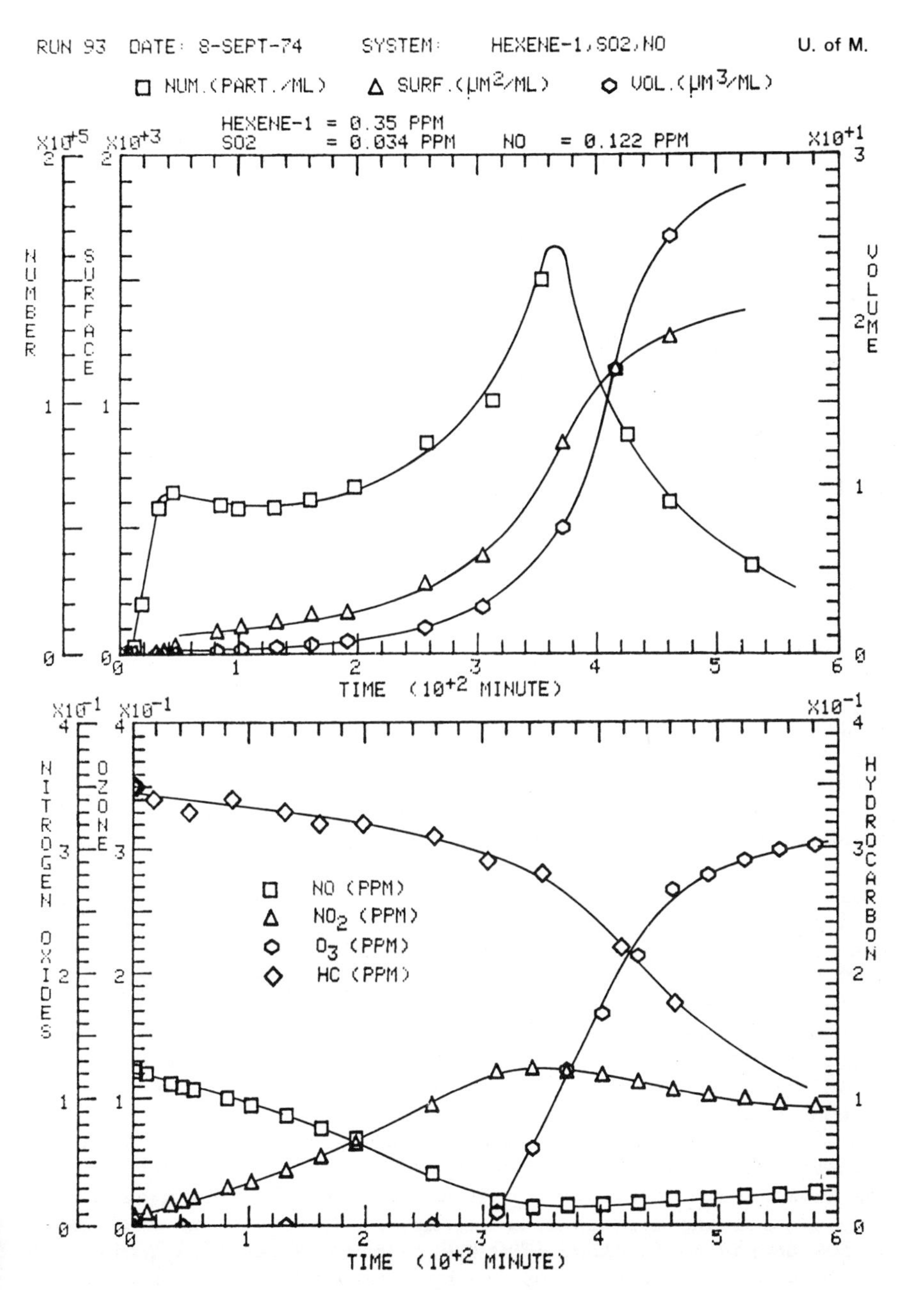

FIGURE 5. Aerosol development and chemistry data for a 1-hexene + SO_2 + NO system.

oxidized would have to be converted to nonvolatile species to produce a significant aerosol yield. In the first 500 minutes of University of Minnesota experiment 92 (Figure 2), about 0.24 ppm or 890 $\mu g/m^3$ of 1-hexene disappeared from the gas phase, during the same period about 0.18 $\mu m^3/cm^3$, or assuming an aerosol density of unity, 0.18 $\mu g/m^3$ of aerosol is formed. This represents only about 0.02 percent of the mass of 1-hexene removed.

In run 93 with SO_2 present (Figure 5), the aerosol yield is raised to about 3.5 percent of the mass of 1-hexene removed. Such a change could easily occur as a result of either scheme 1 or 2. If scheme 3 is important, it should be easy to verify experimentally because in order to explain the observed aerosol formation a significant fraction, that is, more than 20 percent of the SO_2 initially present, would have to be removed from the gas phase. This quantity could be detected by gas phase sulfur monitoring. Analysis of aerosol sulfate content would provide more definitive results.

The m-xylene + NO + SO_2 systems investigated were very reactive in terms of chemical and aerosol behavior. The times to reach $[NO_2]_{max}$ were shorter in every case than the analogous toluene or hexene experiments. As is the case for the other hydrocarbons, two phases of aerosol growth are evident as shown by the data in Figures 6 and 7. The SO_2 photooxidation rates during the initial growth phase were 0.14 and 0.21 percent hr^{-1} for the University of Minnesota tests and 0.32 percent hr^{-1} in the Calspan experiments. m-Xylene and SO_2 appeared to interact only weakly during the initial phase of aerosol growth.

By contrast, during the second growth phase, very rapid aerosol production was observed as the data in Figure 6 show. Here the addition of SO_2 to the m-xylene + NO produces a great deal more aerosol surface and substantial additional volume. The much greater increase in surface area than in volume results from the formation of many more small particles than in the m-xylene + NO system. Comparing the aerosol data with and without SO_2 for the other HC + NO systems reveals the same effect. Smaller particles are present with the added SO_2, since the SO_2 generates a large concentration of tiny nuclei early in the experiment. These particles then serve as sites for subsequent condensation in the second growth phase.

One cyclohexene + NO + SO_2 system was investigated at Calspan (run 9) and one at the University of Minnesota (run 96). The Calspan and University of Minnesota data for this system are very similar and are compared in Figures 8 and 9. The addition of SO_2 to the cyclohexene + NO system produced effects qualitatively similar to those produced by addition of SO_2 to the other hydrocarbon + NO systems studied. Thus the NO, NO_2, O_3, and

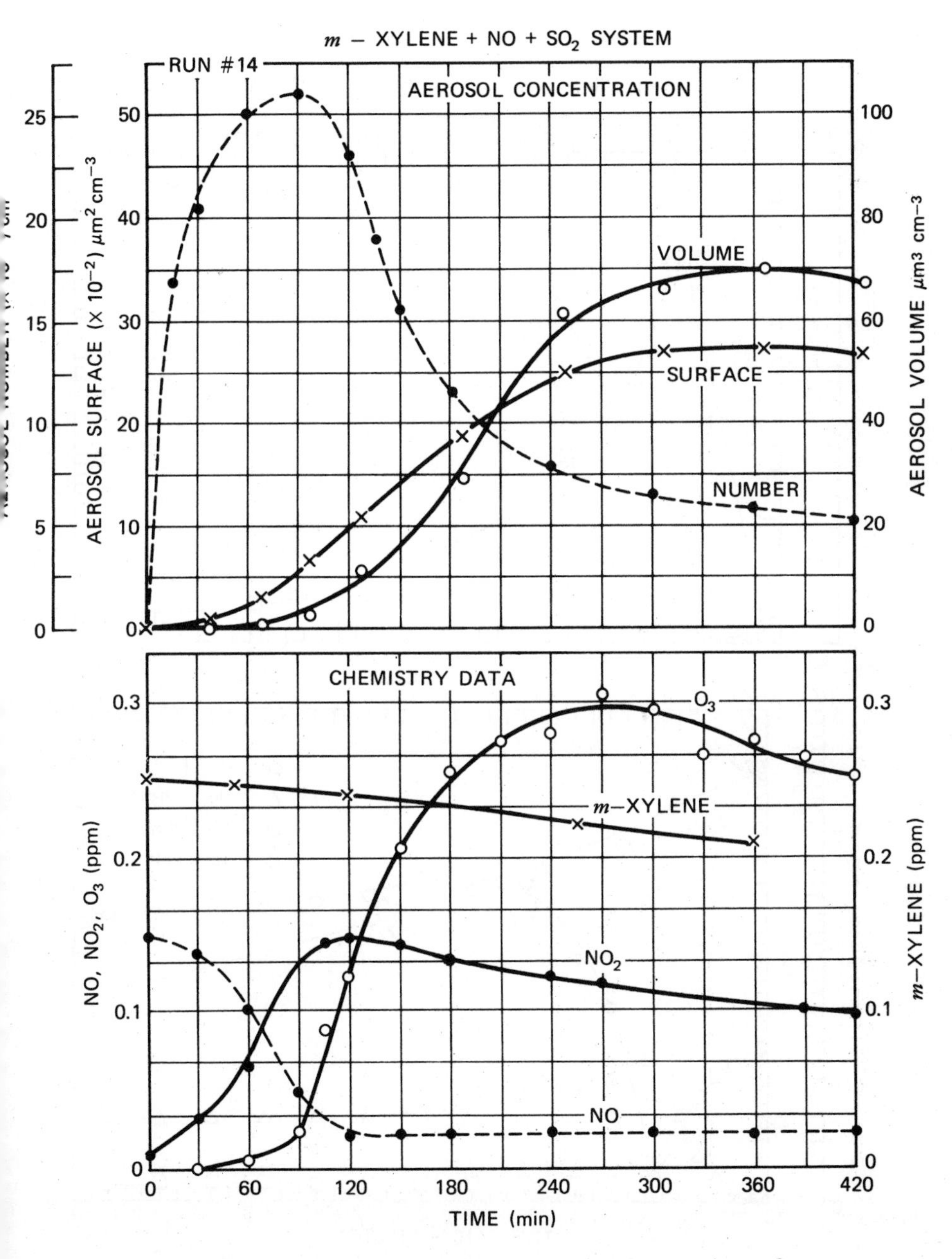

FIGURE 6. Aerosol development and chemistry data for an m-xylene + NO + SO_2 system.

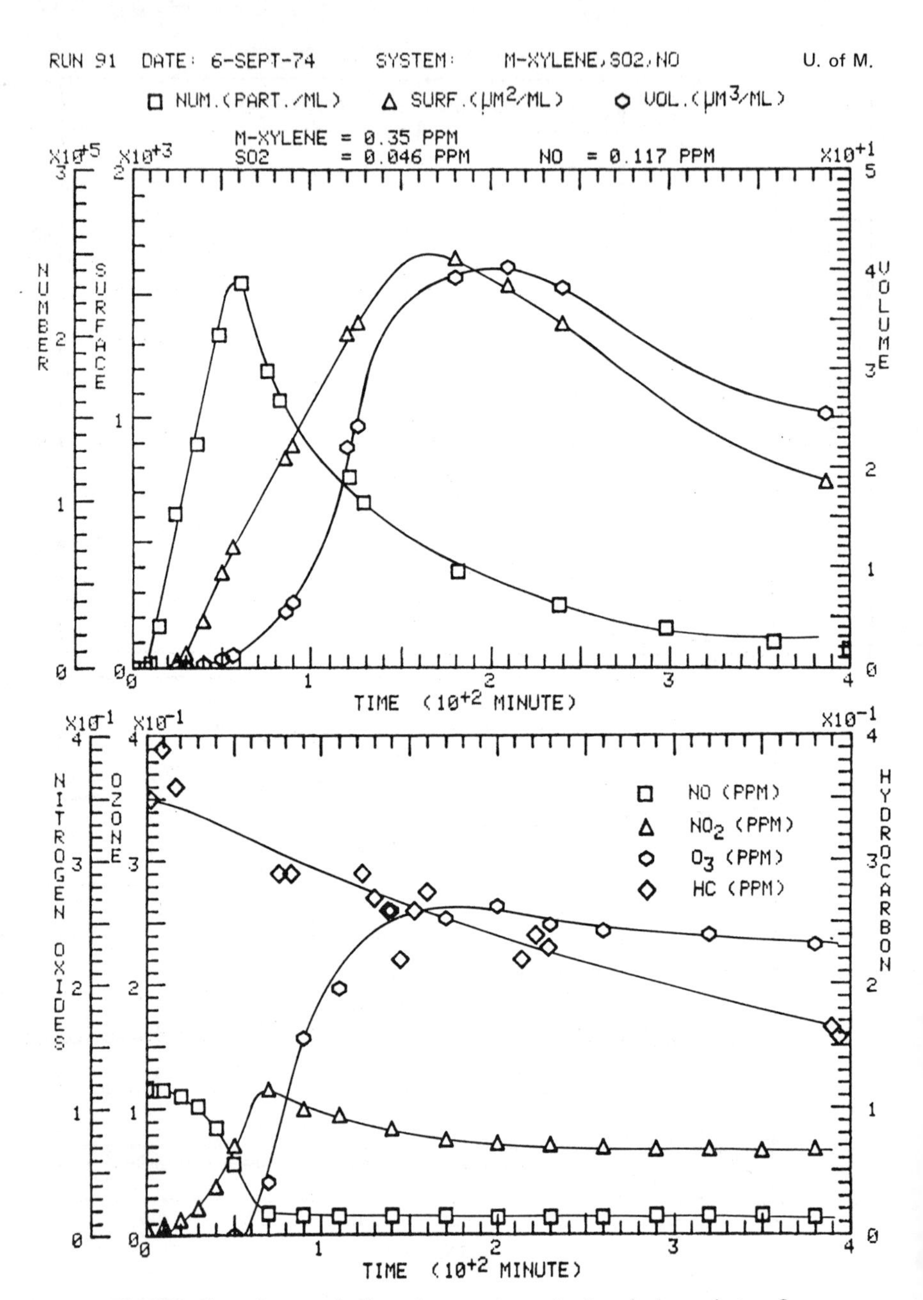

FIGURE 7. Aerosol development and chemistry data for an m-xylene + SO_2 + NO system.

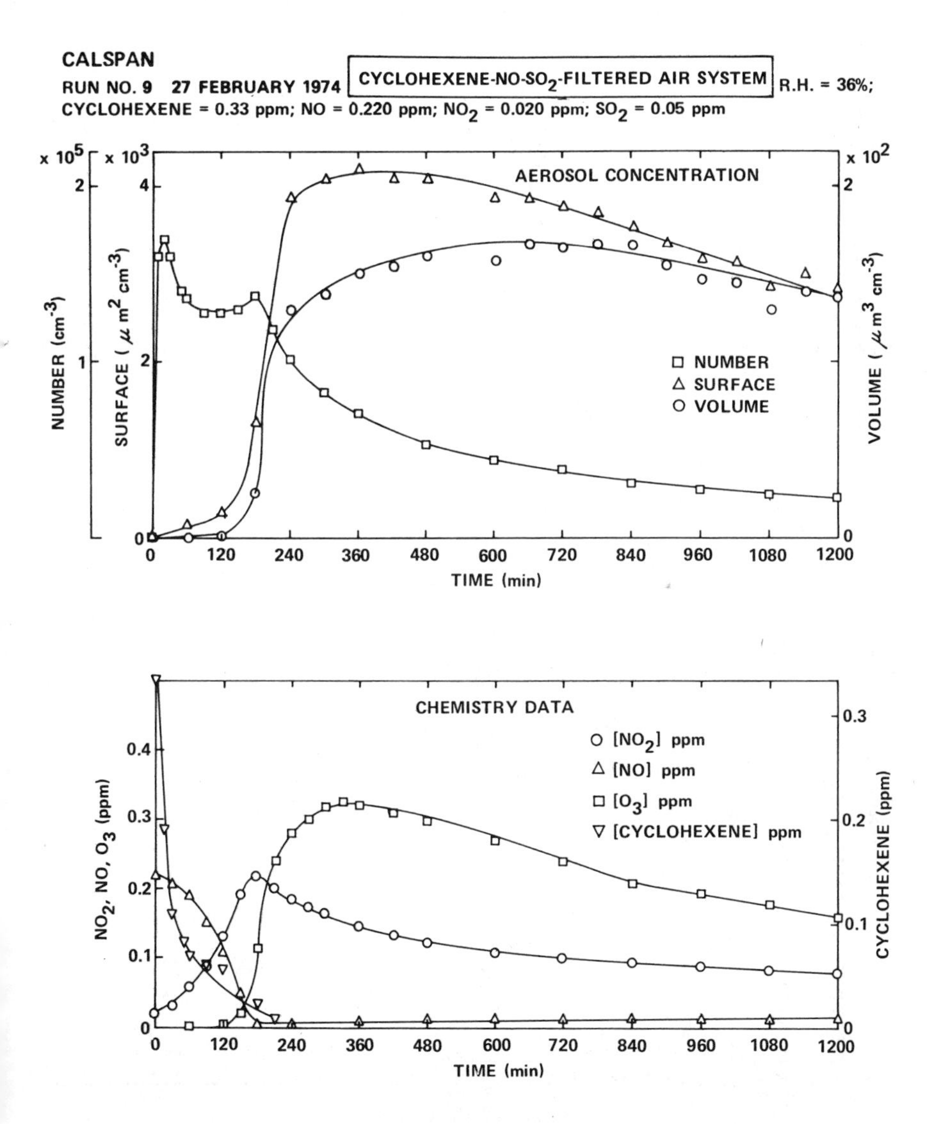

FIGURE 8. Aerosol development and chemistry data for a cyclohexene + NO + SO_2 filtered air system.

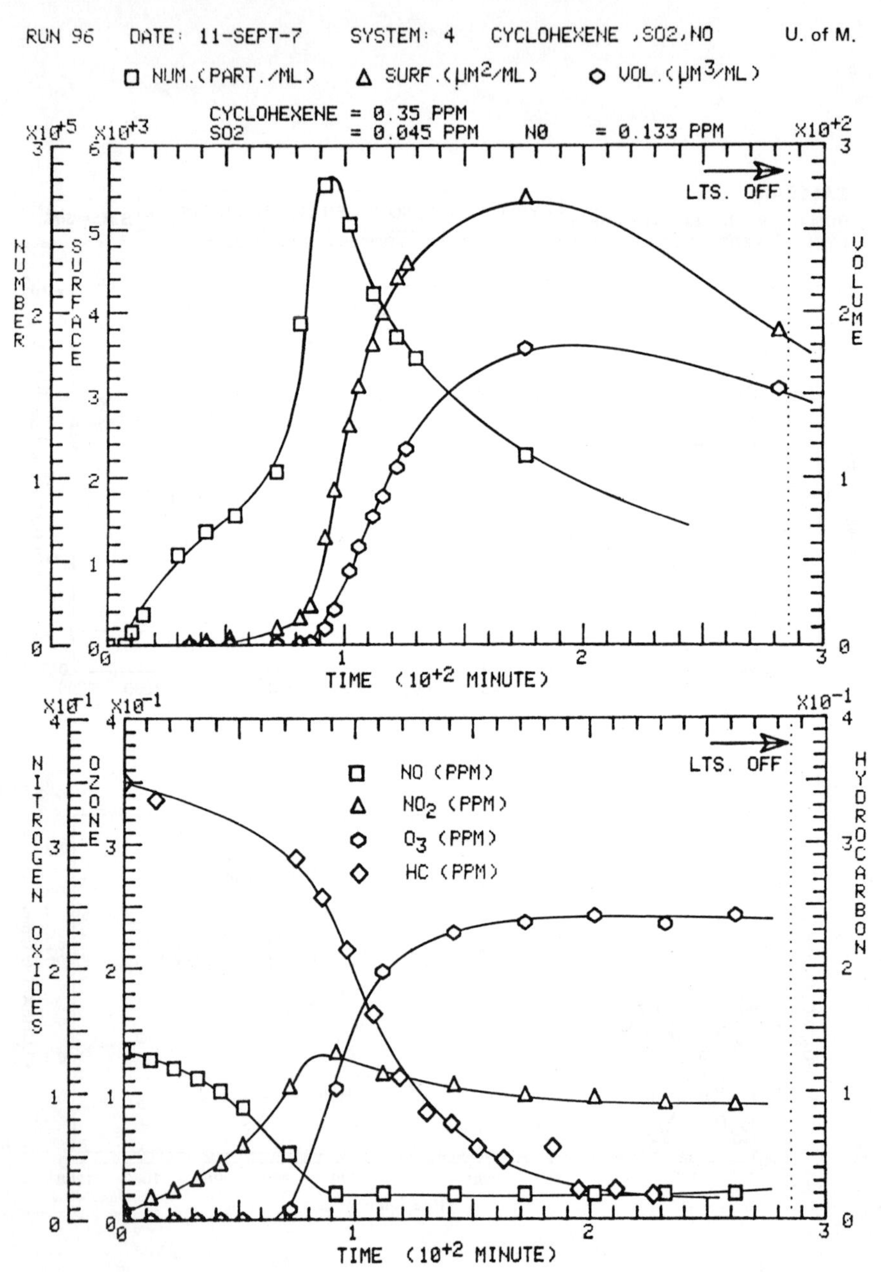

FIGURE 9. Aerosol development and chemistry data for a cyclohexene + SO_2 + NO system.

and hydrocarbon profiles were essentially unchanged by SO_2 addition. At the same time, aerosol behavior was markedly changed, with two-phase aerosol growth resulting.

The changes in aerosol behavior produced by SO_2 addition to this system were dramatic and informative. During the first phase of growth, aerosol production occurred as though a slightly contaminated SO_2 + clean air system were being irradiated. The apparent SO_2 photooxidation rate was 0.20 percent/hr in the University of Minnesota case and 0.28 percent/hr in the Calspan experiment, about the same as that observed for other hydrocarbon + NO + SO_2 systems. With this system, evidence of the second phase of aerosol growth appears before NO has been completely oxidized out of the system. Especially in the University of Minnesota case, the volume against time data becomes nonlinear with marked upward curvature appearing as early as 50 minutes after lights on. However, in both cases, rapid aerosol volume production does not occur until about the time of complete NO oxidation and rapid formation of ozone.

In terms of volume production, the second phase growth is similar to the cyclohexene + NO system in the absence of SO_2. The differences that are observed are not considered significant because of the difficultues associated in measuring $(dv/dt)_{max}$ for the cyclohexene + NO system, that is, rapid growth beyond the upper particle size limit of the EAA. The change in surface and number concentrations associated with the addition of SO_2, however, are very significant. Both number and maximum surface concentration are much larger. However, since the total aerosol volumes produced in all cyclohexene cases are about the same, the addition of SO_2 must lead to a dramatic decrease in particle size.

A relative humidity effect is also apparent in the hydrocarbon + NO + SO_2 experiments summarized in Table 2. Comparison of runs 77 and 88 for toluene, 78 and 93 for 1-hexene, and 82 and 91 for _m_-xylene reveal a marked increase in both the equilibrium surface area and the maximum aerosol production rates at higher relative humidities. Again a higher water content in the aerosol formation is the probably cause.

The effect of SO_2 on particle size can be seen in Figure 10, in which mean surface diameter has been plotted as a function of time for several HC + NO and HC + NO + SO_2 systems studied. Data for a typical SO_2 irradiation are also shown in the figure. The almost explosive growth in the cyclohexene + NO case occurs at about the time of complete oxidation of NO and appearance of appreciable ozone. In the case shown, particle diameters of nearly 0.5 μm were produced in less than an hour from the time of initial nucleation. With SO_2 in the system, the particle sizes are much smaller. Under these conditions, initial nuclei

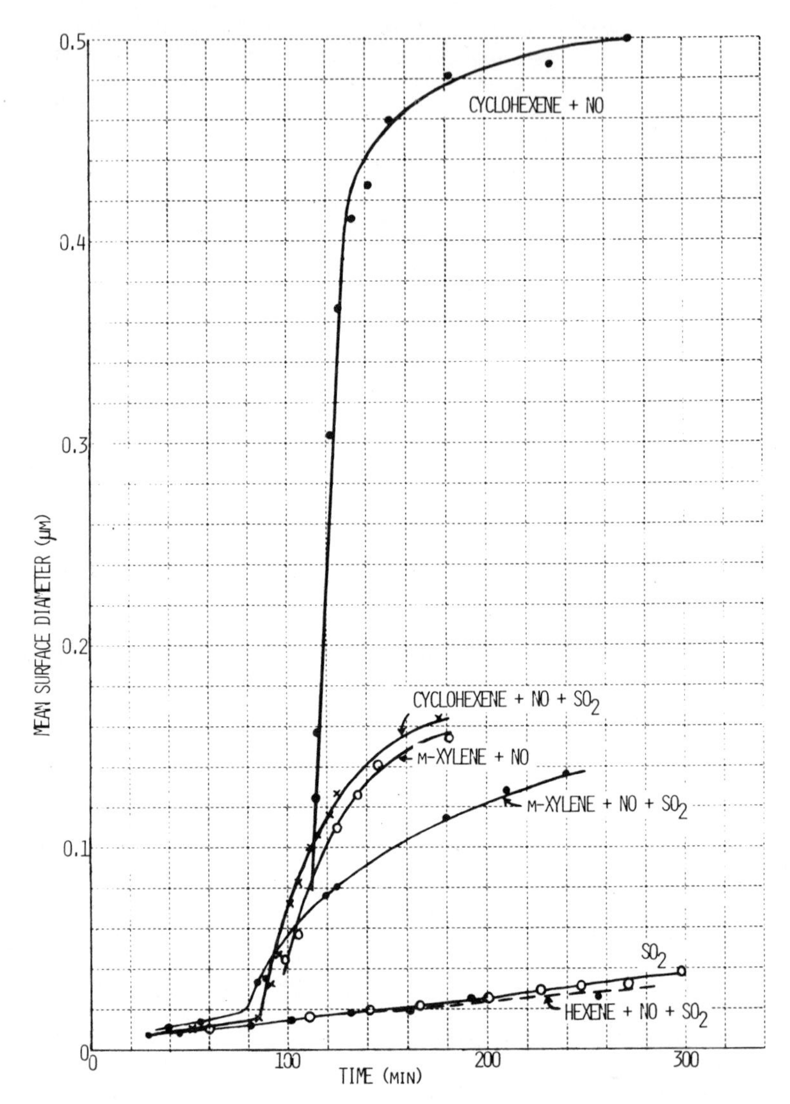

FIGURE 10. Mean surface diameter versus time for several HC + NO, HC + NO + SO_2, and SO_2 experiments.

formation results from SO_2 oxidation. These nuclei provide numerous sites for condensation during the second phase of growth. Consequently, although the addition of SO_2 does not significantly change the total volume of aerosol formed, the aerosol consists of a much larger number of smaller particles.

m-Xylene + NO behaves like the cyclohexene + NO system but to a lesser extent. The addition of SO_2 results in the same effect; initial particle formation and growth at a rate similar to that for SO_2 alone followed by accelerated growth during the second phase. By comparison, the much less reactive hexene + NO + SO_2 system was observed to essentially follow the SO_2 particle growth curve for the duration of the experiment.

The implications of these data relative to the production of light-scattering aerosol can be seen in Figure 11. In the figure

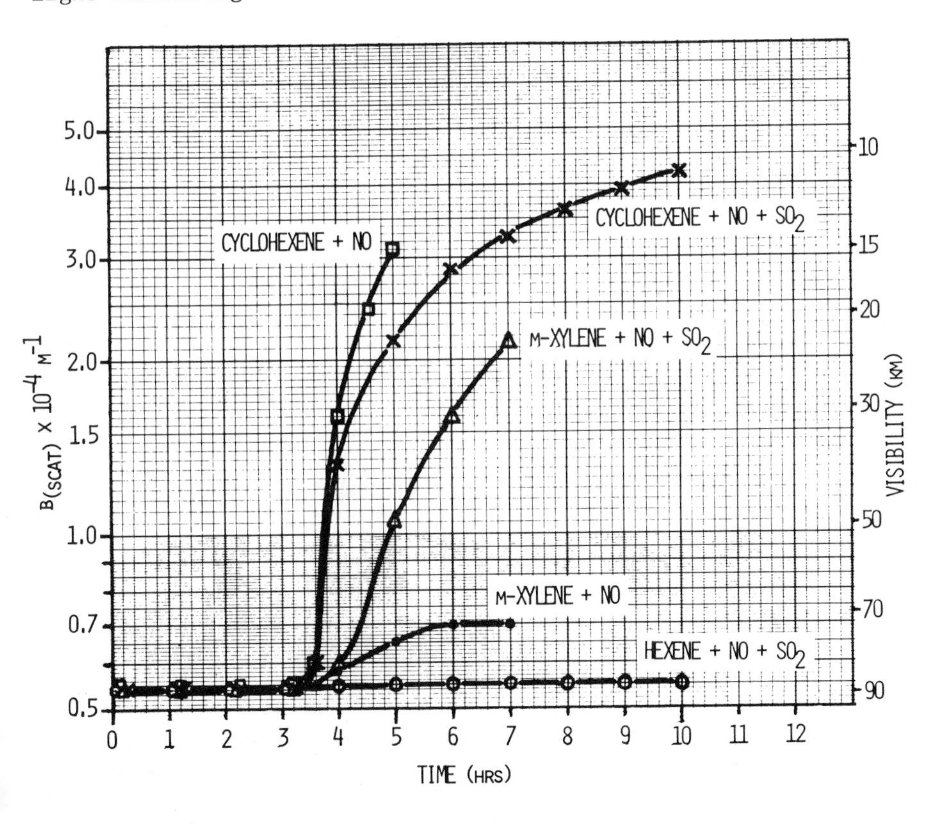

FIGURE 11. Light-scattering coefficient (b_{scat}) of photochemical aerosols versus time.

the light-scattering function b_{scat} is plotted as a function of time for several HC + NO, HC + NO + SO_2 systems in the Calspan chamber. The production of very large particles in the cyclohexene + NO case results in the rapid formation of light scattering aerosol. Although the particle diameters are much larger than in the comparable cyclohexene + NO + SO_2 experiment, b_{scat} is nearly the same because fewer particles are produced. The addition of SO_2 results in smaller but more numerous particles whose effect on light scattering is about the same.

The addition of SO_2 to the m-xylene + NO case can also be seen to produce substantial light-scattering aerosol during the second phase of aerosol growth. For several hours there is no effect; but once NO oxidation is complete and second stage growth begins, large visibility losses are observed. The fact that the concentration of particles is much smaller in the m-xylene + NO case (compared to m-xylene + NO + SO_2) accounts for the lack of appreciable light-scattering aerosol.

By contrast, the 1-hexene system, even in the presence of SO_2, did not produce significant light-scattering aerosol or visibility losses. The same result was obtained in the Calspan tests for the toluene + NO + SO_2 system. The somewhat accelerated particle growth observed in the University of Minnesota tests may have resulted in light-scattering aerosol; however, there was no opportunity to make this measurement. Finally, the SO_2 alone, while producing very large number concentrations of particles, did not produce significant light-scattering aerosol over the duration of these tests.

V. CONCLUSIONS

From the data generated in these experiments, the following points can be made:

1. Each HC + NO experiment can be divided into two phases. In the first phase, NO is converted to NO_2 and some oxidation of hydrocarbon occurs. Ozone starts to appear near the end of this period. The second phase, accompanied by substantial aerosol formation, begins as soon as NO is oxidized out of the system and NO_2 reaches its maximum; ozone grows rapidly and approaches a maximum.

2. The addition of SO_2 to the HC + NO system was generally found to exert a synergistic effect on aerosol formation. At Calspan the effect was greatest for m-xylene, while at the University

of Minnesota the largest effect on aerosol behavior was observed in the 1-hexene + NO + SO_2 system. Possible synergistic effects in the cyclohexene system were masked by the explosive growth of aerosol with and without the addition of SO_2.

3. The addition of SO_2 to the HC + NO system produces a dramatic decrease in the mean particle diameter. This results from the initial formation of very high concentrations of nuclei during the initial stages of the experiment. During the second stage of aerosol growth, condensation proceeds on the existing particles rather than forming fewer but larger particles typical of the HC + NO system.

4. Of the hydrocarbons studied, cyclohexene was the most reactive, both in terms of aerosol and chemical behavior, followed by m-xylene, 1-hexene, and toluene. The main difference observed in the duplicate experiments at the University of Minnesota was that 1-hexene was the least reactive hydrocarbon.

5. SO_2 photooxidation rates of a few tenths of a percent per hour are typically observed in clean filtered air for a light intensity of about 50 percent noonday sun. In the presence of hydrocarbon contamination, accelerated rates are generally observed.

6. It is possible to characterize system reactivity in terms of aerosol behavior. The most important variables are maximum number concentration, equilibrium surface concentration, and volumetric growth rate. These aerosol measures of reactivity have been found to correlate well with other conventional parameters, such as time to $[NO_2]_{max}$ and $[O_3]_{max}$.

7. In all cases involving SO_2, aerosol formation rates were enhanced at high relative humidities, probably as a result of the higher water content in the aerosols.

8. The data generated in these experiments in chambers of widely different physical dimensions show a high degree of correlation. The main difference in the results is indicated in the significantly greater reactivity of toluene found in the University of Minnesota chamber.

ACKNOWLEDGMENTS

The work reported here was sponsored jointly by the U. S. Environmental Protection Agency, Laboratory of Chemistry and Physics, and the Coordinating Research Council for the United States vehicle manufacturing and petroleum industries.

REFERENCES

1. Bray, W. H. 1970. Water Vapor Pressure Control at Aqueous Solutions of Sulfuric Acid, J. Mater. 5, 233-248.

2. Clark, W. E. 1972. Measurement of Aerosol Produced by the Photooxidation of SO_2 in Air, Ph.D. dissertation, University of Minnesota.

3. Clark, W. E. and Whitby, K. T. 1975. Measurement of Aerosols Produced by the Photochemical Oxidation of SO_2 in Air, J. Colloid. Interf. Sci. 51, 477.

4. Davis, D. D. 1974. Absolute Rate Constants for Elementary Reactions of Atmospheric Importance; Results from University of Maryland Gas Kinetics Lab, Report No. 1, Chemistry Department, University of Maryland.

5. Doyle, G. J. 1961. Self Nucleation in the Sulfuric Acid Water System, J. Chem. Phys. 35(3), 795.

6. Friend, J. P., Leifer, R., and Trichon, M. 1973. On the Formation of Stratospheric Aerosols, J. Atmos. Sci. 30, 465-479.

7. Kocmond, W. C., Kittelson, D. B., Yang, J. Y. and Demerjian, K. L. 1973. Determination of the Formation Mechanisms and Composition of Photochemical Aerosols, First Annual Summary Report, Calspan Report No. NA5365-M-1, Calspan Corp., Buffalo, New York.

8. Kocmond, W. C., Kittelson, D. B., Yang, J. Y., and Demerjian, K. L. 1975. Study of Aerosol Formation in Photochemical Air Pollution, Calspan Report No. NA5365-M-2, Calspan Corp., Buffalo, New York.

9. Liu, B. Y. H. and Pui, D. Y. H. 1974. A Submicron Aerosol Standard and the Primary, Absolute Calibration of the Condensation Nuclei Counter, J. Colloid. Interf. Sci. 47, 155.

10. Liu, B. Y. H. and Pui, D. Y. H. 1977. On the Performance of the Electrical Aerosol Analyzer, J. Aerosol Sci., to be published.

11. Reiss, H. 1950. J. Chem. Phys. 18, 840.

12. Takahashi, K., Kasahara, M., and Masayuki, I. 1975. A Kinetic Model of Sulfuric Acid Formation from Photochemical Oxidation of Sulfur Dioxide Vapor, J. Aerosol Sci. 6(1), 45.

13. Whitby, K. T. and Clark, W. E. 1966. Electrical Aerosol Particle Counting and Size Distribution Measuring System for the 0.015 to 1 μm Size Range, Tellus 18, 573.

The Photomodification of Benzo[a]Pyrene, Benzo[b]Fluoranthene, and Benzo[k]Fluoranthene Under Simulated Atmospheric Conditions

DOUGLAS A. LANE* and MORRIS KATZ
Department of Chemistry
York University, Toronto
Ontario, Canada

I. INTRODUCTION

In recent years, there have been a number of attempts to study the modifications of polycyclic aromatic hydrocarbons (PAH) (Falk et al., 1956; Falk et al., 1960; Tebbens et al., 1966; Thomas et al., 1968) under simulated atmospheric conditions. Falk et al. (1956, 1960) dissolved various PAH in lipid solvents and applied the PAH to several sites on Watman filter papers. The PAH on the filter papers were exposed to various conditions including a synthetic smog that had an oxidant content of approximately 30 ppm as determined by the potassium iodide method. Tebbens et al. (1966) and Thomas et al. (1968) produced a smoke through the incomplete combustion of propane and passed the smoke through a 22-foot long, 6-inch diameter Pyrex pipe, the first half of which was irradiated with banks of fluorescent lamps. The particulate smoke was analyzed, both at the entrance to and the exit from the pipe, for its PAH content and, as a consequence of their studies, the half lives for the destruction of benzo[a]-

*Present address: Sciex Limited, Thornhill, Ontario, Canada L3T 1P2 pyrene (BaP) and the PAH, in general, were implied to be of the order of hours or days in the atmosphere.

These experimental conditions, while yielding interesting information, were not representative of actual atmospheric conditions. Oxidant levels of 30 ppm do not occur and fluorescent lamps, sunlamps, and the black light fluorescent lamps, recently advocated by Laity (1971), are poor substitutes for solar radiation in the important (photochemically) spectral range of 295 to 400 nm.

It was, therefore, the aim of this research to develop a reaction system that would approximate as closely as possible the factors contributing to the atmospheric decomposition (namely oxidant and sunlight) of the PAH. A dynamic flow reaction chamber was employed to expose the PAH to constant oxidant (ozone was chosen as the oxidant since it normally accounts for about 90 percent of the normal atmospheric oxidant) levels of 0 to 2.28 ppm and to radiation. The system permitted the heterogeneous gas-solid decomposition reactions to be followed as a function of time, and consequently, half lives for the various reactions were determined.

II. EXPERIMENTAL METHODS

A. Apparatus

The apparatus consisted of an air source, air purification train, ozone generator, reaction chamber, and irradiation source as shown schematically in Figure 1.

The flow of air from a cylinder (Matheson Zero Gas) was regulated by a precision needle valve to a constant flow of 100 ml/min. A Gelman 4.7-cm diameter glass fiber filter (in a Gelman filter holder) was installed in the line to remove any oil droplets or particles that might be introduced into the airstream by the air tank regulator valve. A purification train consisting of absorbing towers of activated charcoal, magnesium perchlorate, and ascarite further served to remove both organic compounds and water moisture from the air stream. A calibrated rotameter was employed to monitor the flow rate of the air through the ozone generator and then to the reaction chamber. All connections downstream of the activated charcoal absorbing tower were borosilicate glass to glass or glass to Teflon to glass.

The ozone generator was fabricated in this laboratory after the design of Hodgeson et al. (1971) and was calibrated by the neutral-buffered potassium iodide method with an air flow of 100 ml/min through the generator. A constant voltage transformer

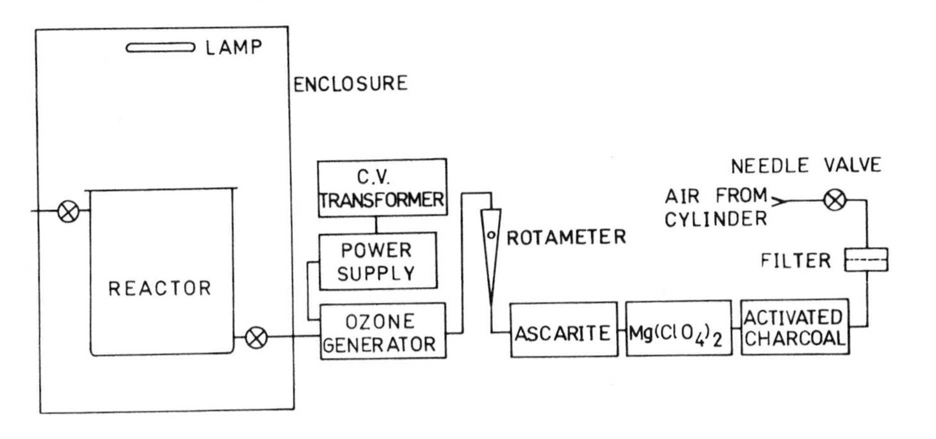

FIGURE 1. Schematic diagram of the reactor system.

regulated the power supply which powered the Pen Ray ultraviolet lamp used in the generator.

The reaction chamber was a 20-liter glass jar with an air inlet stopcock at the bottom and an air exit stopcock diametrically opposite but at the top of the jar. Extruded silicone rubber (6.4 x 6.4 mm) was sealed to the top of the jar with General Electric RTV silicone glue to serve as a gasket between the jar and the glass disc cover (to which it was not sealed) of the reactor. After 12 months of use, there has been no detectable reaction between the silicone rubber gasket or the silicone rubber glue and the ozone used in the reaction chamber. A 0.3-cm thick, 30-cm diameter Pyrex disc covered the top of the reactor and acted as a cut-off filter for the light from the irradiation source. Nine 5-cm diameter petri dishes containing the PAH samples (each individual sample was used to produce one experimental point on a decomposition curve) to be reacted were placed on a support constructed of glass rods. The glass support was placed 6.5 cm above the air inlet and provided ample support for the petri dishes, while creating minimal resistance to the air flowing through the reactor.

Both the reaction chamber and irradiation source were placed inside an enclosure to exclude laboratory fluorescent illumination from the reaction. A temperature build-up was minimized by installing a muffin fan at the top of the enclosure.

A 500-watt General Electric Quartzline Lamp (Q500T3/CL) mounted in a standard photographic housing, 46 cm above the dishes in the chamber, was used to simulate natural sunlight. The spectral output of the lamp was obtained using a Heath 701 modular spectrophotometer, which was equipped with a 1P28 photomultiplier tube. The spectral intensity of the lamp between wavelengths of 290 and 400 nm was determined by actinic irradiance measurements using the standard uranyl nitrate-oxalic acid method and by the photoelectric current induced in a 935 photodiode tube. The phototube was enclosed in a light-tight black box, which had a slit opening 1.6 x 3.18 mm. Both actinic irradiance and photoelectric current measurements were conducted for the lamp, the background irradiation, and natural sunlight. When lamp and background irradiation tests were carried out, a 0.3-cm thick Pyrex filter was placed over the uranyl nitrate-oxalic acid absorbing solutions and over the slit in the phototube housing. No Pyrex filter was used when taking sunlight measurements.

To our knowledge, this is the first time that a Quartzline Lamp has been used to simulate sunlight in photochemical reactions. Figure 2 shows the spectral output of the sun at solar zenith angles of 40° and 60° as reported by Leighton (1961). The spectral distribution of the Quartzline lamp using a 0.3-cm thick Pyrex disc as a filter is also shown but not to the same absolute intensity scale. It should be noted that the spectral distribution of the Quartzline lamp is a continuous spectrum, not a line spectrum as obtained from sun lamps and fluorescent lamps, and that the spectral distribution very closely approximates the solar distribution between 295 and 400 nm.

Actinic irradiance and current measurements were made at a point 46 cm below the lamp to compare the lamp emission with background radiation, both inside and outside the enclosure, and with sunlight. As can be seen from Table 1, fluorescent lighting in the laboratory was between 2 and 10 percent the intensity of the lamp and, therefore, would contribute significantly to the reaction if it were not excluded. Background radiation inside the enclosure was insignificant. This lamp is thus considered to be an excellent, inexpensive laboratory source for the simulation of natural sunlight. Wavelengths down to 230 nm (but at very low intensity) may be obtained if the Pyrex filter is omitted.

B. Sample Preparation, Reaction, and Analysis

In order to simulate the exposure of PAH to atmospheric conditions, less than 500 ng of a given PAH were distributed as

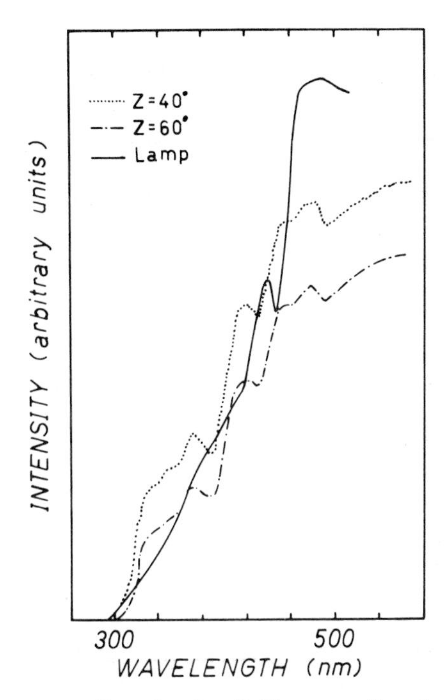

FIGURE 2. Spectral output of the sun for two solar zenith angles (Z) and for the Quartzline Q500T3/CL Lamp after passing through a 0.30-cm Pyrex plate.

a thin dispersion in each of a number of petri dishes. The individual PAH could then be exposed to the desired radiation and ozone conditions.

Aldrich 98 percent pure BaP was used, without further purification, to produce a stock solution of 245 ng/ml BaP in spectrograde n-hexane. A 250 μl aliquot of the stock solution was transferred by Eppendorf pipet to each of twelve 5-cm diameter petri dishes, and the aliquot in each dish was made to coat the bottom of each dish as evenly as possible while carefully evaporating the solvent. Three dishes were selected at random for time t = 0 determinations and to check the reproducibility of the sample preparation procedure. A 250-μl aliquot of the stock solution was also transferred to each of three 10-ml volumetric flasks to enable the preparation of a calibration graph and to check the efficiency of recovery of the BaP from the petri dishes. The nine remaining dishes were placed in the reactor by sliding the glass disc back just enough to admit the

TABLE 1. Actinic Irradiance and Phototube Data for the Various Radiation Sources Tested

Irradiation Tested	Actinic irradiance[a] (photons/sec/cm^2)	Current[b] (amperes)
Direct sunlight	1.20×10^{15} [c]	2.95×10^{-5}
Quartzline lamp output filtered by the 0.30-cm Pyrex filter	9.36×10^{13}	2.10×10^{-6}
Background—laboratory fluorescent irradiation	9.56×10^{12}	4.4×10^{-8}
Background—inside reactor enclosure	---	$\sim 4 \times 10^{-11}$

[a] All actinic irradiance measurements were made on August 28, 1973.

[b] The direct sunlight current reading was taken on March 5, 1974, with a solar zenith angle of approximately 50°.

[c] Measurement at noon with a temperature of 35°C.

dishes, one at a time. The cover was replaced and clamped in position to effect an airtight seal.

The same procedure was employed for BbF and BkF, which were used without further purification, as supplied by J. L. Monkman of the Air Pollution Control Division, Canada Department of the Environment.

The reactor was conditioned overnight with the appropriate gas mixture for the experiment to be performed and the air and air-ozone flow rates were always maintained at 100 ml/min. When required, the lamp was turned on for 1 hour before the dishes containing the particular PAH under study were placed in the reactor, but it was turned off for the brief period during which the dishes were actually being transferred to the reactor.

The course of the reaction was followed by removing dishes from the reactor at specific times and quantitatively transferring the contents of each dish to separate 10-ml volumetric flasks using several washings of spectrograde n-hexane. Subsequently, each flask was made up to volume with spectrograde n-hexane. All preparations and recoveries of samples were, of course, carried out in the absence of laboratory fluorescent lighting. Quantitation of standard and reacted samples was accomplished by the use of standard fluorescence techniques using a Farrand MK I spectrofluorometer.

Each PAH was exposed to the following conditions:

1. Lamp radiation in the absence of ozone,

2. Various ozone concentrations in the dark, and

3. Lamp radiation and various ozone concentrations.

Duplicate runs were made to check the reproducibility of the system and the resulting decomposition plots were combined to produce the oxidation curves presented in this paper.

Reproducibility checks performed for each run showed that the recovery efficiency from the dishes was better than 98 percent and the sample variation was $\pm$ 1.5 percent.

III. RESULTS AND DISCUSSION

The decomposition curves for BaP, BbF, and BkF under various experimental conditions are shown in Figures 3 to 9. Since the decomposition curves were either entirely linear, or had distinctly linear initial slopes, linear regression analyses of the data were performed to determine the initial slope of each decomposition curve. From the slope of the curve, the half life for each experimental condition was determined. The results are tabulated in Table 2.

A blank run performed on BaP (the most reactive of the three PAH tested) in dry, ozone-free air in the dark indicated a 1.8 percent loss after a 24-hour exposure. Such a loss was considered to be negligible and was not taken into account when determining the half lives for the subsequent decomposition experiments.

Under illumination and in the absence of ozone (Figure 3), the three PAH were found to have half lives of 5.3, 8.7, and 14.1 hours for BaP, BbF, and BkF, respectively.

In the dark and in the presence of ozone, however, BaP reacted much more rapidly than either BbF or BkF (Figures 4, 6, and 8). BbF and BkF appeared to react in a similar manner

TABLE 2. The Half Lives of BaP, BbF, and BkF Under Various Reaction Conditions

Reaction Conditions		Half Lives in Hours[a]		
Irradiation	Ozone Concentration (ppm)	BaP	BbF	BkF
None	0.19	0.62	52.7	34.9
	0.70	0.4	10.8	13.8
	2.28	0.3	2.9	3.3
Quartzline Q500T/CL lamp	0.0	5.3	8.7	14.1
	0.19	0.58	4.2	3.9
	0.70	0.2	3.6	3.1
	2.28	0.08	1.9	0.9

[a]The half life was determined from the linear regression analysis of the initial linear portion of each decomposition curve.

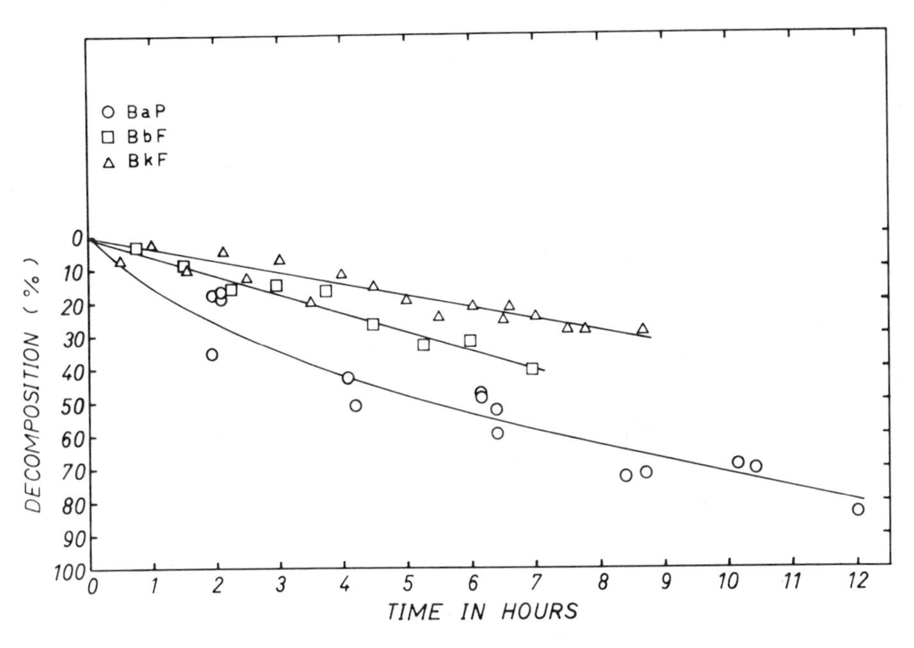

FIGURE 3. Decomposition as a function of time of BaP, BbF, and BkF at 245 ng each under lamp irradiation and in the absence of ozone.

but showed increased differences in their oxidation rates as the ozone concentration was increased.

Under both illumination and ozone, the reactions were, in all cases, accelerated (Figures 5, 7, and 9). BbF and BkF still reacted similarly, but BaP reacted much more rapidly. At 0.19-ppm ozone, BaP had a half life of 0.62 hours and at 0.70 ppm ozone, a half life of only 0.2 hours.

While BbF and BkF were fairly resistant to oxidation in the dark at low ozone concentrations, BaP was exceedingly reactive. Under light and ozone conditions, BbF and BkF reacted much more rapidly than they did in the dark. BaP, however, reacted only slightly faster under illumination and in the presence of ozone than it did in the dark under the same ozone concentrations. These unexpected results indicate that there may be significant losses of PAH (and BaP in particular) through decomposition by oxidant during the normal 24-hour high volume sampling procedure employed to collect suspended particulate matter. A closer examination of the results obtained for BaP will indicate the reasons for this concern.

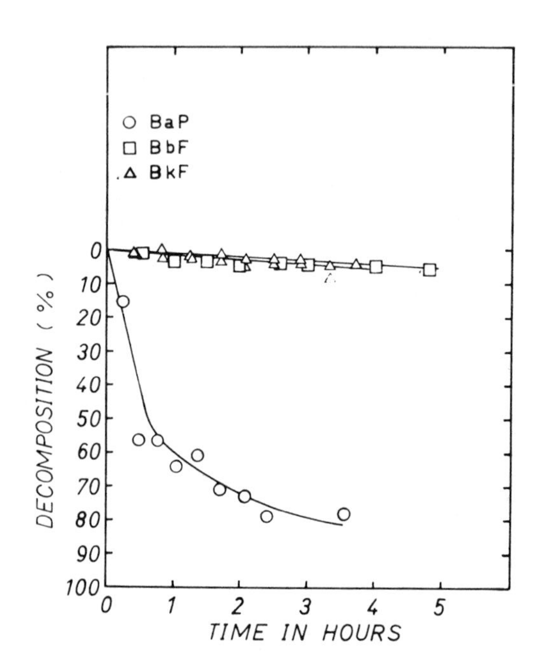

FIGURE 4. Decomposition as a function of time of BaP, BbF, and BkF at 245 ng each in the dark and exposed to 0.19-ppm ozone.

In Figure 6, the decomposition of BaP is shown at an ozone concentration of 0.70 ppm. With an initial sample size of 245 ng per dish, the initial slope indicated a half life of 0.4 hours. The tailing portion of the curve also appeared to be linear and was determined to have a loss rate of 1.5 percent/ hour. Very similar results were obtained for BaP at 0.19-ppm and 2.28-ppm ozone. A similar effect was noted for BbF and BkF.

The nature of the decomposition curves obtained for BaP suggested a rather complex gas-solid reaction that depended upon the extent of the multilayering of the PAH molecules within each sample. For the 5-cm diameter petri dishes used, 245-ng BaP corresponded to about one-third the weight of a monolayer, whereas 490-ng BaP represented approximately two-thirds the weight of a monolayer. The preparation procedure was such that most of the PAH deposited in each dish was in the surface layer (or layers) and that only a relatively small percentage of the PAH existed in an essentially protected state beneath the surface material.

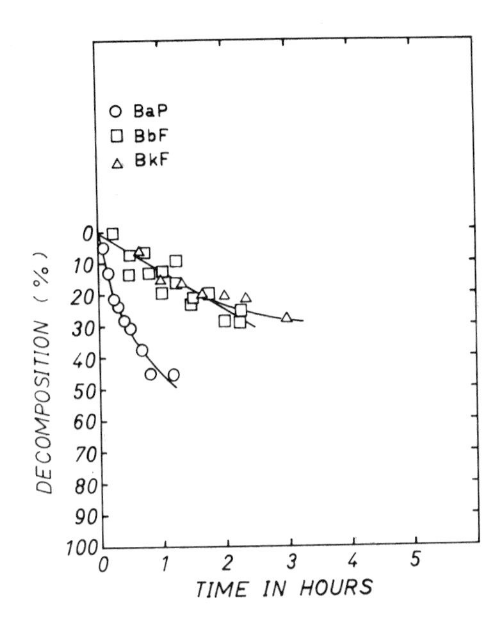

FIGURE 5. Decomposition as a function of time of BaP, BbF, and BkF at 245 ng each under lamp irradiation and exposed to 0.19-ppm ozone.

The oxidation curves obtained for the PAH may be interpreted as indicating that the surface-exposed layer reacted very rapidly with the ozone in the reaction chamber to form oxidized species which, in effect, acted to hinder the oxidation of the subsurface material. Since a slower oxidation was observed after the initial rapid oxidation, one of, or perhaps a combination of, several effects were indicated:

1. The initial oxidized products (e.g., peroxides, epoxides, and quinones) were further oxidized and eventually fragmented to such an extent that the product fragments left the surface, thus exposing the subsurface material to oxidation.

2. The products were mobile on the surface and simply hindered the access of the ozone to the subsurface PAH molecules.

3. There was an upwelling of the subsurface molecules or an upward percolation of the molecules, thus exposing the subsurface molecules to the ozone.

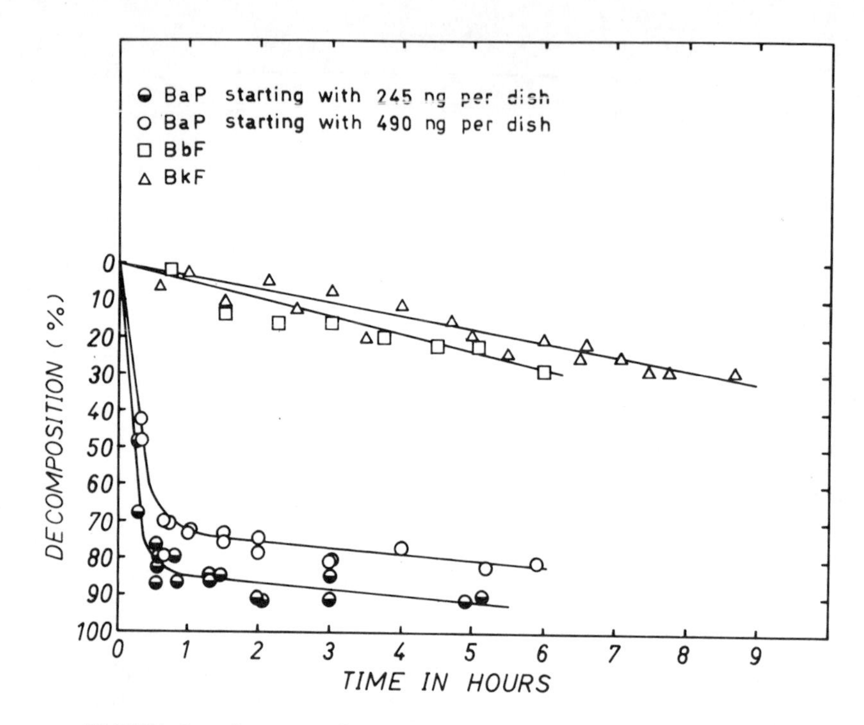

FIGURE 6. Decomposition as a function of time of BaP, BbF, and BkF at 245 ng each and BaP at 490 ng in the dark and exposed to 0.70-ppm ozone.

In order to test this hypothesis, experimental runs were carried out to expose BaP to 0.70-ppm ozone but commencing with 490-ng BaP per dish. The resulting decomposition curve may be seen in Figure 6 (open circles) alongside the curve for the 245-ng BaP per dish decomposition curve (half-filled circles). The two experimental runs had very similar initial slopes (they may, in fact, have been the same) but leveled off at different percent decompositions. The tailing slope was again approximately 1.5 percent/hour. In the higher loading experiments (open circle), a greater percentage of the BaP would have constituted subsurface material, and thus the tailing portion of the curves indicated not only the oxidant penetration rate, but also the proportion of the PAH protected in the subsurface layers. The scatter of the experimental points may be taken as an indication of the efficiency of the production of uniform dispersions from sample to sample.

If much larger quantities of BaP had been employed (e.g., 10 μg) the initial surface reaction would not have been detected

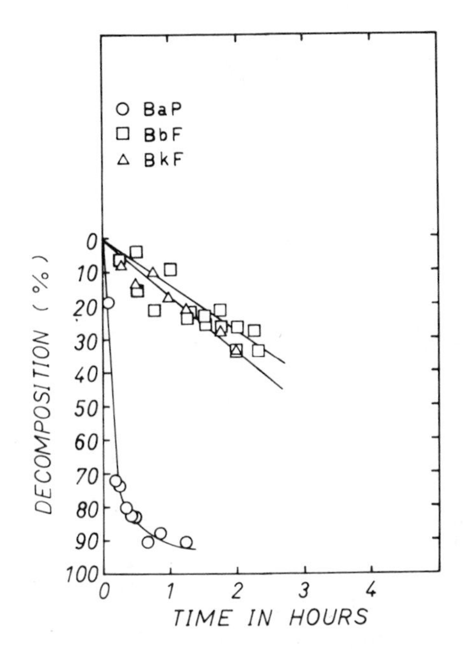

FIGURE 7. Decomposition as a function of time of BaP, BbF, and BkF at 245 ng each under lamp irradiation and exposed to 0.70-ppm ozone.

and only the final penetration rate would have been observed. This may well account for the decomposition described by Falk et al (1956) who used 10-μg samples at each of several sites on filter discs.

IV. APPLICATION OF THE RESULTS TO THE ATMOSPHERIC SYSTEM

It is well known that PAH are produced during the combustion of fossil fuels, tobacco, and refuse; that PAH are present in the resulting emissions; and that the PAH are associated with the atmospheric particulate matter. From the recent review of nucleation processes by Lahaye and Prado (1974), it is reasonable to assume that the PAH exist on the surface of, and in the interstices of soot particles in a multilayered form.

In view of the fact that the simulated atmospheric reactions indicated a rapid surface reaction followed by a slower penetration-type reaction, it is suggested that an analogous process

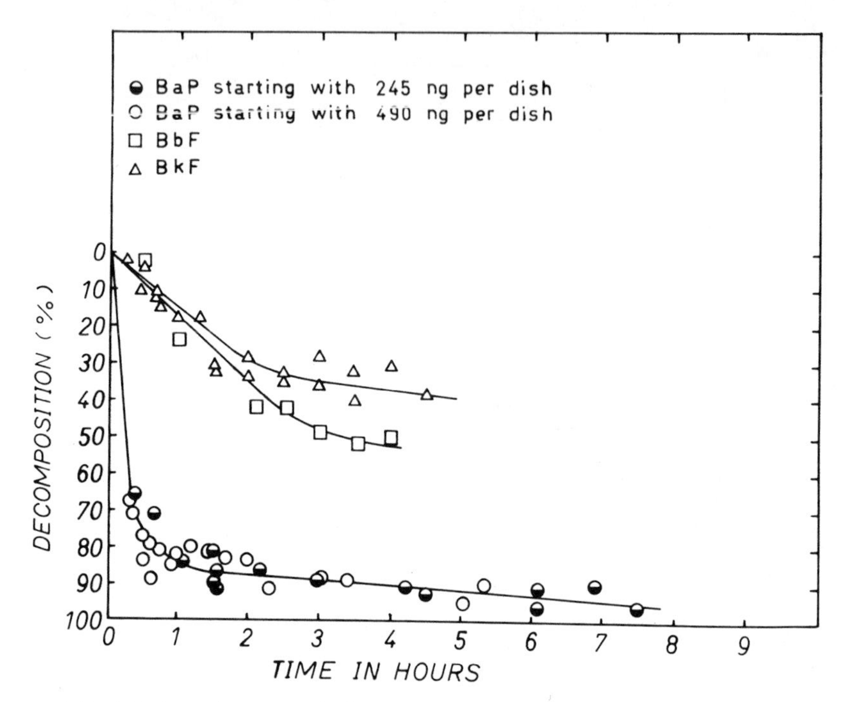

FIGURE 8. Decomposition as a function of time of BaP, BbF, and BkF at 245 ng each and BaP at 490 ng in the dark and exposed to 2.28-ppm ozone.

will affect the PAH adsorbed on the surface of the particulate matter, that is, that the surface-exposed PAH will be oxidized rapidly by the prevailing oxidant and light conditions in the atmosphere, and this will be followed by a slower penetration-type reaction. Subsurface PAH, and PAH physically isolated from reaction by the agglomeration of the soot particles, would have a fairly long survival time.

The oxidation of the PAH will, of course, continue as long as the particle is subjected to the oxidizing conditions. Since particles with diameters below 10 μm are known to have mean atmospheric residence times of 10 to 100 hours (Esmen and Corn, 1971) and since PAH are primarily associated with particles which have diameters below about 3 μm (Pierce and Katz, 1975a), on the average, the PAH will experience a wide variety of oxidant and light conditions. Decomposition of the PAH must certainly occur.

When a particle is trapped on the surface of a Hi-Vol sampler filter, the effect of radiation is removed. However,

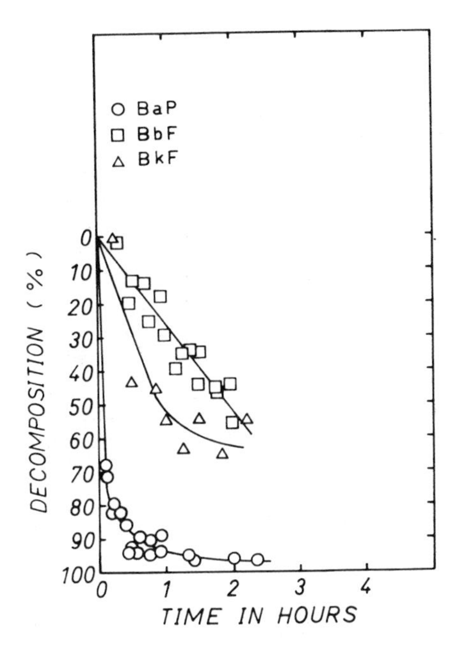

FIGURE 9. Decomposition as a function of time of BaP, BbF, and BkF at 245 ng each under lamp irradiation and exposed to 2.28-ppm ozone.

there are still many oxidants in the air, which is being drawn through the filter. BaP was shown to react only marginally slower in the absence of light when exposed to ozone than in the presence of both light and ozone and it is, therefore, anticipated that PAH oxidation will continue for as long as the PAH on the particulate matter are exposed to the flow of air through the filter. That period of exposure may be from 0 to 24 hours.

Data, both direct and indirect, in the literature support the concept that different PAH are oxidized in the atmosphere (and presumably also on the collection filter) at rates charactersitic of the individual PAH. In most pyrolysis experiments (see, e.g., Badger et al., 1962) and in the exhaust gases from the combustion of liquified gas and gasoline (Del Vecchio et al., 1970), BaP is generated in larger quantities than either BbF or BkF. Yet atmospheric analyses very often indicate either equal BaP, BbF, and BkF mass concentrations or indicate that the BaP mass concentration is less than the BbF and BkF mass concentrations (Bravo et al., 1970; Gordon and Bryan, 1973; Sawicki et al., 1962). This, however, constitutes only indirect evidence for the reaction of the PAH under atmospheric conditions. Direct proof of PAH oxidation necessitates the isolation of known oxidation products of the PAH from air extracts.

Recently Pierce and Katz (1975b) have detected not only the 1,6-, 3,6-, and 6,12-quinones of BaP, but also the quinones of anthracene, benz[a]anthracene, and dibenzo[b,def]chrysene in the air extracts from Toronto, Canada. Thus it appears that the quantity of PAH determined by the standard Hi-Vol filtration process represents only a residual mass determination as far as the PAH are concerned. A significant portion of the PAH injected into the air may be lost, not only through oxidation while the particulate matter is in the atmosphere, but also during the collection procedure while the particle is trapped on the Hi-Vol sampler filter.

The factors affecting the mass concentration of each PAH determined by the standard high volume filtration procedure may be summarized as:

1. The degree of multilayering of the PAH on the surface and in the interstices of the soot particles,

2. The extent to which agglomeration of the soot particles physically isolates PAH from reaction with the ambient oxidant and light,

3. The concentration of atmospheric oxidants,

4. The length of time that the PAH are exposed to sunlight,

5. The residence time of the particle in the atmosphere before collection on the sampling filter.

6. The length of time that the PAH are exposed to the air flow through the sampling filter,

7. The oxidant flux through the Hi-Vol sampler filter,

8. The individual reactivity of each PAH to the ambient atmospheric conditions, and

9. The possible synergistic effects between the PAH on the particle surface or between other oxidants, light, and the PAH.

As a consequence, it appears that a reassessment of the utility of the Hi-Vol sampler for the collection of particulate matter for the subsequent elution of reactive organic compounds, and PAH in particular, is urgently required.

Over the past 20 years or so, BaP has come to be accepted as an indication of atmospheric pollution in general and as a measure of the carcinogenic potential of atmospheric polycyclic organic matter to man (National Academy of Sciences, 1972). In fact, several health studies have attempted to correlate the BaP mass concentrations to the incidence of cancer in the society. In view of the demonstrated high reactivity of BaP under all oxidizing conditions, it appears that BaP was a poor choice for such an indicator. A better indicator (if a single PAH must be chosen) would be BbF since it is, itself, a potent carcinogen and is much more resistant to atmospheric oxidation than is BaP.

REFERENCES

1. Badger, G. M., Donnelly, J. K., and McL. Spotswood, T. 1962. The Formation of Aromatic Hydrocarbons at High Temperature. XV. The Pyrolysis of 2,2,4-Trimethylpentane ("iso-Octane"), Aust. J. Chem. 15, 605-615.

2. Bravo, H., Nulman, A., Nulman, R., Monkman, J. L., and Stanley, T. 1970. Concentrations of Lead, BaP, and BkF in the Atmosphere of Three Mexican Cities, Proc. Second Int. Clean Air Congress, 118-121.

3. Esmen, N. A., and Corn, M. 1976. Residence Time of Particles in Urban Air, Atmos. Environ. 5, 571-578.

4. Falk, H. L., Kotin, P., and Miller, A. 1960. Aromatic Polycyclic Hydrocarbons in Polluted Air as Indicators of Carcinogenic Hazards, Int. J. Air Pollut. 2, 201-209.

5. Falk, H. L., Markul, I., and Kotin, P. 1956. Aromatic Hydrocarbons. IV. Their Fate Following Emission into the Atmosphere and Experimental Exposure to Washed Air and Synthetic Smog, A.M.A. Arch. Ind. Health 13, 13-17.

6. Gordon, R. J. and Bryan, R. J. 1973. Patterns in Airborne and Polynuclear Hydrocarbon Concentrations at Four Los Angeles Sites, Environ. Sci. Technol. 7, 1050-1053.

7. Hodgeson, J. A., Stevens, R. K., and Martin, B. E. 1971. A Stable Ozone Source Applicable as a Secondary Standard for Calibration of Atmospheric Monitors, presented at the Analysis Instrumentation Symposium, Instrument Society of America, Houston, Texas, April.

8. Lahaye, J. and Prado, G. 1974. Formation of Carbon Particles from a Gas Phase: Nucleation Phenomenon, Water Air Soil Pollut. 3, 473-481.

9. Laity, J. L. 1971. A Smog Chamber Study Comparing Blacklight Fluorescent Lamps with Natural Sunlight, Environ. Sci. Technol. 5, 1218-1220.

10. Leighton, P. A. 1961. Photochemistry of Air Pollution, Academic, New York, p. 29.

11. National Academy of Sciences, Washington, D. C. 1972. Particulate Polycyclic Organic Matter, pp. 69-73.

12. Pierce, R. C. and Katz, M. 1975a. Dependency of Polynuclear Aromatic Hydrocarbon Content on Size Distribution of Atmospheric Aerosols, Environ. Sci. Technol. 9, 347-353.

13. Pierce, R. C. and Katz, M. 1975b. Chromatographic Isolation and Spectral Analysis of Polycyclic Quinones: Application to Air Pollution Analysis, Environ. Sci. Technol., December.

14. Sawicki, E., Hauser, T. R., Elbert, W. C., Fox, F. T., and Meeker, J. E. 1962. Polynuclear Aromatic Hydrocarbon Composition in the Atmosphere in Some Large American Cities, Amer. Ind. Hyg. Assoc. J. 23, 137-144.

15. Tebbens, B. D., Thomas, J. F., and Mukai, M. 1966. Fate of Arenes Incorporated with Airborne Soot, Amer. Ind. Hyg. Assoc. J. 27, 415-422.

16. Thomas, J. F., Mukai, M., and Tebbens, B. D. 1968. Fate of Airborne Benzo[a]pyrene, Environ. Sci. Technol. 2, 33-39.

17. Vecchio, V. del, Valori, P., Melchiori, C., and Grella, A. 1970. Polycyclic Aromatic Hydrocarbons from Gasoline-Engine and Liquified Petroleum Gas Engine Exhausts, Pure Appl. Chem. 24, 739-748.

The Oxidation of Chloroethylenes

EUGENIO SANHUEZA*
Universidad de Oriente, Escuela de Ciencias
Departamento de Quimica
Cumana, Venezuela

and

I. C. HISATSUNE and JULIAN HEICKLEN
Department of Chemistry and Center for Air Environment Studies
The Pennsylvania State University
University Park, Pennsylvania

I. INTRODUCTION

In recent years there have been a number of studies of the oxidation of haloethylenes. Recently we have reviewed these studies in detail (Sanhueza et al., 1976) and the reader is

---

*Present address: Instituto Venezolano de Investigaciones Cientificas, Centro de Ingenieria, Apartado 1827, Caracas 101, Venezuela.

referred to that review for specific information and references to the original studies. Here we present the highlights and the general conclusions of this work and indicate the implications of these studies for urban atmospheres.

The studies performed in our laboratory were undertaken because of the concern regarding the fate of chlorinated species in the atmosphere. The chloroethylenes are used extensively as solvents in both chemical manufacture and dry cleaning plants. They are also quite reactive, and thus the chemistry of their oxidation, particularly under atmospheric conditions, is of interest.

The oxidation of the haloethylenes has been initiated in five different ways:

1. Chlorine atom initiation,
2. H_g $6[^3P]$ sensitization,
3. Reaction with $O[^3P]$,
4. Reaction with $O[^3P]$ in the presence of O_2,
5. Reaction with O_3.

We shall discuss in succession these five pathways to oxidation.

II. CHLORINE ATOM-INITIATED OXIDATION

The chlorine atom oxidation mechanism in the presence of O_2 in excess of 100 torr can be represented as:

$$Cl + C_2X_3Cl \rightarrow ClCX_2CXCl \tag{1}$$

$$ClCX_2CXCl + O_2 \rightarrow ClCX_2CXClO_2 \tag{2}$$

$$2ClCX_2CXClO_2 \rightarrow 2ClCX_2CXClO + O_2 \tag{3a}$$

$$\rightarrow (ClCX_2CXClO)_2 + O_2 \tag{3b}$$

$$ClCX_2CXClO \rightarrow ClCX_2CX(O) + Cl \tag{4a}$$

$$\rightarrow ClCX_2 + CXClO \tag{4b}$$

$$ClCX_2 + \frac{1}{2} O_2 \rightarrow CX_2O + Cl \quad (5)$$

where X is either a hydrogen or halogen atom, and Reaction 5 represents an overall process that proceeds through

$$ClCX_2 + O_2 \rightarrow ClCX_2O_2 \quad (6)$$

$$2ClCX_2O_2 \rightarrow 2ClCX_2O + O_2 \quad (7)$$

$$ClCX_2O \rightarrow CX_2O + Cl \quad (8)$$

In general, the chlorine atom adds preferentially to the less chlorinated carbon atom of the olefin in Reaction 1, and this can be attributed to steric rather than polar effects. A long chain oxidation with chain lengths >150 occurs if the exothermicity of either Reaction 4a or 4b exceeds 11 kcal/mol. If the exothermicity = 11 kcal/mol, then the chain length is about 20.

Whether Reaction 4a or 4b is favored is determined by the relative C-Cl and C-C bond strengths in the $ClCX_2CXClO$ radical:

1. For $D(C\text{-}Cl) - D(C\text{-}C) \geq 6$ kcal/mol, reaction 4b predominates.

2. For $D(C\text{-}Cl) - D(C\text{-}C) < -3$ kcal/mol, Reaction 4a predominates, and

3. For $D(C\text{-}Cl) - D(C\text{-}C) = -3$ to 6 kcal/mol, both Reactions 4a and 4b take place.

There are some special cases that do not necessarily conform to the general rules outlined here. These are:

1. With C_2H_3Cl, the radical produced is $CClH_2CHClO$, which decomposes exclusively via Reaction 4b to give CH_2Cl, which is a terminating radical and does not oxidize to regenerate chlorine atoms. Thus there is no chain process with C_2H_3Cl.

2. Radicals of the type $ClCX_2CXHO$ never decompose to give H atoms and $ClCX_2CX(O)$.

3. Radicals of the type CX_3CH_2O do not decompose via either Reaction 4a or 4b.

4. All perfluorochloroethylenes undergo a long chain oxidation.

III. HG-SENSITIZED OXIDATION

When the oxidation is initiated by Hg $6[^3P]$, produced by absorption of radiation at 2537 Å by Hg vapor, the same free-radical longchain oxidation is observed as when initiation is by chlorine atoms. However, the chain lengths are shorter and increase with the olefin concentration to upper limiting values. With chlorine atom initiation, there was no dependence of the chain length on the olefin, the absorbed intensity (I_a), or the O_2 pressure at high O_2 pressures. Also in the Hg-sensitized oxidation, CO was always produced, whereas this was not so with some chloroethylenes in the chlorine atom-initiated oxidation.

The chloroethylene pressure dependence and the production of CO are considered as evidence for additional initiation and/or terminating reactions. The reactions suggested are

$$C_2X_2Cl_2 + Hg\ 6[^3P] \rightarrow C_2X_2Cl + \frac{1}{2} Hg_2Cl_2 \tag{9}$$

$$C_2X_2Cl + O_2 \rightarrow C_2X_2ClO_2 \tag{10}$$

$$C_2X_2ClO_2 \rightarrow CO \text{ via termination} \tag{11}$$

$$C_2X_2ClO_2 + C_2X_2Cl_2 \rightarrow C_2X_2Cl_3 + (CXO)_2 \tag{12}$$

where Reactions 9, 10, and 12 represent overall rather than fundamental reactions. In particular, Reaction 11 cannot be complete since another free radical must ultimately be involved to lead to termination. However, Reactions 9 through 12 do lead to the proper kinetic rate laws.

IV. $O[^3P]$ Atom Reaction

The reaction of $O[^3P]$ with C_2X_4 proceeds by three routes:

$$O(^3P) + C_2X_4 \rightarrow CX_2 + CX_2O \text{ (or } CO + X_2) \qquad (13a)$$

$$\rightarrow CX_2CX_2O^* \qquad (13b)$$

$$\rightarrow CX_3CX(O) \text{ (or } CX_3 + XCO) \qquad (13c)$$

The major paths are Reactions 13a and 13b and the relative importance of these paths are:

Haloethylene	C=C Cleavage	Excited Molecule
Chloroethylenes (no F)	0.19 - 0.31	0.55 - 0.81
C_2F_4, CFClCFCl	0.80 - 0.85	0.15 - 0.20
CF_2CCl_2 (*cis* and *trans*)	0	1.00

Reaction 13c occurs to a minor extent with CCl_2CH_2, C_2H_3Cl, and possibly *cis*- and *trans*-CHClCHCl. This is to be contrasted with C_2H_4, where Reaction 13c is the major, if not exclusive, process.

In Reaction 13, the $O[^3P]$ atom exclusively attacks the less chlorinated carbon atom of the ethylene. The rate coefficients are 0.1 through 1.0 of that for C_2H_4 at 25°C.

V. REACTION WITH $O[^3P]$ in the Presence of O_2

In the presence of O_2 the CXCl fragment produced in Reaction 13a (presumably triplet CXCl) reacts with O_2 as follows:

$$CXCl + O_2 \rightarrow XO + ClCO \qquad (14)$$

This is followed by the same longchain free radical oxidation observed with Cl or Hg $6[^3P]$ initiation, except that the chain length depends on $[C_2X_3Cl]/I_a^{1/2}$. This suggests that termination proceeds by radical-radical reactions; one radical must be a chain carrier, and the other must be one not present in the chlorine atom or Hg sensitized initiated oxidations (since there was no I_a dependence in those systems). Possible terminating reactions are:

$$Cl + XCO \rightarrow ClX + CO \qquad (15)$$

$$ClO + ClCO \rightarrow Cl_2O + CO \qquad (16a)$$

$$\rightarrow Cl_2 + CO_2 \qquad (16b)$$

VI. REACTIONS WITH O_3

The chlorinated ethylenes can react with O_3 to produce a π-complex or in some cases a molozonide. For those ethylenes (C_2H_4 and C_2H_3Cl) that form a molozonide, as verified by observing its infrared spectrum in a low temperature matrix, the reaction rate is first order in both the olefin and O_3 concentration. The initial steps in the mechanism can be represented as

$$O_3 + \gt C{=}C\lt \rightarrow \gt C(O{-}O{-}O)C\lt \rightarrow \gt C{=}O + \gt\dot{C}{-}O{-}\dot{O} \qquad (17)$$

The diradical (or zwitterion) then initiates a long chain oxidation, which is strongly inhibited in the presence of O_2.

The more chlorinated ethylenes do not form molozonides but do produce π-complexes. The rate laws for the ozonolysis are quite complex and for CCl_2CH_2, C_2Cl_4, and <u>cis</u>- and <u>trans</u>-CHClCHCl, can be represented by

$$\text{Rate} = k[\text{Olefin}]^n[O_3]^m \qquad (18)$$

with both n and m varying between 1 and 2 depending on the reactant pressures. Again a diradical mechanism must be involved, since the presence of O_2 strongly inhibits the ozonolysis (O_2 would promote a mono-free radical chain). The details of the initial steps are not clear, but the formation of the π-complex is reversible, and the indicated reaction steps are

$$O_3 + C_2X_4 \rightleftarrows C_2X_4 \cdot O_3 \qquad (19)$$

(π-complex)

$$C_2X_4 \cdot O_3 + C_2X_4, O_3 \rightarrow \text{products} \qquad (20)$$

As an illustration of the complexity of the reaction mechanism, Figures 1 through 4 show how the rate differs under different conditions. Figure 1 shows the time dependence for reactants and products in the O_3—C_2Cl_4 reaction. Initially, the reaction is rapid. However, O_2 is a product, and as it accumulates, the reaction is markedly inhibited, the rate becoming immeasurably slow after enough O_2 accumulates. If excess O_2 is present initially, the reaction likewise does not proceed.

Figure 2 is a pseudo first-order semilog plot of the CHClO production as a function of reaction time in the reaction of O_3 with excess <u>cis</u>-CHClCHCl. Initially, the reaction is markedly inhibited as the O_2 accumulates. Later, however, the reaction

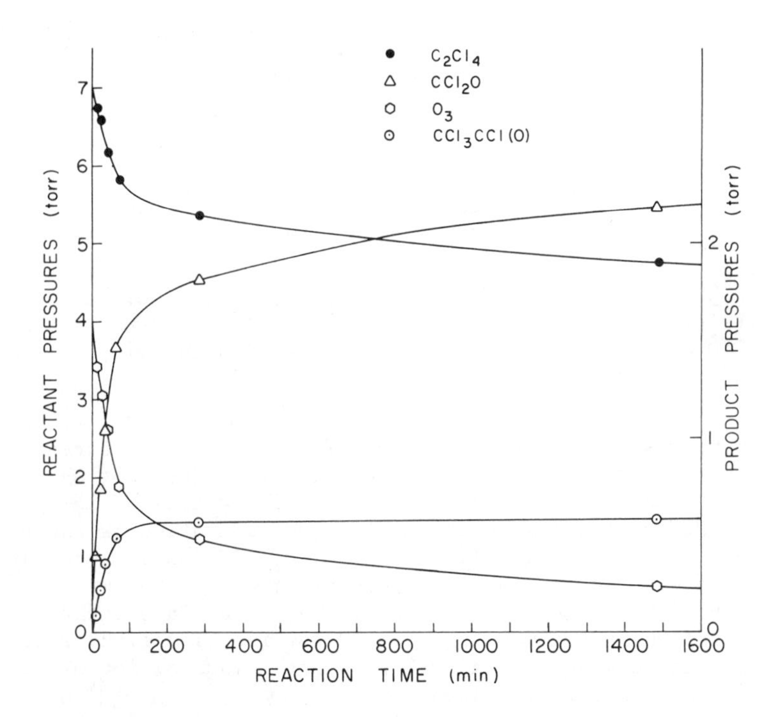

FIGURE 1. Time dependence of the composition of C_2Cl_4 ozonolysis reaction at 24°C: $[C_2Cl_4]_0$ = 6.9 torr, $[O_3]_0$ = 4.1 torr. After Mathias et al., 1974, by permission of the National Research Council of Canada.

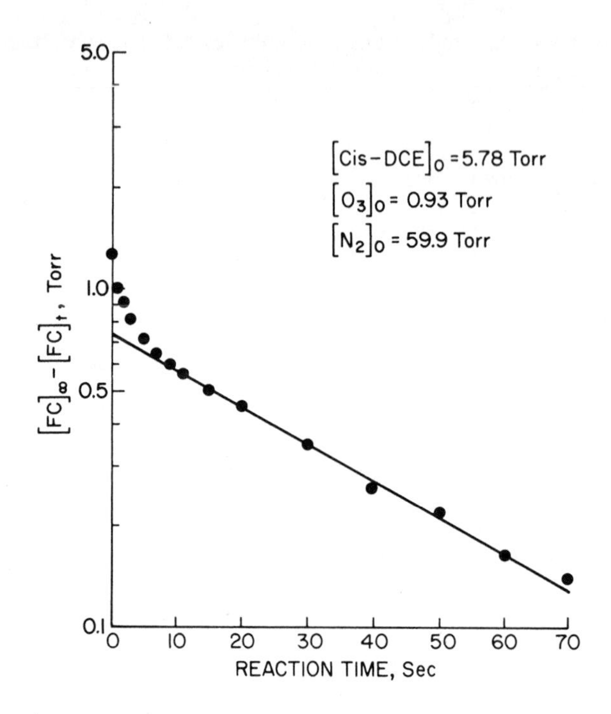

FIGURE 2. First-order kinetic plot of cis-CHClCHCl reaction with ozone at 23°C in N_2 buffer (after Blume et al. (1976) with permission of John Wiley and Sons.)

can be represented nicely as pseudo first order. If excess O_2 is present initially (Figure 3), the reaction does not show the initial deviation from linearity. An illustration of a reaction that is second order in each reactant concentration is shown in Figure 4. This reaction is for the O_3-cis-CHClCHCl system in excess O_2 with equal starting concentrations of the two reactants. Since the CHClCHCl and O_3 are removed at equal rates, their concentrations are always equal, and the reaction fits pseudo fourth-order dependence in each reactant. Thus a plot of $[O_3]^{-3}$ versus reaction time is linear.

As an aside, we point out that the ozonation of C_2F_4 is different from the ozonation of either C_2H_4 or any of the chlorinated ethylenes studied, since the addition of O_2 promotes, rather than inhibits, the oxidation.

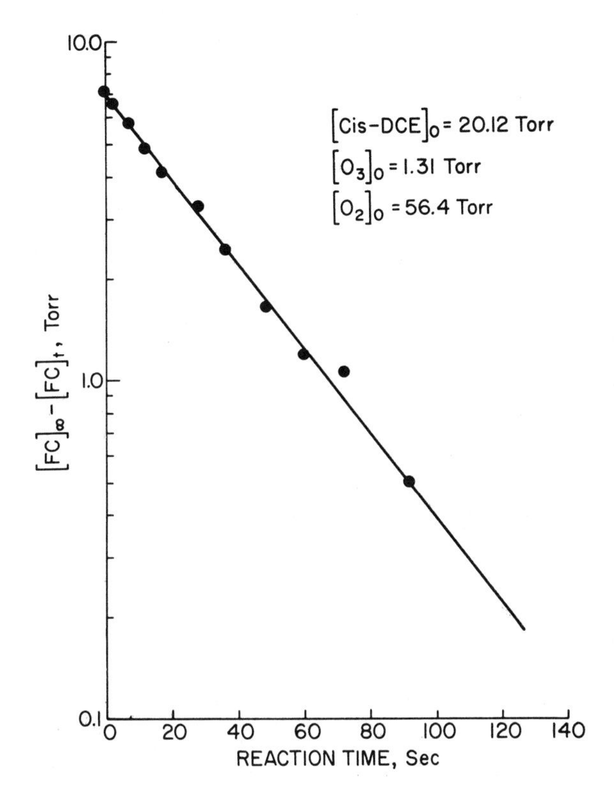

FIGURE 3. First-order kinetic plot of cis-CHClCHCl reaction with ozone at 23°C in O_2 buffer (after Blume et al. (1976) with permission of John Wiley and Sons, New York).

VII. IMPLICATION FOR URBAN ATMOSPHERES

The implications of the oxidation of chloroethylenes for urban atmospheres is the following:

1. Free radical, long chain oxidation carried by Cl atoms,

2. Products are CXClO and $CX_2ClCX(O)$,

3. $O[^3P]$ attack at 0.1 through 1 times rate of $O[^3P] + C_2H_4$ at 25°C, and

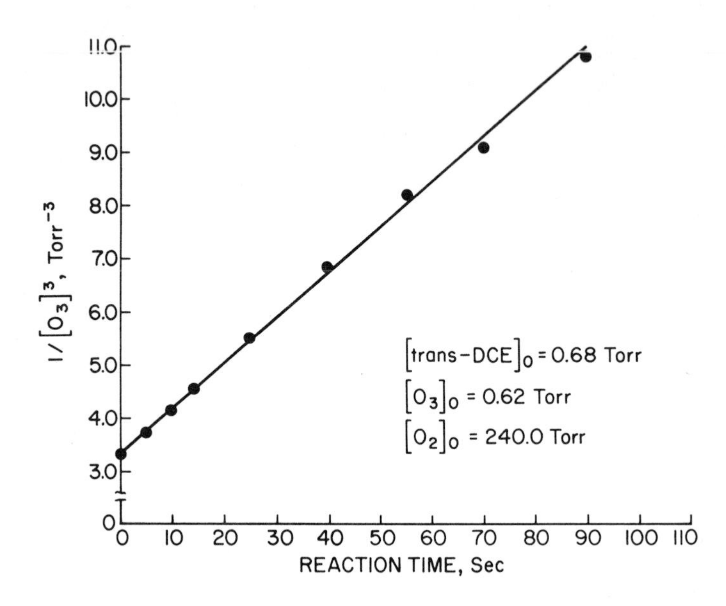

FIGURE 4. Fourth-order kinetic plot of trans-CHClCHCl reaction with ozone at 23°C (after Blume et al. (1976) with permission of John Wiley and Sons, New York).

4. O_3 reactions in presence of O_2 are too slow to be important.

The question then is: Do the chloroethylenes or their oxidation products in the atmosphere pose a health hazard? Table 1 lists the emissions of the chloroethylenes and the recommended threshold limit values (TLV) for them and some of their oxidation products. Since atmospheric concentrations of chloroethylenes have never exceeded a few ppm, chloroethylenes do not pose a direct health problem. However, near industrial or dry cleaning plants that emit C_2Cl_4, and possible C_2Cl_3H, the oxidation of these chloroolefins under conditions of high radical concentrations (such as in photochemical smog) could oxidize them with reaction half lives of a few hours. Thus CCl_2O levels could exceed its TLV of 0.1 ppm. Unfortunately no measurements of chloro-ethylene concentrations or those of its oxidation products now exist in the vicinity of the emission source.

TABLE 1. Emissions and Toxicity

Molecule	1968 Production (10^8 lb)	Oxidation Products (%)	TLV[a] (ppm)
C_2Cl_4	6.4	$CCl_3CCl(O)$ (75%), CCl_2o (25%)	100
C_2Cl_3H	5.2	$CHCl_2CCl(O)$ (90%), CO and CCl_2O	100
$C_2Cl_2H_2$	48.0	CHClO (71%) from CHClCHCl	200
C_2ClH_3	14.6	CHClO (74%), CO (25%)	200[b]
HCl	--	--	5
CCl_2O	--	--	0.1

[a]40-hour week, 8-hour day industrial standard.

[b]Carcinogenic at 50 ppm.

REFERENCES

1. Blume, C., Hisatsune, I. C., and Heicklen. J. 1976. Int. J. Chem. Kinet., 8, 235.

2. Mathias, E., Sanhueza, E., Hisatsune, I. C., and Heicklen, J. 1974. Can. J. Chem. 52, 3852.

3. Sanhueza, E., Hisatsune, I. C., and Heicklen, J. 1976. Chem. Rev. 76, 801.

The Chemical Fate of Hydrides of Boron and Phosphorus from Industrial Processing

EDWARD J. SOWINSKI*
Occupational Health and Safety
Western Electric Company
Allentown, Pennsylvania

and

IRWIN H. SUFFET
Environmental Studies Institute
Department of Chemistry
Drexel University
Philadelphia, Pennsylvania

*Present address: Uniroyal Chemical, Naugatuck, Connecticut 06770.

I. INTRODUCTION

The significance of characterizing specific chemical species that exist in the environment has been demonstrated by pollutants that are generated in the environment and/or persist in the environment. For example, the importance of environmental changes and persistence has been described in the case of mercury, pesticides, and polychlorinated biphenyls (PCBs) (Giblin and Massaro, 1973; Edwards, 1975; Metcalf et al., 1971). However, an additional critical area that can contribute to harmful environmental pollutants has been neglected heretofore. This involves chemical changes that may occur in industrial manufacturing processes that can result in secondary pollutants having a potential environmental hazard. This concerns a concept that is part of chemical fate. The potential environmental impact of any new industrial chemical should concern a consideration of this concept. It can best be evaluated by industrial users of new chemicals, and it should concern the question of whether or not specific by-products are formed in a manufacturing process and whether they and the parent chemical are stable throughout the manufacturing process and the environment. For this purpose, pilot processes may be monitored, and/or laboratory procedures can be employed. The use of laboratory procedures to follow environmental fate has been reported in other areas. For example, a laboratory ecosystem has been developed to study the biogradability of pesticides (Metcalf et al., 1971). Also, an analogous procedure for determining the physical fate (dispersion) of chemicals has been outlined in the area of air pollution (Hanna, 1973).

From the practical viewpoint, an evaluation of chemical fate of a new industrial chemical is best initiated in early periods of industrial research and later in stepwise coordination with further programs of research and development. In this manner if stable and toxic industrial by-products are detected, substitute procedures can be initiated or more adequate controls can be designed than would otherwise be the case. Specifically, process modification can be more effectively utilized as a desirable primary source control choice and abatement methods used as secondary control choices.

A difficulty exists in these areas due to potential differences in chemical stability and reactivity at trace concentrations as compared to high concentrations. For example, it has recently been shown that a particular compound (dichloroacetylene), which is unstable at high airborne concentrations (>200 ppm), is stable at low concentrations and exhibits toxic effects (Saalfield et al., 1971). Consequently, the stability of potentially harmful air pollutants must be determined at or near potential environmental levels and not at high concentrations, as for example, near the explosion limits of a compound.

Metal hydrides of boron and phosphorus represent two classes of potentially hazardous occupational and environmental air pollutants that require an improved understanding of chemical fate (U. S. Department of Health, Education, and Welfare, 1969). The toxicity, industrial use, and chemistry of the metal hydrides are discussed to indicate the research needed to assess the fate of boron and phosphorus hydrides and thereby the potential environmental impact of their industrial use. Then studies to determine the potential environmental impact of these chemicals will be described as these chemicals are used in the electronics industry.

II. TOXICOLOGY

Boron hydrides can present a hazard to man by inhalation from acute, subacute, and chronic aspects. However, there is uncertainty as to specific compounds and the amount of exposure necessary to produce signs of overexposure in man. Table 1 outlines the threshold limit values (TLV) of these compounds (American Conference of Governmental Industrial Hygienists, 1975).

TABLE 1. Threshold Limit Values of Boron and Phosphorus Hydrides

	TLV (ppm)[a]
B_2H_6	0.1
B_5H_9	0.005
B_6H_{10}	Not determined
$B_{10}H_{14}$	0.05
PH_3	0.3

[a]From American Conference of Governmental Industrial Hygienists, 1975.

Diborane presents a hazard by the route of inhalation. Its principal effect has been reported to be caused by an exothermic reaction in the lungs during hydrolysis to boric acid and hydrogen (Levinskas, 1955). Beside inhalation, pentaborane and decaborane present an additional hazard because they can be absorbed through the skin and cause systemic effects. It has been suggested that several subacute exposures to pentaborane may be potentially more serious than a single exposure (Landez and Scott, 1971). Mechanisms for systemic boron hydride toxicity are not known. However, it has been reasoned that the reducing ability of these compounds might account at least in part for their toxicity (Levinskas, 1955). More recently it has been indicated that the boranes act primarily by inactivating pyridoxal enzymes (Hughes et al., 1967). Since the higher boranes hydrolyze slowly with water, it has also been reasoned that their physiological action may be due to their combination with body substances such as amines, alcohols, or free SH groups (Adams, 1964).

Phosphine is a highly toxic gas. A concentration of 2000 ppm can be lethal within a few minutes (American Conference of Governmental Industrial Hygienists, 1975). The TLV of phosphine is shown in Table 1.

III. INDUSTRIAL USES

In the past, major uses of boron hydrides involved diborane and its conversion to higher boron hydrides, which were used as high energy fuels and propellants. More recently, boron hydrides have been used in the manufacture of amine complexes, as catalysts in polymerization reactions, as semiconductor dopant sources, as synthetic rubber vulcanizers, and generally as reducing agents (Callery Chemical Company, 1972; Anon., 1972). Potential industrial applications for the boron hydrides appear to be numerous. These involve the synthesis of carborane siloxane polymers from decaborane, applications as wire insulators, compression molded seals, furnace parts, and electrical components (Anon., 1971).

In the electronics industry, diborane and phosphine are used as boron and phosphorous dopant sources for slices of silicon crystals that are used as semiconductors. The industrial use of diborane and phosphine has significance as an environmental hazard and control problem, since these compounds are toxic, and moreover they may yield environmentally stable and toxic by-products under industrial applications. Although a great deal of effort has been expended on studies concerning the use of inorganic hydrides as dopants in the electronic industry under pyrolysis conditions, very little is known about the mechanisms and products involved (Bloem, 1972; Ghandhi, 1968).

IV. BORON HYDRIDE CHEMISTRY

Boron hydrides can be divided into two groups of formulas: B_nH_{n+4} (e.g., B_2H_6, B_5H_9, B_6H_{10}, and $B_{10}H_{14}$) and B_nH_{n+6} (e.g., B_5H_{11}, B_6H_{12}, and B_8H_{14}). The members of the first group tend to be more stable thermally, while those of the second group tend to be unstable and readily convertible to the more stable series. The discovery and characterization of most B_nH_{n+4} boron hydrides was accomplished by Alfred Stock (1935). Stock also reported the occurrence of nonvolatile polymeric hydrides which result from heating lower molecular weight hydrides. More recently, boron hydrides containing eight and nine boron atoms, octaborane and nanoborane, have been reported (Shapiro and Keilin, 1959; Kotlensky and Schaeffer, 1958). The boron hydrides are generally prepared from the pyrolysis of diborane.

A. Pyrolysis of Diborane

A review of pyrolysis investigations since the time of Stock reveals that two fundamental areas have been explored. These are kinetic and mechanistic studies and studies aimed at optimizing pyrolysis product yields for commercial purposes. The bulk of pyrolysis investigations in both areas has involved diborane.

Kinetic and mechanistic studies of B_2H_6 pyrolysis have been conducted primarily at low pressures and temperatures in the range 85 to 191°C in static systems (Clarke and Pease, 1951; Bragg et al., 1951; Morey, 1958). The appearance of B_2H_6 pyrolysis products in static systems at this temperature range has been measured in minutes (Fehlner, 1965). Synthetic pyrolysis reactions, on the other hand, are generally done at high temperatures and pressures and short residence times (seconds) in flow reactors (Adams, 1964; Hughes et al., 1967). Generally, low pressure and temperatures favor unstable intermediates and products, while high pressures, short residence times, and high temperatures favor higher yields of stable products. It is of interest that the order of reaction of B_2H_6 pyrolysis changes with temperature (Morey, 1958; Fehlner, 1965).

The quantitative formation of specific pyrolysis products from an industrial application of B_2H_6 can be expected to depend on conditions of temperature, pressure, and pyrolysis residence time. These factors must be evaluated in order to determine specific B_2H_6 pyrolysis products from a manufacturing process. No specific data have been found on the products that may occur from industrial flow pyrolysis operations using B_2H_6 in the parts-per-million concentration range at atmospheric pressure. Similarly, no data are available on the products that may occur from industrial flow pyrolysis operations using PH_3 in the parts-per-million concentration range at atmospheric pressure.

B. Oxidation of Boron Hydrides

While Stock reported the boron hydrides to be unstable and sensitive to air and moisture, he also reported that they do not ignite spontaneously on contact with air. Data on the oxidation of boron hydrides have been reported primarily in terms of explosion limits and flammability limits for diborane at temperatures above those corresponding to slow decomposition. When diborane is ignited in air, a persistent odor characteristic of boron hydrides has been observed (Price, 1950). This report indicates that, in the heat of ignition, some diborane is converted to higher, more stable hydrides. No specific data have been found on the stability of boron hydrides at trace parts-per-million concentrations in the presence of excess oxygen as found in the atmospheric environment. However, the existence of lower flammability limits and the reported incomplete combustion of diborane-oxygen mixtures indicates that hydride stability at trace concentrations in the presence of excess oxygen may occur. In addition, the reported formation of higher hydrides from the pyrolysis B_2H_6 and oxygen atoms indicates an additional route for industrial by-product formation (Fehlner and Strong, 1962). This is important since some industrial operations involve the pyrolysis of diborane in the presence of molecular oxygen.

C. Hydrolysis of Boron Hydrides

Stock reported that hydrolysis at room temperature proceeded quite differently with the different hydrides. Diborane hydrolyzes very rapidly (in a few seconds) with the formation of boric acid and hydrogen, while decaborane hydrolyzes very slowly. Overall, Stock reported that reactivity with water decreases with rising molecular weight of hydrides (Stock, 1933).

No data have been found on the effect of atmospheric levels of water vapor (near 10,000 ppm H_2O) on the stability of boron hydrides in the air environment. However, water vapor may be expected to influence boron hydride stability in the air environment by hydrolysis. Although the extent of this effect is unknown, the reported slow hydrolysis of higher hydrides in aqueous solution indicates that some degree of boron hydride stability in the presence of water vapor may occur.

In summary, information is not available on concentrations of boron, particularly the more toxic boron hydrides, in the atmosphere. The stability of boron hydrides in the ambient air environment at trace concentrations is unknown. Studies on the stability of boron hydrides have been done at low pressure and at explosion limits, but not at atmospheric pressure and trace concentrations where implications to the environment are critical.

This work concerns an evaluation of some factors which can influence the occupational and air pollution hazard potential of diborane (B_2H_6). It involves the possible formation of more stable and toxic boron hydrides in industrial applications of diborane under flow pyrolysis conditions at atmospheric pressure and the stability of diborane and its pyrolysis products at trace parts-per-million concentration in the presence of excess oxygen and water vapor as may be found in the ambient air environment. Although not used as extensively as diborane, phosphine (PH_3) represents a similar situation. For comparison purposes, the atmospheric pressure flow pyrolysis of phosphine has been studied in addition to the studies involving diborane.

The experimental objectives of this study include a determination of pyrolysis and stability.

1. The conditions that can influence the qualitative and quantitative nature of pyrolysis products generated at atmospheric pressure from parts-per-million concentration are evaluated. These include: (a) different reactor geometrics, (b) variable contact time in a flow reactor, and (c) variable pyrolysis temperature.

2. The stability (half life) of B_2H_6 and its pyrolysis products at parts-per-million concentration are evaluated. The following atmospheres have been used: (a) nitrogen, (b) 20 percent O_2, and (c) 20 percent O_2 and water vapor.

V. EXPERIMENTAL APPARATUS AND PROCEDURES

The experimental laboratory apparatus has been designed as a self-contained, flexible system with component parts to do the following: (a) produce pyrolysis products representative of a manufacturing process, (b) identify and quantitate pyrolysis products, and (c) determine the stability of pyrolysis products. The apparatus has been designed for evaluating the chemical fate of gaseous inorganic hydrides and of chemicals related to the hydrides, for example, metal hydrides and organometallics. It consists of (a) a pyrolysis system, (b) a gas-handling system including a static reactor, (c) a gas chromatograph (GC), (d) three detectors [flame photometric (FPD), electron capture (EC), and microcoulometry (MC)], and (e) a mass spectrometer (MS) interfacing system. A block diagram of the experimental apparatus is shown in Figure 1.

Previous publications by the authors describe the use of the gas-handling system, programmed temperature chromatography, the use of flame photometric, electron capture, microcoulometric and mass spectrometric detection, and confirmatory techniques for the boron hydrides (Sowinski and Suffet, 1971, 1972, 1973, 1974, 1975). This system was also applied to the pyrolysis of phosphine in the present case.

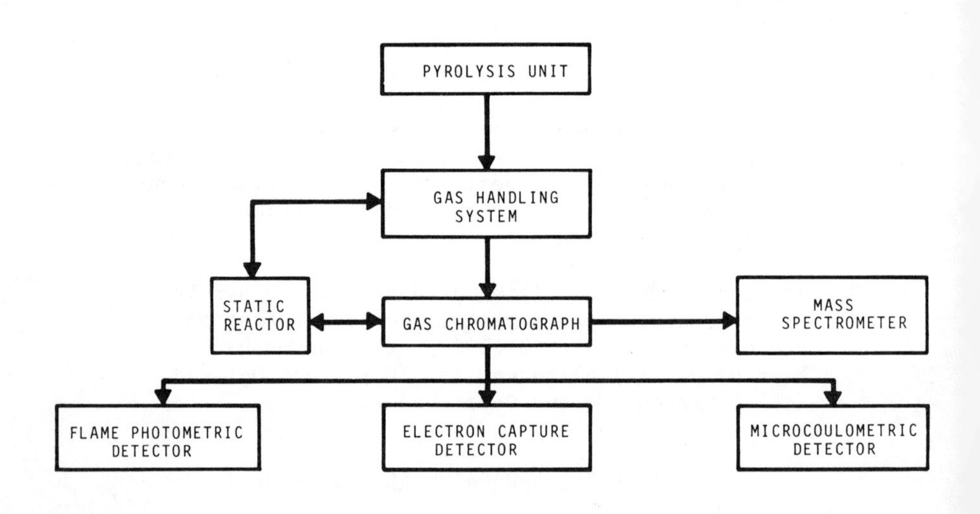

FIGURE 1. Block diagram of experimental apparatus.

VI. EXPERIMENTAL RESULTS

A. Pyrolysis

The first experiment was to pyrolyze B_2H_6 at different temperatures and determine if any products could be observed. The conditions employed were: (two microreactors, gold helix reactor, or quartz tube) a flow rate of B_2H_6 (1000 ppm in N_2) of 40 ml/min and atmospheric pressure in the flow reactor. A maximum of three products in addition to B_2H_6 were observed with the FPD and MC detectors at the same retention times with the experimental programmed temperature GC procedure used (Sowinski and Śuffet, 1972). The three products were subsequently identified as B_5H_9, B_6H_{10}, and $B_{10}H_{14}$ with multiple GC retention times, Kovats retention indices, and GC/MS.

Figure 2 shows the experimentally determined relative responses observed for the parent B_2H_6 and its pyrolysis products as a function of pyrolysis temperature. The relative response for the boron hydrides is reported in Figure 2 as log percent response of initial boron weight. This procedure served to eliminate variances in relative response due to the hydride moiety and provides a common representative scale for all four boron hydrides.

The data in Figure 2 reveal several important facts. First, at temperatures less than 210°C, no diborane appeared to be decomposed as products were not detected. Second, pyrolysis products were observed only in a temperature range of 210 to 270°C with a maximum yield of products at a temperature of approximately 260°C. At this temperature, approximately 30 percent of the original B_2H_6 was decomposed and approximately 12 percent of the original B_2H_6 was detected in the form of B_5H_9, B_6H_{10}, and $B_{10}H_{14}$. Third, no volatile boron hydrides, within the defined detector limits, were observed at pyrolysis temperatures greater than 270°C. Fourth, B_2H_6 was not observed at pyrolysis temperatures greater than 350°C. The data was reproduced in both types of flow reactor. A white solid was observed also in the quartz tube reactor after the pyrolysis experiments.

The disappearance of chromatographically detectable boron hydrides at elevated pyrolysis temperatures is of pertinent interest to air pollution control of the boron hydrides. An explanation for the absence of volatile boron hydrides at high pyrolysis temperatures may relate to the formation of nonvolatile higher hydrides in the pyrolysis system. This would be in accord with the observations of Stock and other investigators. The white solid observed in the quartz tube reactor after pyrolysis could be these higher hydrides.

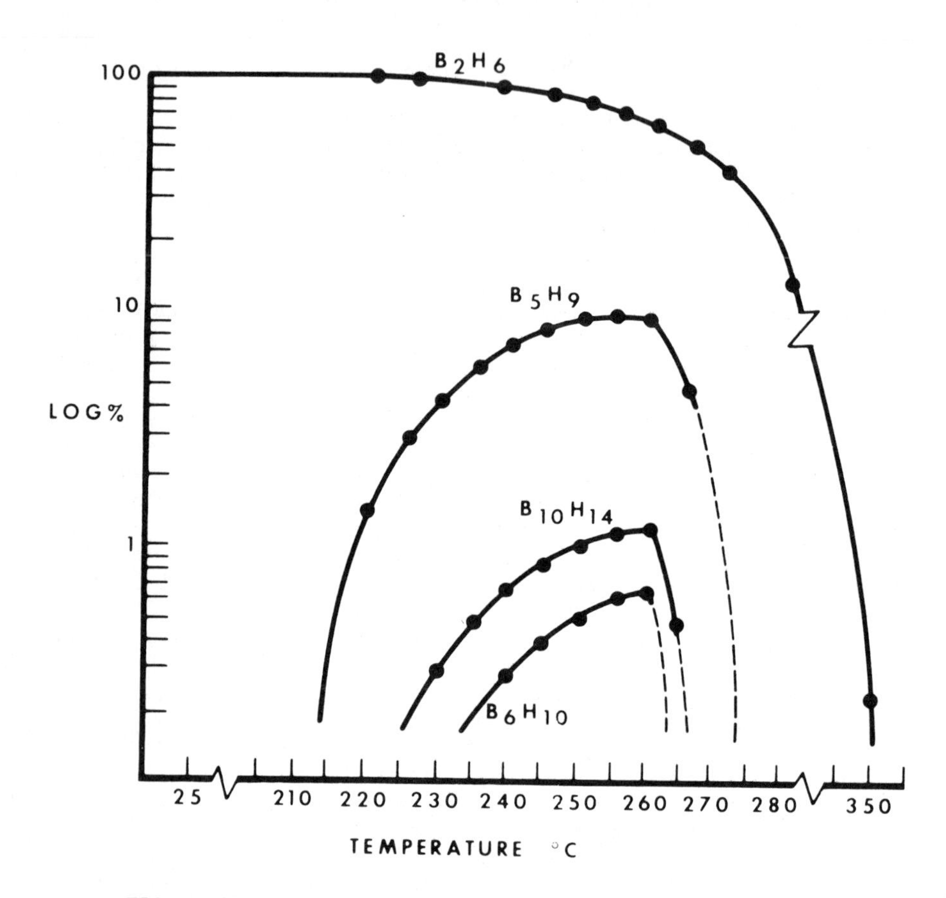

FIGURE 2. Boron content of pyrolysis sample: Log percent of initial boron weight in terms of B_2H_6, B_5H_9, B_6H_{10}, and $B_{10}H_{14}$ as a function of pyrolysis temperature.

Emission, mass, X-ray, and infrared spectroscopic analyses of the white solid were made. The emission spectrographic analysis indicated the presence of boron as expected. Additionally, silicon was also detected in the emission spectra. This may have been introduced into the sample from the quartz tube during sample preparation. An X-ray diffraction pattern of the white solid indicated the presence of B_2O_3. X-ray diffraction has been used to identify polymeric boron hydrides (ASTM, 1971). No previously characterized boron hydride was detected. An infrared spectrum of the white solid also indicated the presence of B_2O_3; however, a characteristic BH band at 2500 cm^{-1} was additionally observed. Mass spectra of the white solid were obtained using a solids probe heated to 300°C. Background mass spectra at 300°C up to 400 m/e were clean; however, mass units up to 180 m/e were observed for the white solid. This indicates the presence of boron compounds higher in molecular weight than $B_{10}H_{14}$.

Although higher hydride products were not confirmed in analysis of the white solid, their presence was indicated by mass spectrometry and infrared spectrometry. The large proportion of boron oxide detected by X-ray and infrared analysis is puzzling, since B_2H_6 was not pyrolyzed in the presence of oxygen, and higher boron hydrides are reported to be relatively stable in air. However, oxygen was introduced into the closed pyrolysis system closely downstream of the pyrolysis unit in oxidation studies and some back diffusion, especially at low B_2H_6 flow rates could have occurred. Also, the white solid was exposed to air during sample preparation and analysis. Oxidation of boron hydride polymers could have occurred as a result of these two factors.

B_2H_6/N_2 was pyrolyzed in the gold reactor with variable flow rates giving contact times in the range 10 seconds to 1 minute. Contact times in this range are representative of conditions in semiconductor manufacturing processes. Approximate atmospheric pressure in the flow reactor and a pyrolysis temperature of 255° C was maintained. The variation in contact time of 10 seconds to 1 minute resulted in no qualitative variation and little or no quantitative variation in B_2H_6 pyrolysis products.

The uniqueness of the pyrolysis experiments concerns the flow conditions at atmospheric pressure and concentration of B_2H_6 used. These conditions yield three stable boron hydride products. No short-lived intermediates were observed. This reaction is apparently complete in less than 10 seconds at constant temperature since no quantitative variation in product formations occurred with contact times in the range 10 seconds to 1 minute. These results contrast with data obtained from the only previously reported chromatographic study of B_2H_6 pyrolysis (Borer et al., 1960). That study, performed at 112°C and atmospheric pressure, reported the formation of B_5H_9, B_5H_{11}, and B_4H_{10} in a time span of minutes. However, the data were obtained from a static system and a lower temperature than this study. Additionally, the chromatographic conditions that were used in the previous study were apparently not capable of detecting hydrides higher than B_5. These contrasting results indicate the importance of analytical methodology for the detection of specific pyrolysis products and the influence of pyrolysis conditions on the qualitative nature of products and time required for their formation.

B. Stability of Boron Hydrides

The static reactor was used to evaluate the stability of B_2H_6 and its pyrolysis products as they were generated from pyrolysis in the laboratory system. Three basic conditions were evaluated by following time dependent stability (decay rates) of the boron hydrides: (a) B_2H_6 and the pyrolysis products in nitrogen, (b) B_2H_6 and the pyrolysis products in 20 percent oxygen, and (c) B_2H_6 and the pyrolysis products in 20 percent oxygen and water vapor (approximately 10,000 ppm). Two experiments were run for each of these conditions. First, the stability of B_2H_6 alone was tested, and second, a mixture of the pyrolysis products in the proportions obtained from the optimum pyrolysis temperature (260°C), Figure 2 was tested. This phase of the study was intended to determine the stability of the boron hydrides in the presence of excess oxygen and water vapor as found in the air environment.

The static reactor was sampled periodically using the gas-handling system. Concentration changes were monitored as a function of time by the GC temperature-programming procedure (Sowinski and Suffet, 1972). A solution of $B_{10}H_{14}$ (1.0 μl of 10^{-4} M $B_{10}H_{14}$ in cyclohexane) was used as an external reference and the FPD was used to monitor the experiments. At least two runs were made for each set of concentration time data. Plots of log concentration versus time were fit with a least-squares technique using a Hewlett-Packard desk top computer.

The stability of B_2H_6 was evaluated independently of the pyrolysis products using nitrogen diluted concentrations of B_2H_6 at 1, 25, and 800 ppm in the static reactor. The stability of B_2H_6 was evaluated in the absence of the pyrolysis products for two reasons. First, it was difficult to observe B_2H_6 and the first pyrolysis product (B_5H_9) with the same chromatographic attenuation as they occur quantitatively under pyrolysis conditions (Figure 2). Second, it was desirable to determine if higher hydrides could be formed from nitrogen diluted concentrations of B_2H_6 in the presence of oxygen. This reaction has been indirectly indicated to yield higher hydrides by a study in which higher hydrides were formed in the presence of molecular oxygen. Data were obtained with the static reactor at two temperatures: 30 and 100°C. A 10-cm^3 static reactor sample loop was used for sampling the pyrolysis products (B_5H_9, B_6H_{10}, and $B_{10}H_{14}$) and B_2H_6 concentrations of 1 and 25 ppm. A 5-cm^3 sample loop was used for sampling B_2H_6 at 800 ppm. The static reactor was purged at 100°C with nitrogen between runs until no volatile boron hydrides were observed with the FPD under programmed temperature chromatographic conditions.

1. Stability of B_2H_6 and Pyrolysis Products in Nitrogen. B_2H_6 at 1000 ppm in nitrogen showed less than a 5 percent change in concentration at 30°C over a time period of 100 hours. Also, the B_2H_6 pyrolysis products (B_5H_9, B_6H_{10}, and $B_{10}H_{14}$) showed less than a 5 percent change in concentration at 30°C over a time period of 100 hours (Figures 3, 4, and 5).

2. Stability of B_2H_6 in 20 Percent Oxygen at Atmospheric Pressure. The decay of 1, 25, and 800 ppm B_2H_6 at 30°C and 800 ppm B_2H_6 at 100°C in the presence of 20 percent oxygen plotted linearly on log concentration versus time plots. Figure 3 shows the plot for 800 ppm. This indicates the rate of decay to be first order. In the case of a first-order reaction, the half life is dependent on the initial concentration. Consequently, the half life of B_2H_6 at different initial concentrations in the presence of excess oxygen may be used as an index of relative stability.

The observed half lives for 1, 25, and 800 ppm B_2H_6 plus oxygen at 30°C were 21, 28, and 30 hours, respectively (Table 2). The relatively large difference between the half life of 1 ppm and the two higher concentrations is probably due to experimental error associated with adsorption of B_2H_6 in the static reactor and associated transfer lines. Also, the 1-ppm B_2H_6 concentration is the closest of the three concentrations to the detectability limit for a 10-cm^3 sample with the FPD and quantitative error may be a factor in affecting an observed half life at the 1-ppm level.

Chromatograms representing the decay of 800 ppm B_2H_6 with 20 percent oxygen at 30°C were temperature programmed with the technique used to isolate the pyrolysis products observed. Neither B_5H_9, B_6H_{10}, or $B_{10}H_{14}$ were observed after 40 hours and approximately 60 percent of the B_2H_6 has disappeared. However, $B_{10}H_{14}$ was observed in temperature programmed samples extracted from the static reactor having an initial concentration of 800 ppm B_2H_6 plus 20 percent oxygen at 100°C after 20 hours from initial filling of the static reactor and approximately 60 percent of the B_2H_6 reacted. No B_5H_9 or B_6H_{10} was observed under these conditions.

The lack of formation of B_5H_9, B_6H_{10}, and $B_{10}H_{14}$ from B_2H_6 plus 20 percent oxygen at 30°C indicates that B_2H_6 will not

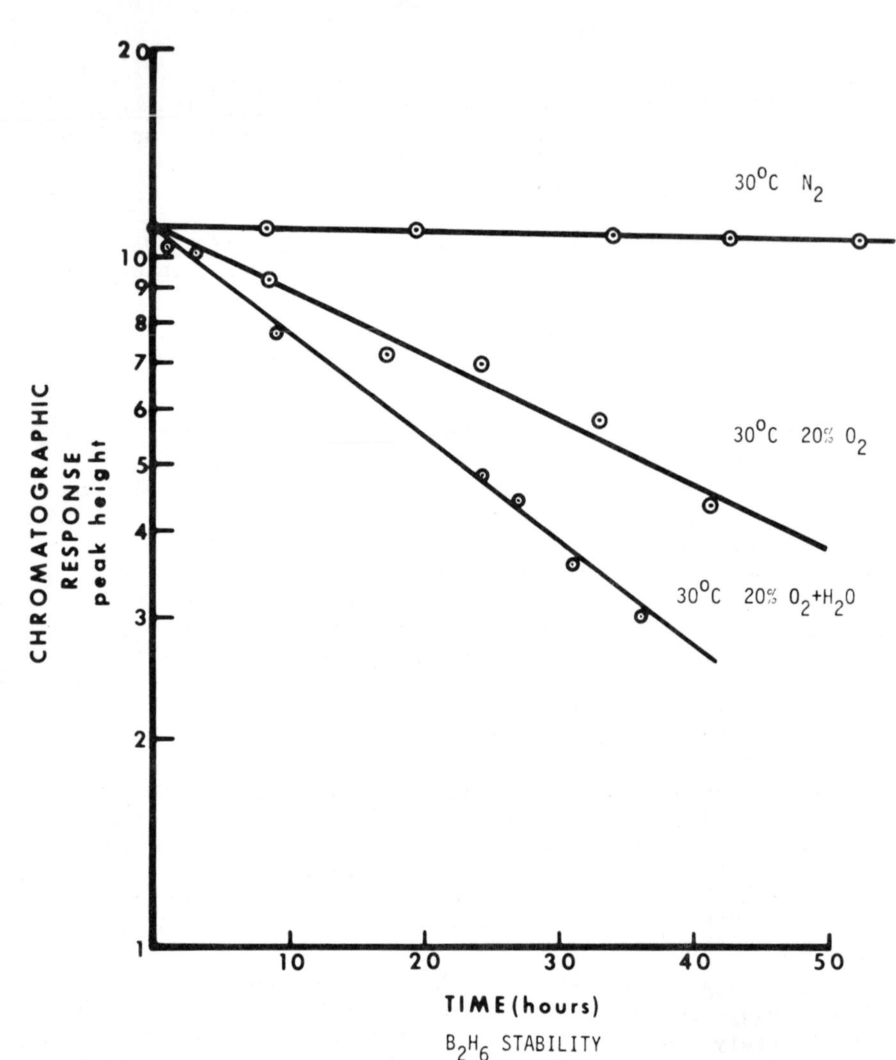

B_2H_6 STABILITY

FIGURE 3. B_2H_6 stability at 30°C in the presence of N_2, 20 percent O_2, and 20 percent O_2 and water vapor in the range of 10,000 ppm. Initial concentration of B_2H_6 is 800 ppm.

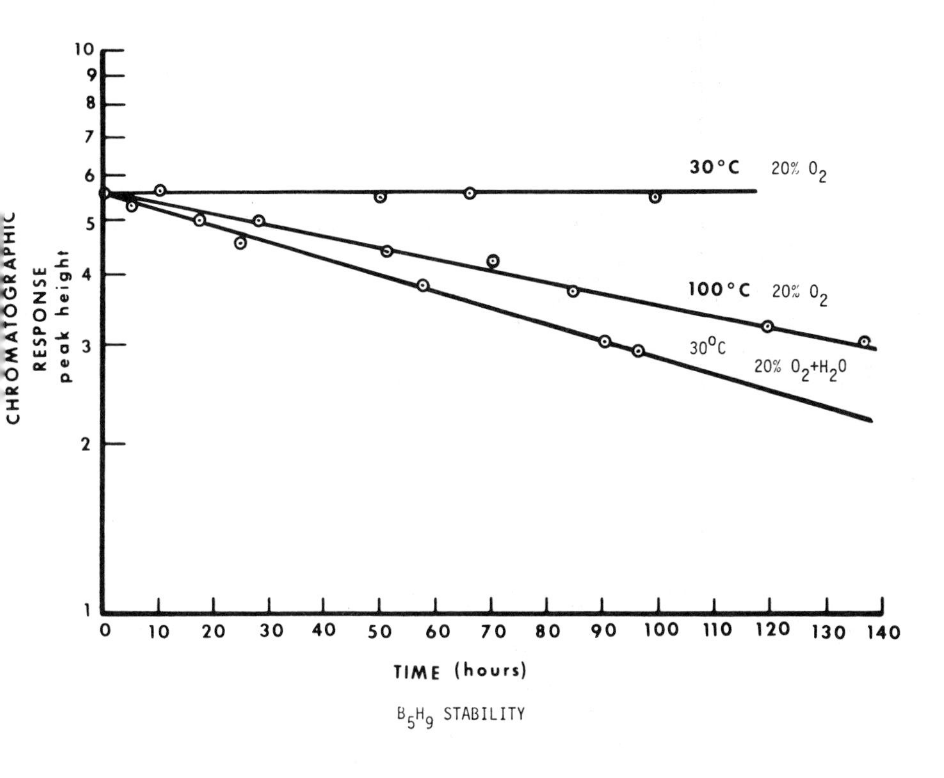

FIGURE 4. B_5H_9 stability at 30 and 100°C in the presence of 20 percent O_2 and at 30°C in the presence of 20 percent O_2 and water vapor in the range of 10,000 ppm.

interfere with the oxidation of mixtures of B_2H_6 and its pyrolysis products via competing formation reactions in the static reactor. However, the formation of $B_{10}H_{14}$ from B_2H_6 plus oxygen at 100°C indicates a potential competing formation reaction which may interfere with the oxidation of B_2H_6 and its pyrolysis products in the static reactor at elevated temperature.

3. Stability of the B_2H_6 Pyrolysis Products in 20 Percent Oxygen at Atmospheric Pressure. The next experiment consisted of filling the static reactor with B_2H_6 and the pyrolysis products in the proportions generated from pyrolysis of 800 ppm B_2H_6 at 260°C (Figure 2). Resultant concentrations in the static reactor

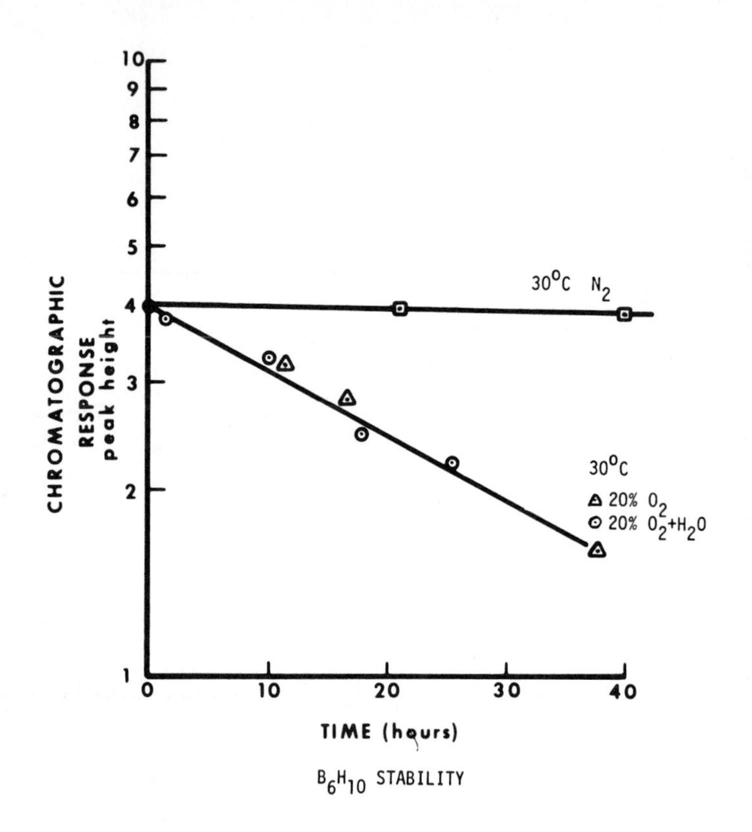

FIGURE 5. B_6H_{10} stability at 30°C in the presence of N_2, 20 percent O_2, and 20 percent O_2 and water vapor in the range of 10,000 ppm.

were approximately B_2H_6 = 500 ppm, B_5H_9 = 50 ppm, B_6H_{10} = 5 ppm, and $B_{10}H_{14}$ = 5 ppm. These concentrations were calculated from the relative proportions of pyrolysis products generated from B_2H_6 (1000 ppm) at 260°C, Figure 2. The rate of decay was followed at 20 and 100°C.

The first pyrolysis product, B_5H_9, showed less than 5 percent change in 100 hours at 30°C in the presence of 20 percent oxygen. However, B_5H_9 decayed linearly on a log concentration versus time plot in the presence of 20 percent oxygen at 100°C, Figure 4. Table 2 shows the first-order rate constant and half life observed.

The decay of the second pyrolysis product, B_6H_{10}, in the presence of 20 percent oxygen at 30°C also plotted linearly on a

TABLE 2. Observed First-Order Kinetic Data for Boron Hydride Oxidation at Atmospheric Pressure

	30°C		100°C	
	½ Life(hrs)	$k(hr^{-1})$	½ Life(hrs)	$k(hr^{-1})$
B_2H_6 (1 ppm)	21	3.3×10^{-2}	-	-
B_2H_6 (25 ppm)	28	2.45×10^{-2}	-	-
B_2H_6 (800 ppm)	30	2.3×10^{-2}	15	4.6×10^{-2}
B_5H_9(∿50 ppm)	stable>100	-	140	4.9×10^{-3}
B_6H_{10} (∿5 ppm)	28	2.45×10^{-2}	Not Detectable	
$B_{10}H_{14}$ (∿5 ppm)	stable>100	-	15	4.6×10^{-2}

log concentration versus time plot, Figure 5. However, B_6H_{10} was not detected in samples taken from the static reactor containing the B_2H_6 pyrolysis effluent plus 20 percent oxygen at 100°C. Table 2 shows the first-order rate constant and half life observed for B_6H_{10} at 30°C.

The third pyrolysis product, $B_{10}H_{14}$, showed less than 5 percent change in 100 hours at 30°C in the presence of 20 percent oxygen. However, the decay of $B_{10}H_{14}$ in the presence of 20 percent oxygen at 100°C plotted linearly on a log concentration versus time plot, Figure 6. Table 2 shows the first-order rate constant and half life observed. An interference may occur in the decay of $B_{10}H_{14}$ with oxygen at 100°C due to the observed formation of $B_{10}H_{14}$ from B_2H_6 plus oxygen at 100°C. However, the observed half life (15 hours) for decay of $B_{10}H_{14}$ with oxygen at 100°C is less than the time (20 hours) required for formation of $B_{10}H_{14}$ at 100°C from B_2H_6 in the presence of oxygen. This observation negates a potential positive interference of the observed decay of $B_{10}H_{14}$ plus oxygen at 100°C.

The apparent first-order decay for the boron hydrides with excess oxygen at atmospheric pressure is in accord with first-order kinetics where one reactant is present in large excess over another so that during a run there is essentially no change in concentration of one reactant. In this case, oxygen is present in large excess as it would be in the ambient air environment.

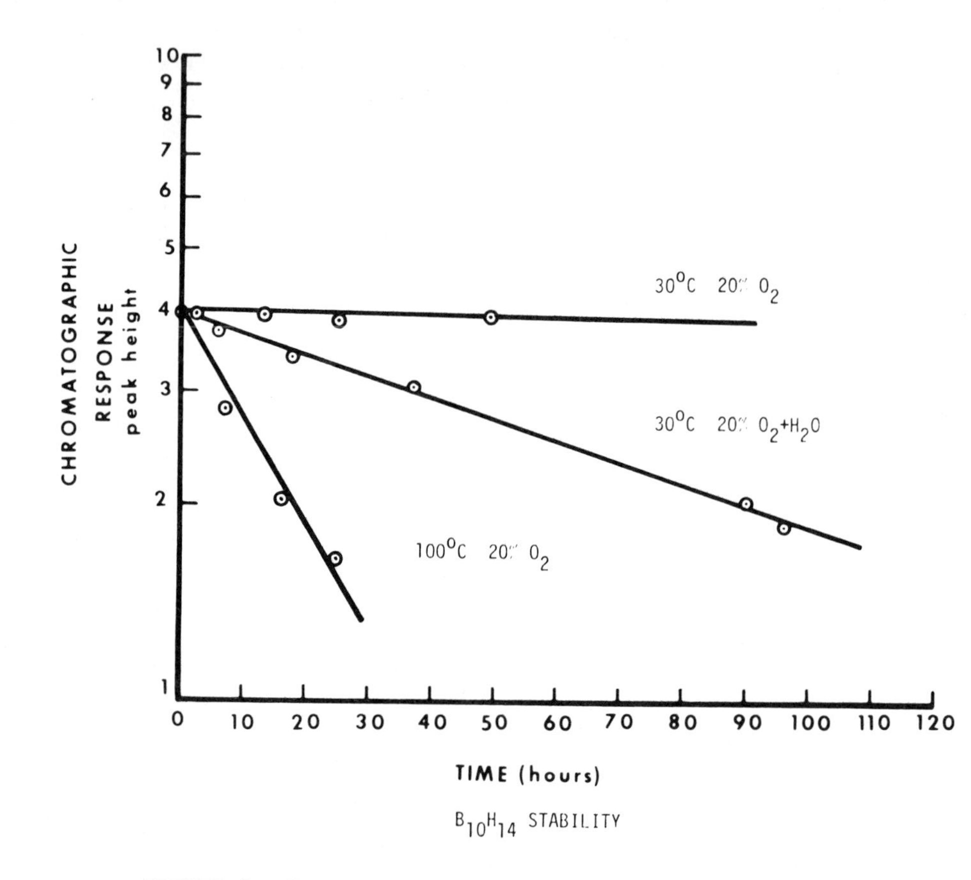

FIGURE 6. $B_{10}H_{14}$ stability at 30 and 100°C in the presence of 20 percent O_2 and 30°C in the presence of 20 percent O_2 and water vapor in the range of 10,000 ppm.

No volatile boron hydride oxidation products were detected with programmed temperature GC analysis. Consequently, the products of the observed decompositions for B_2H_6 and the B_2H_6 pyrolysis products (B_5H_9, B_6H_{10}, and $B_{10}H_{14}$) are probably non-volatile boron oxides which deposit in the static reactor.

4. Stability of B_2H_6 in 20 Percent Oxygen and Water Vapor at Atmospheric Pressure. The decay of 800 ppm at 30°C in the presence of 20 percent oxygen and water vapor (∿10,000 ppm) in the static reactor was followed. This plotted linearly on a log concentration versus time plot, Figure 3. Table 3 shows the

TABLE 3. Observed First-Order Kinetic Data for Boron Hydride Oxidation and Hydrolysis at Atmospheric Pressure

	30°C	
	½ Life(hrs)	k (hr^{-1})
B_2H_6(800 ppm)	20	3.45×10^{-2}
B_5H_9(∿50 ppm)	100	6.9×10^{-3}
B_6H_{10}(∿5 ppm)	28	2.45×10^{-2}
$B_{10}H_{14}$(∿5 ppm)	90	7.6×10^{-3}

first-order rate constant and half life observed. This is in accord with pseudo first-order kinetics as for the oxidation experiments discussed previously, except in this case water vapor, in addition to oxygen, is present in large excess as it would be in the environment.

5. Stability of the B_2H_6 Pyrolysis Products in 20 Percent Oxygen and Water Vapor at Atmospheric Pressure. The static reactor was filled with a mixture of B_2H_6 and the pyrolysis products in the same manner as for the oxidation experiment. The decay of the pyrolysis products (B_5H_9, B_6H_{10}, and $B_{10}H_{14}$) in the presence of 20 percent oxygen and water vapor (10,000 ppm) was in accord with pseudo first-order kinetics as for the oxidation experiments except water vapor is also present in large excess as it would be in the environment, Figures 4, 5, and 6, respectively. Table 3 shows the first-order rate constants and half lives observed.

The significance of the oxidation and hydrolysis experiments performed in the static reactor concerns the potential stability of the boron hydrides in the ambient air environment at trace concentrations. The apparent decay rates and half lives observed as a result of these experiments indicates that boron hydrides can be potentially harmful air pollutants. This is apparent from the observed time intervals required to reduce the boron hydrides observed in this study to nonvolatile products. B_5H_9 and $B_{10}H_{14}$ are the most stable as indicated by the observed half lives of 100 and 90 hours, respectively (Table 3). B_2H_6

and B_6H_{10} are less stable as indicated by the observed half lives of 20 and 28 hours, respectively (Table 3).

Previous investigations concerning the oxidation and hydrolysis of boron hydrides have been carried out primarily at the explosion limits and at low pressures. In none of these cases was any report made which characterized the oxidation and hydrolysis of the boron hydrides in the parts-per-million concentration range in the presence of excess oxygen and water vapor as found in the ambient air environment. Consequently, the study of the stabilities reported here contributes to a better understanding of the air pollution potential of the boron hydrides.

C. Phosphine Pyrolysis

Phosphine was pyrolyzed in a manner analogous to diborane. For this purpose, the laboratory system that has been previously described was used (Sowinski and Suffet, 1975). This gas chromatographic-programmed temperature procedure previously reported for boron hydrides was used to separate PH_3 products (Sowinski and Suffet, 1972). Also the flame photometric detector previously reported for boron hydride detection was fitted with an interference filter at 526 nm and successfully used as a phosphorous specific detection system.

The pyrolysis of phosphine (1000 ppm/N_2) was carried out in a temperature range of 100 to 800°C. One predominant pyrolysis product was observed in a temperature range of 400 to 800°C (Figure 7). An example chromatogram of the chromatographic separation for PH_3 and the pyrolysis product is shown in Figure 8. The product identity has been confirmed as P_4 using GC/MS.

VII. SUMMARY

A study of diborane flow pyrolysis in a modeled industrial process at atmospheric pressure has determined:

1. At least three volatile toxic products (B_5H_9, B_6H_{10}, and $B_{10}H_{14}$) are formed in the flow pyrolysis of B_2H_6 at atmospheric pressure,

2. Formation temperature ranges for the diborane pyrolysis products,

3. Stability data for diborane and its atmospheric pressure pyrolysis products in the presence of excess oxygen and water vapor as can be found in the ambient air environment.

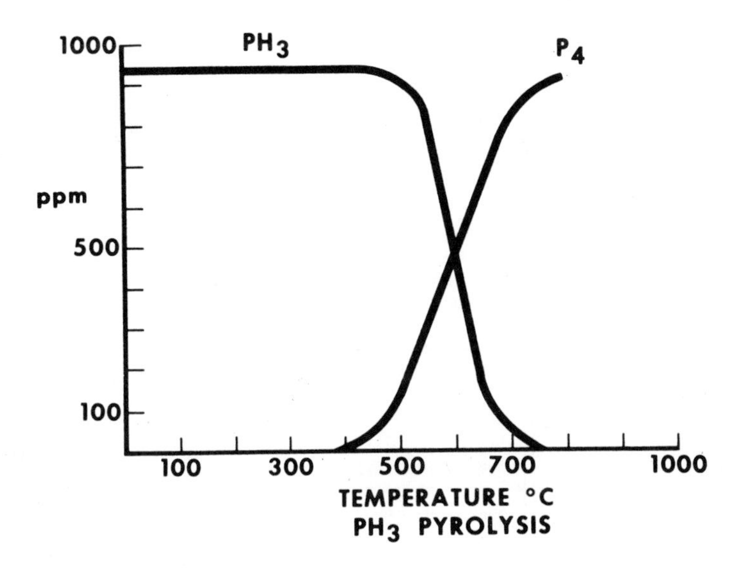

FIGURE 7. PH_3 pyrolysis: ppm compound (PH_3 and P_4) as a function of pyrolysis temperature.

The formation of toxic boron hydride pyrolysis products and the stability of these chemicals and parent B_2H_6 in the presence of excess oxygen and water vapor indicates that these chemicals can be harmful air pollutants. Consequently, controls may be desirable for industrial operations using diborane under pyrolysis conditions.

Three conditions (reactor geometry, diborane flow rate, and pyrolysis temperature) were considered as they may affect the generation of diborane pyrolysis products in the laboratory system. Of these, only variable pyrolysis temperature was found to influence product formations. As a result, a source control technique for the diborane pyrolysis products is indicated by the temperature conditions required for product formations from the pyrolysis of diborane (Figure 2). For example, the products may be eliminated at their industrial source if the formation temperatures are avoided in a manufacturing process. Also, the parent compound, B_2H_6, can be apparently eliminated from atmospheric pressure pyrolysis in a flow system at temperatures exceeding 350°C. Modifications of manufacturing processes using these guidelines are desirable as primary control choices, since they can avoid the need for expensive exhaust treatment methods.

There always can be errors in data obtained from laboratory systems used to predict the formation and stability (chemical

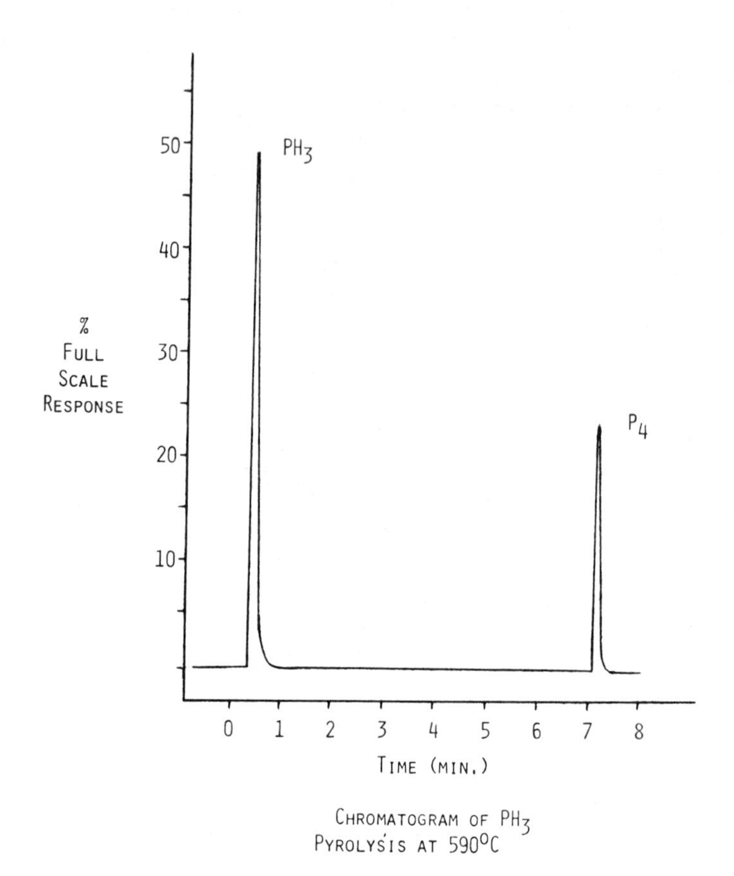

FIGURE 8. Separation of PH_3 and its pyrolysis product P_4 after pyrolysis at 590°C on 6 percent OV-17 liquid phase by programmed temperature gas chromatography with a flame photometric detector. Gas chromatography condition as Sowinski and Suffet (1971).

fate) of potentially harmful environmental pollutants. Two basic reasons apply: first, there may be inadequacies in the laboratory system, and second, there may be parameters that are not represented in the laboratory system. In the case of the laboratory system developed for the boron hydrides, it appears that relatively few shortcomings exist. One problem may relate to surface adsorption of the boron hydrides in the static reactor and associated transfer line. This may be particularly relevant for low concentrations (1 ppm) of B_2H_6 and it might account, at least in part, for the half life of 1 ppm B_2H_6 plus oxygen being shorter by 10 hours as compared to 800 ppm B_2H_6 plus oxygen at 30°C (Table 2).

Parameters that are not represented in the laboratory system may also influence the chemical fate of the boron hydrides. For example, the presence of silicon in industrial applications of diborane in the electronics industry may provide a reactive site for the boron hydrides. Also, boron hydrides used in industry are commonly exhausted through ductwork in which subsequent reaction with other exhausted materials such as Lewis base compounds like amines may occur. Exhaust treatment methods such as water scrubbing and burn-offs on the chemical fate of the boron hydrides is uncertain since higher boron hydrides have been reported to show considerable stability in water, and the combustion of B_2H_6, as may be found in a burn-off, has been reported to yield higher hydrides resulting from pyrolysis reactions that occur in the flame (Price, 1950). A similar phenomenon was observed in this study with the formation of $B_{10}H_{14}$ from B_2H_6 and oxygen at 100°C in the static reactor.

Since shortcomings can exist in data obtained from a laboratory system, it is necessary to sample actual industrial processes in order to validate laboratory data and achieve a more reliable model of actual environmental conditions. However, the use of a laboratory system is indispensable in order to obtain meaningful sampling data. Two basic reasons apply. First, selective and sensitive analytical methods which are developed as part of a laboratory system provide the methodology for analyzing environmental samples. Second, a product template with which environmental samples can be compared is provided by a controlled laboratory system. By applying the analytical methods and products template to the analysis of environmental samples, laboratory data on potential air pollutants can be refined, the efficiency of control techniques for removing harmful products can be more reliably determined, and biological and physical fate parameters can be more effectively determined.

REFERENCES

1. Adams, R. M. 1964. Boron, Metallo-Boron Compounds and Boranes, Interscience, New York.

2. Anon. 1971. Carborane Polymers Stable at High Temperature, Chem. Eng. News, March 22.

3. Anon. 1972. New Hydrides Offer Improved Properties, Chem. Eng. News, June 19.

4. American Conference of Government Industrial Hygienists. 1975. Threshold Limit Values for Airborne Contaminants, Cincinnati, Ohio.

5. ASTM. 1971. Inorganic Index to the Powder Diffraction Files, ASTM, Philadelphia, Pa.

6. Bloem, J. 1972. Doping in Chemically Vapour Deposited Epitaxial Silicon, J. Cryst. Growth 13, 302.

7. Borer, K., Littlewood, A. B., and Phillips, C. S. G. 1960. A Gas Chromatographic Study of the Diborane Pyrolysis, J. Inorg. Nucl. Chem. 15, 316.

8. Bragg, J. K., McCarty, L. V., and Norton, F. J. 1951. Kinetics of Pyrolysis of Diborane, J. Amer. Chem. Soc. 73, 2134.

9. Clarke, R. P. and Pease, P. M. 1951. A Preliminary Study of the Kinetics of Pyrolysis of Diborane, J. Amer. Chem. Soc. 73, 2132.

10. Edwards, J. G. 1975. Science 189, 174.

11. Fehlner, T. 1965. On the Mechanism for the Pyrolysis of Diborane, J. Amer. Chem. Soc. 87, 4200.

12. Fehlner, T. and Strong, R. L. 1962. The Reaction Between Oxygen Atoms and Diborane, J. Chem. Soc. 2893.

13. Ghandhi, S. K. 1968. The Theory and Practice of Microelectronics, Wiley, New York.

14. Giblin, F. J. and Massaro, E. J. 1973. Toxicol. Appl. Pharmacol. 24, 81.

15. Gustafson, C. G. 1970. PCBs: Their Presence and Persistence in the Environment, Environ. Sci. Technol. 10, 814.

16. Hanna, S. R. 1973. Atmos. Environ. 7, 803.

17. Hughes, R. L., Smith, I. C., and Lawless, E. W. 1967. Production of the Boranes and Related Research, Academic, New York.

18. Kotlensky, W. and Schaeffer, R. J. 1958. Decomposition of Diborane in a Silent Discharge. Isolation of B_6H_{10} and B_9H_{15}, J. Amer. Chem. Soc. 80, 4517.

19. Landez, J. H. and Scott, W. N. 1971. Effects of Boranes upon Tissues of the Rat, Proc. Soc. Exp. Biol. Med. 136, 1389.

20. Levinskas, G. J. 1955. Comparative Toxicity of Boranes, Amer. Indust. Hyg. Assoc. Quart. 16, 4.

21. Matherson Gas Company. 1974. Gas Handling Handbook.

22. Metcalf, R. L., Sangha, G. K. and Kapoor, I. P. 1971. Environ. Sci. Technol. 5, 709-713.

23. Morey, J. R. 1958. Kinetics and Mechanism of Diborane Pyrolysis, Diss. Abst. 19, 1223.

24. Price, F. P. 1950. First and Second Pressure Limits of Explosion of Diborane-Oxygen Mixtures, J. Amer. Chem. Soc. 72, 4361.

25. Private communication. 1972. Callery Chemical Company, July 26.

26. Saalfield, F. E., Williams, F. W., and Saunders, R. A. 1971. Identification of Trace Contaminants in Enclosed Atmospheres, American Laboratory, July, 8-16.

27. Shapiro, I. and Keilin, B. 1959. Mass Spectrum of Octaborane, J. Amer. Chem. Soc. 76, 3864.

28. Stock, A. 1933. Hydrides of Boron and Silicon, Cornell University Press, Ithaca, New York.

29. Sowinski, E. J. and Suffet, I. H. 1971. Characterization of a Melpar Flame Photometric-Gas Chromatographic System for Application to Boron Hydrides, J. Chromatogr. Sci. 9, 632.

30. Sowinski, E. J. and Suffet, I. H. 1972. Programmed Temperature Gas Chromatography of Boron Hydrides, Anal. Chem. 44, 2237.

31. Sowinski, E. J. and Suffet, I. H. 1973. An Approach for Studying the Air Pollution Potential of Significant Products Resulting from an Industrial Process, Proc. Third Int. Clean Air Congress, Section D, p. 35, Dusseldorf, Germany.

32. Sowinski, E. J. and Suffet, I. H. 1974. Gas Chromatographic Detection and Confirmation of Volatile Boron Hydrides at Trace Levels, Anal. Chem. 46, 1218.

33. Sowinski, E. J. and Suffet, I. H. 1975. A Gas Sampling and Static Reactor System for Use in Evaluating the Environmental Fate of Volatile Chemicals from Industrial Processes at Trace Levels: Application to Volatile Inorganic Hydrides as Used in the Semiconductor Industry, No. 75280, Proc. 21st Int. Instrument Soc. of America Instrumentation Symposium, pp. 463-466, Philadelphia, Pennsylvania, May 19.

34. U. S. Department of Health, Education, and Welfare; Public Health Service; and National Air Pollution Control Administration. 1969. Preliminary Air Pollution Survey of Boron and Its Compounds, Raleigh, North Carolina, October.

SECTION IV
BIOLOGICAL FATE OF POLLUTANTS IN THE AQUATIC ENVIRONMENT

Biological Fate and Transformation of Pollutants in Water

ROBERT L. METCALF
Institute of Environmental Studies and
Department of Entomology
University of Illinois
Urbana-Champaign, Illinois

I. INTRODUCTION

Problems of water quality associated with man-made chemicals have greatly intensified during the past 30 years. In the 1940s, we were told that the solution to aquatic pollution was dilution. Today we are deeply concerned about the seizure by FDA of fish from Lake Michigan contaminated with DDT, dieldrin, and polychlorinated biphenyls (PCBs) that have been biomagnified 10^5- to 10^6-fold from parts-per-trillion residues in the water.

The aqueous environment, river, lake, estuary, and ocean, is the ultimate repository for trace amounts of all the synthetic chemicals made by man. These contaminants together with natural products liberated by industrial society, for example, asbestos, petroleum, metals, terpenoids, tannins, humic acids, comprise an almost endless variety from which to study water pollution. Increasing awareness is focused on the aquatic micropollutants, compounds generally present in parts-per-million to parts-per-trillion levels, yet which have the properties of (a) low water and high lipid solubility, (b) bioconcentration through organisms of trophic webs or directly from water, (c) delayed symptoms of biological activity, and (d) toxicity, carcinogenicity, mutagenicity, or teratogenicity (Warner, 1967).

The inventiveness of the United States' organic chemical industry and its exponential growth are the source of major concern about water pollution. The first edition of the Merck Index (1889) "containing a summary of whatever chemical products are today adjudged as being useful in either medicine or technology" listed 828 chemicals; the eighth edition (1968) listed about 10,000. The Toxic Substances List (National Institute for Occupational Safety and Health, 1973) catalogs 25,043 potentially hazardous chemicals. The United States' production of organic chemicals has grown from 10 x 10^9 lb (4.5 x 10^9 kg) in 1943 to 140 x 10^9 lb (64 x 10^9 kg) in 1972. The rate of growth has been exponential, increasing about 9 percent per year with a doubling time of about 8 years (U. S. Tariff Commission data, 1970). Organic chemicals utilized as intermediates in synthesis, fuels, detergents, pesticides, plasticizers, pharmaceuticals, food and feed additives, brighteners, solvents, paints, lacquers, rubber chemicals, household products, plastics, fibers, and elastomers enter into the total environment *accidentally* from industrial and household effluents, from employment as drugs and feed supplements, and as byproducts of transportation; or *purposefully* from applications of pesticides, preservatives, and protective coatings.

Two facets of this proliferation of aquatic contaminants are especially alarming: the number of new organic chemicals, estimated at 500 through 700, which enter the environment in appreciable trace quantities (Lee, 1964), and the long delay required to understand and appreciate hazardous effects. Thus vinyl chloride (United States production was 5.6 x 10^9 lb in 1974) was produced for 40 years before it was discovered to be a carcinogen (Maltoni and Lefemine, 1975), and phthalate ester plasticizers (United States production was 1 x 10^9 lb in 1970) were produced for 30 years before they were found to be teratogens (Singh et al., 1973). These problems are highlighted by the very recent concern over the carcinogenic effects of *bis*-

chloromethyl ether (Kuschner et al., 1975) and the highly persistent toxicity of the polybromobiphenyl flame retardants (Anon, 1975). The importance of the entire subject of chemical contamination of food and water is indicated by the suggestion that 80 percent of all human cancers are caused by chemicals encountered during our daily lives (Anon, 1972, 1974a).

II. CONTAMINATION OF WATER RESOURCES

The extent of contamination of United States water resources by organic chemicals is not well appreciated. The few careful surveys that have been made suggest that such contamination is both extensive and potentially hazardous. The waters of the Kanawha River in West Virginia, draining an extensive manufacturing complex, were found to contain as high as 457 ppb of carbon-chloroform extractives from which more than 100 synthetic organic chemicals were identified including phenols, substituted benzenes, aldehydes, ketones, alcohols, chloroethyl ether, acetophenone, diphenyl ether, pyridine and other nitrogenous bases, nitrates, acids, tetralin, naphthalene, detergents, DDT, and aldrin (Middleton, 1960). The Charles River in Boston was found to contain more than 30 detectable synthetic organic chemical pollutants including naphthalene, anthracene, pyrene, and dialkyl phthalates in concentrations ranging from 0.05 to 1 ppb (Hites and Biemann, 1972). These apparently entered largely from automobile exhaust and sewer runoff. The municipal water supply of Evansville, Indiana, taken from the Ohio River, was found to contain over 40 synthetic organic chemicals including toluene, ethyl benzene, xylene, styrene, hexachlorobenzene, chlorhydroxybenzophenone, bromodichloromethane, chlorodibromomethane, tetrachloroethylene, bromoform, hexachloroethane, *bis*-(2-chloroethyl ether) and *bis*-(2-chloroisopropyl) ether, in concentrations up to 0.8 ppb (Kleopfer and Fairless, 1972).

The most publicized survey of this nature is the EPA (1974) study of the finished municipal water supply of New Orleans, Louisiana, taken from the Mississippi River. At least 66 organic chemicals were characterized and measured including the following with maximum levels in parts per billion: plasticizers—diisobutyl phthalate 0.59, benzyl butyl phthalate 0.81, di-(2-ethylhexyl) phthalate 0.31, dibutyl phthalate 0.19, dimethyl phthalate 0.27, dipropyl phthalate 0.14, triphenyl phosphate 0.12, dioctyl adipate 0.10; pesticides—atrazine 5.1, deethylatrazine 0.5, dieldrin 0.07, heptachloronorbornane (toxaphene) 0.06, endrin 0.004; detergents—alkylbenzenes C_2-C_3 0.33; organohalogens—chloroform 113, bromoform 0.57, dibromodichloroethane 0.33, dibromochloromethane 1.1, 1,2-dichloroethane 8, hexachlorobutadiene 0.27, hexachloroethane 4.4, tetrachloroethylene 0.5, tetrachloroethane 0.11, trichloroethane 0.45, bis-(2-chloroisopropyl) ether 0.18, bis-(2-chloroethyl) ether 0.16; miscellaneous—dihydrocarvone 0.14, isophorone 2.9, limonene 0.03, toluene 0.10, 2,6-di-t-butyl-p-benzoquinone 0.23. Breidenbach et al. (1967) showed nearly ubiquitous contamination of the surface waters of the United States by organochlorine pesticides in concentrations ranging from DDT (total) 0.008 to 0.144 ppb, dieldrin 0.008 to 0.122 ppb, endrin 0.008 to 0.214 ppb, heptachlor epoxide 0.001 to 0.008 ppb, and BHC (T) 0.003 to 0.022 ppb. The paper by Koefler et al (in this volume) reviews the pollutants found in United States drinking water supplies.

III. BIOCONCENTRATION FROM WATER

The ability of living organisms to concentrate and accumulate relatively high concentrations of lipid soluble organic compounds, either directly from water or indirectly through consumption of contaminated food web organisms, is perhaps the most serious problem resulting from water pollution by organic micropollutants. This process is termed bioconcentration or ecological magnification and is of major importance in determining the safety of the human food supply, for example, consumption of tuna fish contaminated with methyl mercury from the ocean; coho salmon, lake trout, and chubs contaminated with DDT, dieldrin, and PCBs from Lake Michigan (Environmental Protection Agency, 1972); milk contaminated with polybromobiphenyls (PBB) (Anon., 1975); or chickens contaminated with dieldrin from soybean sludge (Anon., 1974b). Thus the Environmental Protection Agency has proposed evaluation of bioconcentration as a major criterion in establishing effluent standards for pollution (Quarles, 1973).

The magnitude of bioconcentration in nature is well demonstrated by data for the lake trout, Salvelinus namaycush from Lake Michigan (Environmental Protection Agency, 1972) as summarized in Table 1, where average bioconcentration factors (ppm in trout/ppm in water) were: DDT, 3.13×10^6; dieldrin, 1.35×10^5; PCBs, 3.41×10^6.

Bioconcentrations of this order can have gross biological and social consequences. As an example, coho salmon, Oncorhynchus kisutch, from Lake Michigan formed about 40 percent of the diet used for commercial production of mink, in the Great Lakes region. Mink raised in this area have been suffering from severe reproductive difficulties since about 1967. After a great

TABLE 1. Bioconcentration of DDT, Dieldrin, and PCBs in Lake Trout *Salvelinus namaycush* of Lake Michigan (Environmental Protection Agency, 1972)

Compound	Average Concentration (ppm)		Bioconcentration factor
	Open Lake	Lake Trout	
DDT-(T)	0.0000060	18.80	3.13×10^6
Dieldrin	0.0000020	0.26	1.30×10^5
PCBs	0.0000082	28.0	3.41×10^6

deal of research, this reproductive failure was traced to the presence of PCB residues, now reaching 13 to 15 ppm, in the coho salmon component of the diet (Environmental Protection Agency, 1972; Ringer, Aulerich, and Polin, 1974). PCBs fed to mink are embryotoxic to the developing fetus causing total cessation of reproduction when ingested at levels of 5 ppm or less (Ringer et al., 1974; Aulerich et al., 1973).

The bioconcentration of DDT and dieldrin by fish, particularly those in cold water lakes where detoxication processes are minimal, is essentially a linear function with the age of the fish. Thus Youngs et al. (1972) evaluated total residues of DDT and its degradation products DDE and DDD in lake trout of various ages from Cayuga Lake in New York. As the trout aged from 1 to 12 years, residues of DDT and its metabolites increased proportionately from about 1 ppm DDT-(T) at 1 year of age to 16 to 20 ppm at 11 years of age. The correlation between DDT-(T) residues in parts per million and age ranged from 0.92 to 0.94 over a three-year period, 1968 to 1970. Similar data for residues of DDT and dieldrin and size of lake trout from Lake Michigan are shown in Table 2 (Environmental Protection Agency, 1972). The correlation between size of fish and DDT residues was $r = 0.986$, and for dieldrin residues was $r = 0.962$.

The U. S. Food and Drug Administration has established action levels for residues in fish of DDT, 5 ppm; dieldrin, 0.3 ppm; and PCBs, 5 ppm. Enforcement of the levels essentially will prevent the sale of Lake Michigan fish (Environmental Protection Agency, 1972). The grave nature of the contamination of the Great Lakes with micropollutants is intensified by the low water turnover times, as Rainey (1967) has estimated that it requires

TABLE 2. Bioconcentration of DDT and Dieldrin in Lake Trout *Salvelinus namaycush* from Lake Michigan (Environmental Protection Agency, 1972)

Lake Trout Size (in.)	Whole Body Residues (ppm)	
	DDT-(T)	Dieldrin
2.0 - 5.9	0.89	0.03
6.0 - 9.9	2.24	0.12
10.0 - 15.9	6.00	0.14
16.0 - 21.9	8.00	0.21
22.0 - 26.9	14.62	0.23
27.0 - 32.9	19.23	0.26

about 100 years for 90 percent waste removal in Lake Michigan and >500 years in Lake Superior. Thus the contamination of these lakes is a major disaster for which no apparent solution exists. Rainey has calculated the change in concentration of pollutants with time T as

$$C_2 = C_2^0 \exp\left(-\frac{RT}{V}\right) \left[C_1 + \left(\frac{Q}{R}\right)\right] \left[1 - \exp\left(-\frac{RT}{V}\right)\right] \quad (1)$$

where V is the volume, C_2^0 is the concentration of pollutant at initial time T = 0, R is the flow rate, the concentration C_1 of pollutants entering is constant, and pollutants Q are added at a constant rate so that their concentration C_2 is uniform.

A. Role of Partition Coefficient

The partition of chemicals between lipids and water determines the entry of lipid soluble substances either directly from water through cuticle or from water to blood through gills, and from blood to tissues. The (octanol/H_2O) partition coefficient has been used by Fujita et al. (1964) as a constant to express the distribution of drugs in living systems. The log (octanol/H_2O) partition has been defined as Π, a linear free energy parameter which can be approximated by

$$\Pi_X = \Sigma \text{ substituent constants} + \Pi_H \quad (2)$$

where H is the value for benzene or a benzene derivative and X is a substituted benzene. Neely et al. (1974) correlated log bioconcentration, defined as k_1k_2 (where k_1 is the kinetic rate constant for uptake and k_2 is the kinetic rate constant for elimination) in the muscle of rainbow trout, *Salmo gairdneri*, with log partition coefficient (Fujita et al., 1964) for a variety of organic compounds including 1,1,2,2-tetrachloroethylene, carbon tetrachloride, *p*-dichlorobenzene, diphenyl, diphenyl ether, 2-biphenyl phenyl ether, hexachlorobenzene, and 2,2',4,4'-tetrachlorodiphenyl ether. The correlation coefficient was r = 0.948 and F test indicated a confidence level of 0.999.

Lu and Metcalf (1975) correlated Π determined experimentally with log bioconcentration in the mosquito fish *Gambusia affinis* using the relative ppm in fish/ppm in water after 3 days, for 11 chemicals as shown in Figure 1. The correlation coefficient r = 0.79 indicated a highly significant relationship that might be improved by more accurate determination of partition co-

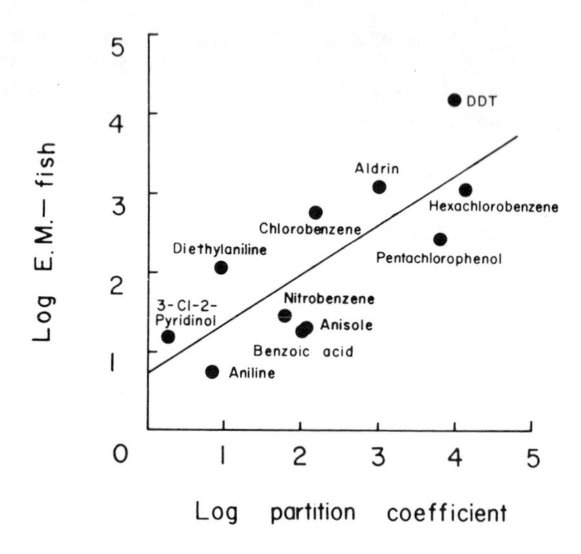

FIGURE 1. Relationship of log octanol/water partition coefficient (Π) to ecological magnification (bioconcentration) of various organic compounds in the fish *Gambusia affinis* (after Lu and Metcalf, 1975. Reprinted with permission from the National Institute of Environmental Health Science, U. S. Department of Health, Education, and Welfare).

efficients as these are readily influenced by traces of octanol which are very difficult to separate from water. It is apparent that the octanol/H_2O partition coefficient for any organic compound can be used as a first approximation of its relative propensity for bioconcentration in living organisms. The end result, however, is clearly a function of the stability of the compound in the water and the rate at which it is degraded to more water partitioning metabolites in the organism.

B. Role of Water Solubility

For organic compounds, at least, there is an inverse relationship between water solubility and bioconcentration. Exchange equilibria between water and tissue lipids were suggested by Hamelink et al. (1971) as explaining the bioconcentration of DDT, dieldrin, toxaphene, and lindane in aquatic organisms, and this was shown to be inversely proportional to water solubility. Lu and Metcalf (1975) demonstrated a high degree of correlation ($r = -0.93$) for the 11 representative organic compounds shown in

Figure 1, between log water solubility in parts per million and log bioconcentration in Gambusia affinis as determined in model ecosystem studies. This relationship has been extended to 38 organic compounds as shown in Figure 2, with a correlation coefficient r = -0.8. This data reflects the general lipid solubility of organic compounds, and water solubility can be used as an initial approximation of the propensity of organic compounds to biomagnify.

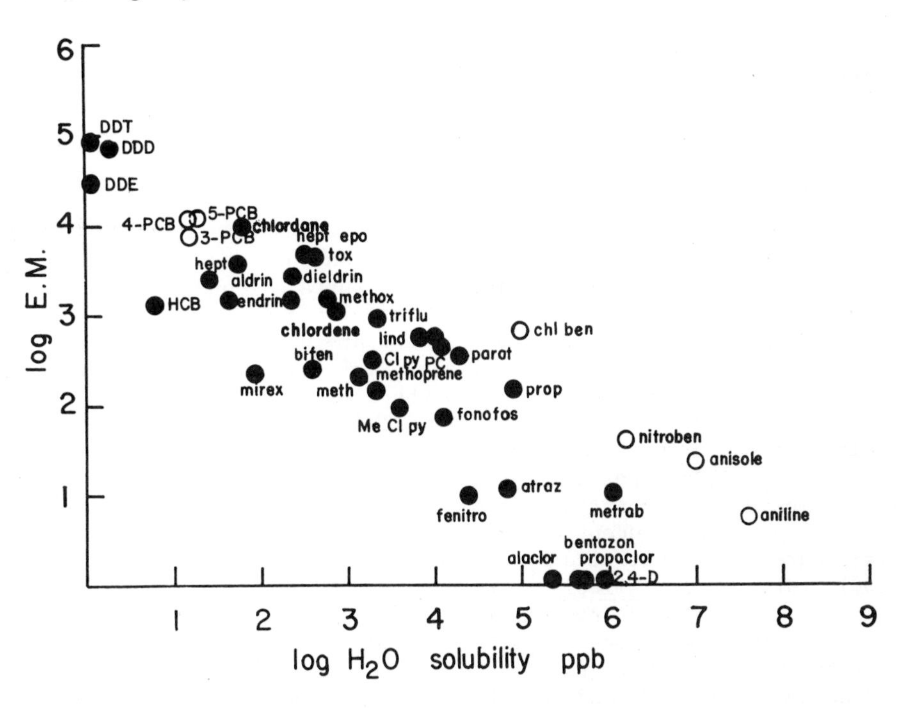

FIGURE 2. Relationship of water solubility to ecological magnification (bioconcentration) of various pesticides and organic compounds by the fish Gambusia affinis. Data from laboratory model ecosystem studies. The PCB's were 2,5,2'-trichlorobiphenyl (3-PCB); 2,5,2',5'-tetrachlorobiphenyl (4-PCB); and 2,4,5,2',5'-pentachlorobiphenyl (5-PCB). Meth is 2,2-bis-(p-methylphenyl)-1,1,1-trichloroethane, tox is toxaphene, hept is heptachlor, methox is methoxychlor, prop is propoxur, Cl py is chloropyrifos, fenitro is fenithrothion, and atraz is atrazine.

C. Quantitative Aspects of Bioconcentration

1. Absorption Through Integument. Most aquatic organisms, including plants and animals and both invertebrates and vertebrates, absorb lipid soluble xenobiotics (man-made compounds foreign to the normal environment of organisms) directly through the integument. This absorption is generally a first-order process, that is, the amount penetrating per unit time is a function of the amount absorbed at the absorbing surface. This absorption process has been expressed as a form of Fick's law (Buerger, 1966):

$$\frac{dS}{dt} = - DA \frac{dc}{dx} \tag{3}$$

where D is the diffusion coefficient, A is the cross-sectional area of surface, x is the thickness of surface (constant), c is the concentration of diffusing substance, and S is the total amount of diffusing substances present at the surface. Buerger (1966) has extended this to the comparable biological situation of penetration through a series of barriers, and concluded that the penetration of the integument will follow first-order kinetics in an integument composed of any number of phases structured in any possible arrangement.

The rate of absorption from water is a function of the rate of exposure, or concentration of xenobiotic, times the time of exposure as demonstrated by Reinbold et al. (1971) with [^{14}C] DDT, for the fish *Tilapia mossambica* and *Lepomis cyanellus*, where the total body residue was directly related to the concentration in water times the time of exposure in days. This relationship, shown in Figure 3, also holds for the relatively sessile snail *Physa* and for the water flea *Daphnia magna* (Metcalf et al., 1973) and is complicated by the rate of flow of contaminated water around the body of the organism or through its gills by differing rates of separation. Similar rates of uptake have been shown for 2,5,2'-trichlorobiphenyl, 2,5,2',5'-tetrachlorobiphenyl, 2,5,2'4',5'-pentachlorobiphenyl (Sanborn et al., 1975) and for di-2-ethylhexyl phthalate in the alga *Elodea*, and mosquito larva *Culex* (Goldstein et al., 1969). The relationship is probably universal for lipid soluble substances.

2. Pharmacodynamics of Bioconcentration. The quantitative aspects of xenobiotic absorption and metabolism by aquatic organisms living in a large pool of constant contamination, for example a lake, represent an example of the compartmental analysis

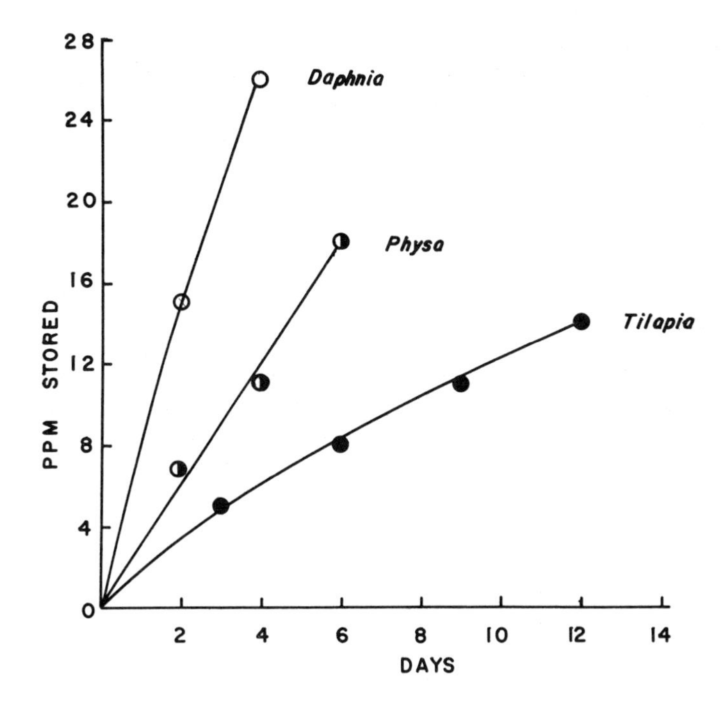

FIGURE 3. Uptake of [^{14}C] DDT from water at 0.003 ppm by the fish Tilapia mossambica, Daphnia magna, and the snail Physa sp. (data from Reinbold et al., 1971).

of the pharmocokinetics of drug absorption and metabolism (Goldstein et al., 1969; Kerr and Vass, 1973). Assume that the aquatic organism is represented by a single homogenous compartment of constant size V (organism mass or volume), and the xenobiotic is absorbed at a constant zero-order rate k_a, determined by the lipid/water partition coefficient, from an aquatic environment contaminated at a constant level C_e. Then the concentration in the organism C_o can be represented as

$$\frac{dC_o}{dt} = \frac{C_e k_a}{V} - k_c C_o \qquad (4)$$

Here k_c is the rate constant for clearance of the drug by the combination of the processes of degradation, elimination, and "growth dilution." This occurs at a rate proportional to the internal concentration in the organism C_o, and has first-order kinetics.

Integration of Eq. 5 gives the concentration in the organism at any time t as

$$C_o = \frac{C_e k_a}{k_c V}(1 - e^{-k_c t}) \tag{5}$$

If the rate of absorption of the xenobiotic into a living organism is constant and the rate of output is exponential, then the concentration in the organism C_o will increase until a steady state is reached where clearance is equal to absorption. Thus as t becomes infinite, $1 - e^{-k_c t}$ becomes unity, and at the steady state,

$$C_o = \frac{C_e k_a}{k_c V} \tag{6}$$

This is called the plateau principle (Goldstein et al., 1969).

However, in poikilothermic aquatic organisms (e.g., fish) where k_c is determined by enzymatic processes influenced by environmental temperatures, the rate of clearance k_c becomes very small in comparison with the rate of absorption k_a largely independent of temperature except as a function of respiratory rate. Therefore, the length of time required to reach a plateau or stable residue level will often exceed the longevity of the organism. This phenomena can become of acute importance for the stable organochlorine xenobiotics where lipid/water partitioning is very high and degradation and elimination very slow, as in the lake trout in Lake Michigan (Table 2) where bioconcentration of DDT, dieldrin, and PCBs is a linear function with age.

Although these equations are oversimplified in the assumption of constant parameters, which seldom are found in most ecological systems, they are useful to describe the rates of accumulation of xenobiotics both in controlled systems and in large abyssal lakes or oceans where constant conditions are more likely to prevail. They are clearly useful in comparing the bioconcentration of various organic compounds in various organisms and in evaluating additional environmental factors that can make them less applicable in a variety of specific situations.

3. Bioconcentration from Food. The rate of xenobiotic accumulation from ingestion of contaminated food or prey can be treated as a modification of Eq. 3 in which the factor for accumulation of the xenobiotic is the daily intake or concentration in prey C_p times the weight of prey W_p times a suitable factor (c) to measure extraction of contaminant from the food. Thus

$$\frac{dC_o}{dt} = \frac{c(C_P W_P)}{V} - k_c C_o \qquad (7)$$

These factors are not very well understood quantitatively for most organisms. However, Reinbold et al. (1971) contaminated *Daphnia magna* with DDT by placing them in water containing 0.003 ppm for 48 hours until the daphnia contained 25 to 26 ppm DDT (dry weight). The daphnia were fed to the guppy *Lebistes reticulatus* in clean water. As shown in Figure 4, the rate of storage of DDT in the guppy was nearly linear with time over a 20-day interval, but methoxychlor, which is readily cleared by the guppy, did not bioconcentrate appreciably when fed at approximately the same rate, 21 to 22 ppm, in daphnia.

The plateau principal seems to be more readily demonstrated when organochlorine compounds are ingested contained in food than when they are absorbed through the integument. Grzenda et al. (1970, 1971) studied the uptake and elimination of DDT and dieldrin by the goldfish *Carassius auratus* fed a contaminated diet for 192 days. The percentage of intake stored in body tissues declined progressively with time of feeding as shown in Figure 5. Similar results were obtained by Gruger et al. (1974) for coho salmon fed a diet contaminated with 3,4,3',4'-tetrachlorobiphenyl, 2,4,5,2',4',6'-hexachlorobiphenyl, and 2,4,5,2',4',5'-hexachlorobiphenyl (Figure 5).

IV. BIOTRANSFORMATIONS

The processes of metabolism and degradation of xenobiotic compounds in aquatic organisms are qualitatively similar to those of terrestrial animals. Basically, the aquatic organism seeks to convert relatively water-insoluble and lipid-soluble xenobiotics into more water-soluble compounds that can be excreted rather than stored in tissue lipids. The enzymes most generally involved in this detoxication process are the microsomal oxidases (National Academy of Sciences, 1972), which effect hydroxylation of organic molecules in which a single atom of molecular oxygen is inserted through an intermediary, the cytochrome P-450 OH radical:

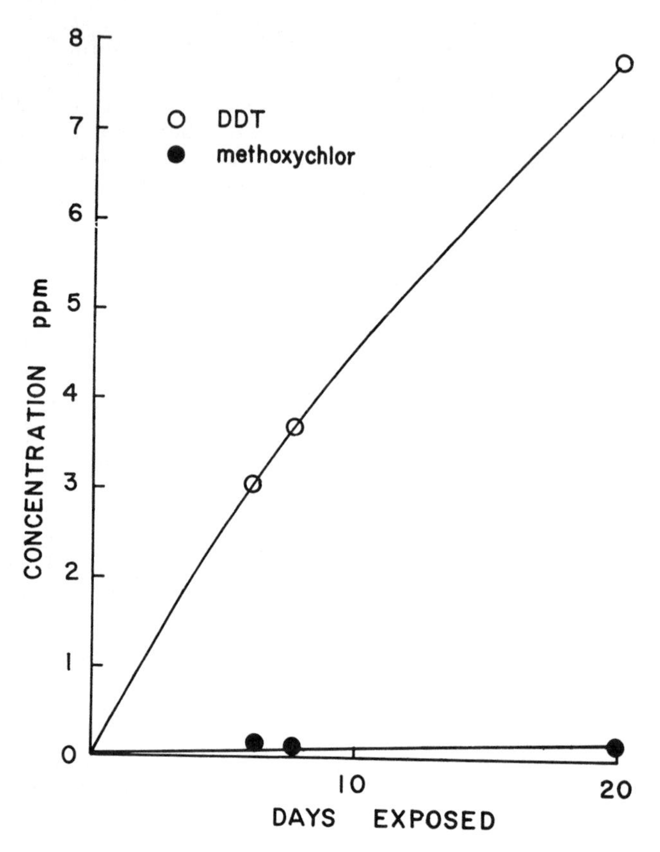

FIGURE 4. Uptake of [^{14}C] DDT and [^{3}H] methoxychlor by the fish fed on Daphnia magna containing 25 ppm DDT and 22 ppm methoxychlor (data from Reinbold et al., 1971).

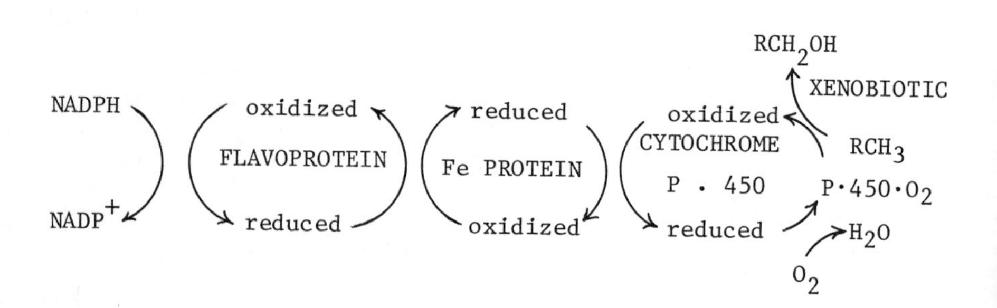

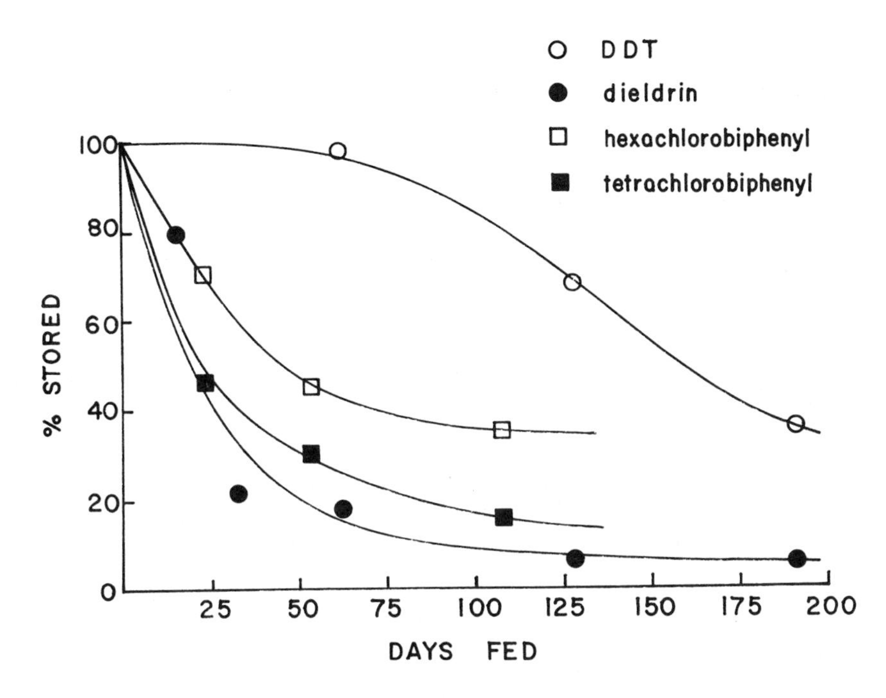

FIGURE 5. Relationship of body storage of DDT and dieldrin fed to goldfish, *Carassius auratus*, and of 3,4,3',4'-tetrachlorobiphenyl and 2,4,5,2',4',5'-hexachlorobiphenyl fed to coho salmon, *Oncorhynchus kisutch* (data from Grzenda et al., 1970, 1971 and Gruger et al., 1975).

The types of reactions mediated by the microsomal oxidase enzymes are those that can be accomplished by direct attack of the ·OH, usually followed by rearrangement or further degradation (Metcalf, 1966):

1. Ring hydroxylation: $C_6H_6 + \cdot OH \longrightarrow C_6H_5OH$

2. Sulfoxidation: $RSCH_3 + \cdot OH \longrightarrow RS(\rightarrow O)CH_3$

3. *O*-Dealkylation: $ROCH_3 + \cdot OH \longrightarrow ROCH_2OH \rightarrow ROH$

4. N-Dealkylation: $RNHCH_3 + \cdot OH \longrightarrow RNHCH_2OH \longrightarrow RNH_2$

5. Side chain oxidation:

$$C_6H_5CH_3 + \cdot OH \longrightarrow C_6H_5CH_2OH \longrightarrow C_6H_5COOH$$

6. Epoxidation:

Cl, Cl, Cl, Cl, ClCCl, HCH (aldrin) + ·OH ⟶ Cl, Cl, Cl, Cl, ClCCl, HCH, O (dieldrin)

7. Desulfuration:

$$O_2N\text{-}C_6H_4\text{-}OP(S)(OC_2H_5)_2 + \cdot OH \longrightarrow O_2N\text{-}C_6H_4\text{-}OP(O)(OC_2H_5)_2$$

It is noteworthy that while these reactions commonly result in more water-soluble, less toxic products, for example, detoxication mechanisms, Reaction 6 may result in a more persistent, more toxic compound as in the formation of dieldrin from aldrin, and Reaction 7 may convert relatively inert phosphorothionates to highly toxic phosphates, for example, formation of paraoxon from parathion. Thus these latter two reactions represent intoxication mechanisms.

The importance of the microsomal oxidases in the clearance of xenobiotics from fish was demonstrated by Reinbold and Metcalf (1976) using the synergist piperonyl butoxide, which is an effective inhibitor of these enzymes presumably by interference with the P-450·O_2 activated complex (Casida, 1970). Green sunfish were exposed to ^{14}C-radiolabeled aldrin, methoxychlor, and trifluralin at 0.01 ppm in water alone or together with 0.1 ppm piperonyl butoxide (3,4-methylenedioxy-6-propylbenzyl butyl diethyleneglycol ether) for periods of 16 days and radioassays were conducted on the amount and nature of degradative products formed as shown in Table 3. The presence of piperonyl butoxide

TABLE 3. Effects of Piperonyl Butoxide at 0.1 ppm on Degradation by Green Sunfish *Lepomis cyanellus* of Pesticides at 0.01 ppm (Reinbold and Metcalf, 1976)

	ppm in Body of Fish							
	Methoxychlor		Trifluralin		Aldrin		Dieldrin	
Days	Alone	p.b.[a]	Alone	p.b.[a]	Alone	p.b.[a]	Alone	p.b.[a]
1	0.80	2.99	1.98	3.63	2.01	3.96	3.18	1.19
2	0.87	4.28	1.61	3.57	1.29	5.18	3.97	1.57
4	0.33	4.65	1.06	4.68	0.40	4.57	3.90	2.42
8	0.07	2.74	0.14	3.10	0.18	3.45	4.94	2.69
16	0.04	0.60	0.005	0.225	0.05	1.08	3.29	3.02

[a]Piperonyl butoxide synergist.

increased the body storage of methoxychlor by 15-fold and substantially shifted the degradative pathway away from O-dealkylation (Reaction 3) toward dehydrochlorination to form 2,2-bis-(p-methoxyphenyl)-1,1-dichloroethylene, which was stored in 17-fold greater amounts. Similarly, piperonyl butoxide reduced the rate of formation of dieldrin from aldrin but increased the storage of the parent ^{14}C-radiolabeled aldrin by 17-fold. With trifluralin (α-trifluro-2,6-dinitro-N,N-dipropyl-p-toluidine), the presence of piperonyl butoxide reduced the rate of N-dealkylation (Reaction 4) and increased the amount of trifluralin in the green sunfish by 45-fold.

A. Comparative Aspects of Biotransformation

There are two schools of thought regarding the evolutionary development of the microsomal oxidases. Brodie and Maickel (1961) suggest that these arose as primitive aquatic life forms moved onto land and became terrestrial feeders encountering a much wider variety of xenobiotics. Adamson et al. (1965) view these enzymes as normally present for metabolism of endogenous products such as steroids and acting on xenobiotics only where structural similarities permit. In any event, the microsomal oxidases are under genetic control, and the evolutionary process can be discerned easily in the selection of insecticide resistant insects by exposure to xenobiotic insecticides. In many instances, survival occurs through enhanced microsomal oxidase activity. However, for life forms in general, evolutionary pressures could scarcely have equipped them to cope with the hundreds of thousands of exotic microcontaminants, many of which are deliberately designed for maximum environmental persistence.

From a relatively small amount of data, it appears that an evolutionary gradient exists in the levels of the microsomal system in life forms and that aquatic organisms are relatively deficient compared to many terrestrial species. Generalizations are complicated by substantial differences in temperature optima for the enzymes and in wide variations in responses of males and females. Thus oversimplifications can be misleading and closely related life forms may have dramatic differences in microsomal oxidase activities.

Comparison of the rates of microsomal epoxidation of aldrin to dieldrin (Reaction 6) in various animals is fairly representative of differences that are found (Terriere, 1968): quail ♂ 108, rat ♂ 84, housefly 41, rat ♀ 21, quail ♀ 9, blowfly 3, and trout 6. This reaction provides an excellent demonstration of the effects of phyllogeny upon biodegradative pathways. Using ^{14}C-labeled aldrin in a 3-day aquatic model ecosystem, Lu

(1973) showed the conversion of aldrin to dieldrin and tissue storage to be in the order alga < daphnia < snail < mosquito < fish as shown in Figure 6.

Not all biotransformations are microsomal. DDT is readily dehydrochlorinated in some organisms by the enzyme DDT'ase, to produce the ethylene DDE or 2,2-bis-(*p*-chlorophenyl)-1,1-dichloroethylene, and is reductively dechlorinated to produce DDD or 2,2-bis-(*p*-chlorophenyl)-1,1-dichloroethane. The relative rates as shown by use of ^{14}C-labeled DDT in typical organisms of the model aquatic ecosystem are shown in Figure 6 (Lu, 1973). It is of interest that the mosquito larva *Culex pipiens quinquefasciatus*, well known to be naturally resistant to the action of DDT, has an unusually high level of DDT'ase enzyme.

B. Degradation of Model Compounds

The environmental fate of simple radiolabeled benzene derivatives in a model aquatic ecosystem was studied by Lu and Metcalf (1975). The compounds were ecologically magnified in the fish *Gambusia affinis* over a short period as determined by their partition coefficients (Section III.A) and water solubilities as suggested in Figures 1 and 2. Degradation in the various aquatic organisms followed generally recognized microsomal oxidative pathways leading toward more polar derivatives and to conjugation and elimination.

Aniline was detoxified by methylation, acetylation, hydroxylation, and conjugations. *Daphnia* and *Physa* (snail) degraded it completely to polar compounds and *N*-methyl and *N*, *N*-dimethylaniline were found in *Oedogonium* (alga) and in *Culex* (mosquito). *Gambusia* (fish) retained small amounts of aniline, the methylated products, acetanilide, and almost equal amounts of *o*-, *m*-, and *p*-aminophenols.

Anisole was stored in all the organisms in fairly substantial amounts but was also degraded by *O*-demethylation to phenol, and hydroxylated to give *o*-, *m*-, and *p*-methoxyphenols, followed by conjugation.

Benzoic acid was stored in the highest quantities in *Physa* and *Daphnia*. Conjugation through glycine to give hippuric acid was an important detoxication mechanism. Hydroxylation to phenolic acids occurred to a limited extent in *Physa*, *Daphnia*, and *Culex*. Catechol was the key degradation product leading to complete degradation.

Chlorobenzene was highly persistent and bioconcentrated, and small amounts of *o*- and *p*-chlorophenol and 4-chlorocatechol were found.

Nitrobenzene because of its high polarity was neither stored nor biomagnified (Figure 1) but was resistant to degra-

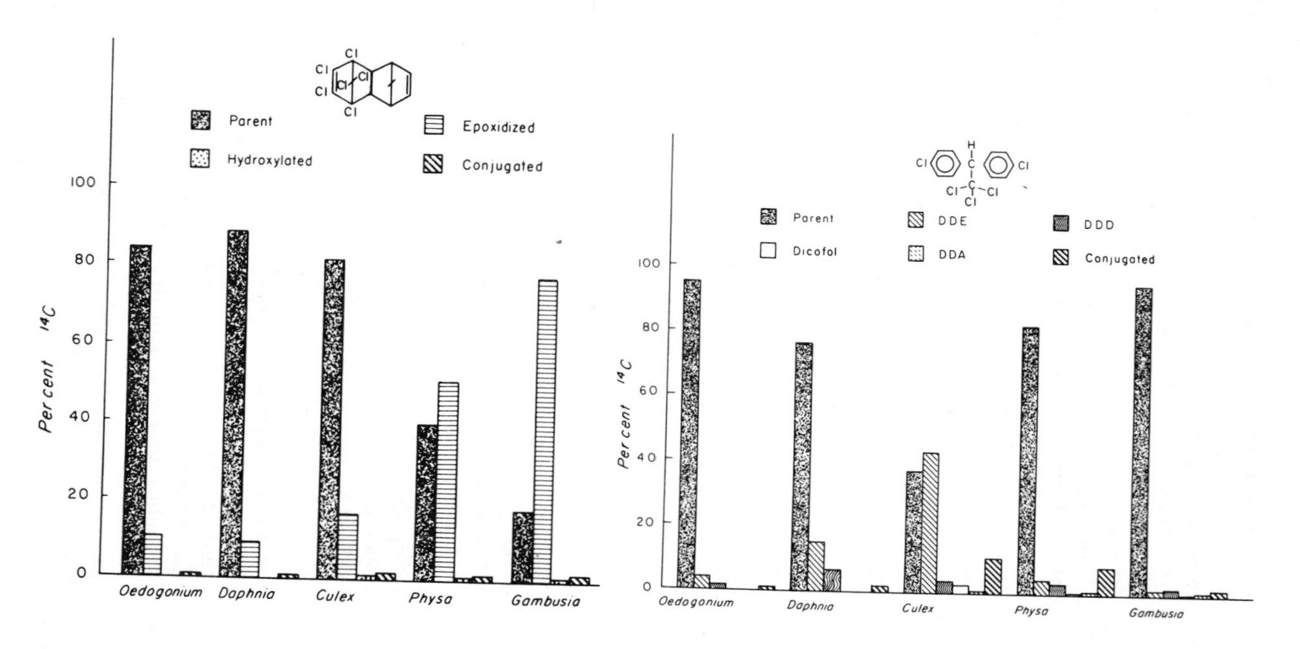

FIGURE 6. Degradative fate of [^{14}C] aldrin and [^{14}C] DDT in organisms of model aquatic ecosystem (data from Lu, 1973).

dation. It was reduced to aniline in all organisms and subsequently acetylated in *Gambusia*. Hydroxylation to give nitrophenols and aminophenols occurred in *Oedogonium*, *Daphnia*, and *Gambusia*.

The relative biodegradability of these simple substances provides patterns that can readily be applied to more complex compounds. The relative susceptibility to degradation and clearance is correlated with the intrinsic polarity of the aromatic nucleus as shown by Lu and Metcalf (1975). In Figure 7 Hammett's σ values for aniline, anisole, chlorobenzene, and nitrobenzene, giving a quantitative measure of electron density around the aromatic nucleus, correlated with the total percentage of hydroxylated and conjugated products in the 5-key organisms ($r = 0.91$) and in *Gambusia* ($r = 0.99$). As shown by the correlation constants, these relations were highly significant (Lu and Metcalf, 1975).

V. DESIGN OF BIODEGRADABLE COMPOUNDS

This chapter has emphasized the problems resulting from the pollution of the aquatic environment by persistent organic compounds that are relatively nondegradable in living organisms, for example, the organochlorines. Recognition of the problem and restriction of contamination and/or usage as with the PCBs are important avenues of environmental quality control. However, for purposeful contaminants, such as pesticides and detergents, redesign of the products themselves to promote biodegradability is equally important. Abelson (1970) has suggested that producers of fat-soluble, nonbiodegradable organic chemicals should carefully consider what they may responsibly set loose on the environment.

Biodegradability can be imparted to lipid soluble organic compounds by incorporating degradophores, that is, molecular groupings that are substrates for attack by the microsomal oxidases. These enzymes convert xenobiotic molecules into more water-partitioning derivatives that are excreted rather than bioconcentrated and stored in lipid tissues.

An excellent example of this sort is the substitution of methoxychlor for DDT. The two *p*-CH_3O groups of methoxychlor (H_2O sol 0.62 ppm) are readily *O*-dealkylated *in vivo*, to form 2,2,-bis-(*p*-hydroxyphenyl)-1,1,1-trichloroethane (H_2O sol 76 ppm) (Kapoor et al., 1970). The relative biodegradability of methoxychlor compared to DDT is shown in Figure 4, and the importance of *O*-demethylation on the degradation and clearance of methoxychlor in the green sunfish in Table 3.

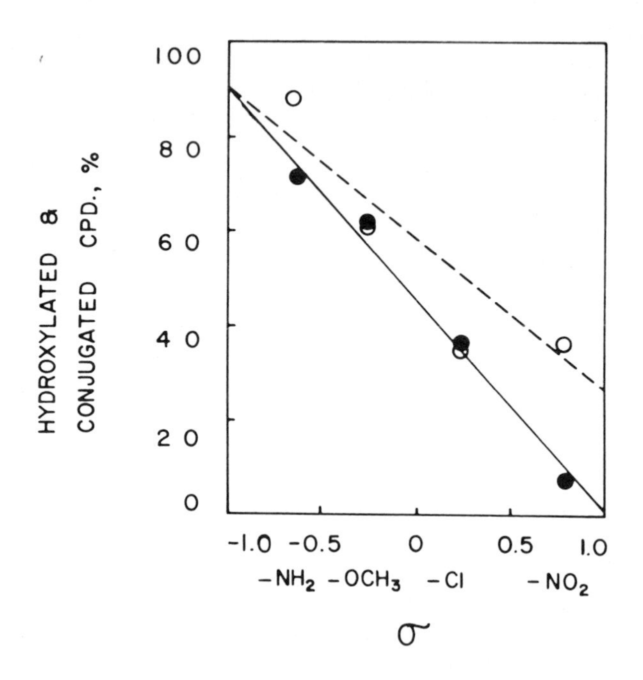

FIGURE 7. Relationship between Hammett's σ values for aniline (NH_2), anisole (CH_3O), chlorobenzene (Cl), and nitrobenzene (NO_2) and total percentage of hydroxylated and conjugated derivatives found in organisms of model aquatic ecosystem. Open circles are averages from five organisms—alga, daphnia, snail, mosquito, and fish; and solid circles are the fish *Gambusia* (after Lu and Metcalf, 1975). Reprinted with permission from the National Institute of Environmental Health Science, U. S. Department of Health, Education, and Welfare.

Similar effects can be achieved with alkyl side chains that are degradophores readily oxidizable to carboxylic acids. Methylchlor or 2,2-bis-(*p*-methylphenyl)-1,1,1-trichloroethane (H_2O sol 2.21 ppm) is oxidized *in vivo* to 2,2-bis-(*p*-carboxyphenyl)-1,1,1-trichloroethane (H_2O sol 50 ppm) (Kapoor et al., 1972). More effective DDT-type insecticides with enhanced biodegradability result from combinations of degradophores, such as 2-(*p*-methoxyphenyl)-2-(*p*-methylphenyl)-1,1,1-trichloroethane (Kapoor et al., 1973). The effects of degradophores upon bioconcentration of DDT-type molecules in the fish *Gambusia*, using the model ecosystem evaluation is shown in Table 4 (Kapoor et al., 1973).

TABLE 4. Effects of Degradophores on Bioconcentration of DDT-type Analogs in the Fish *Gambusia affinis*[a]

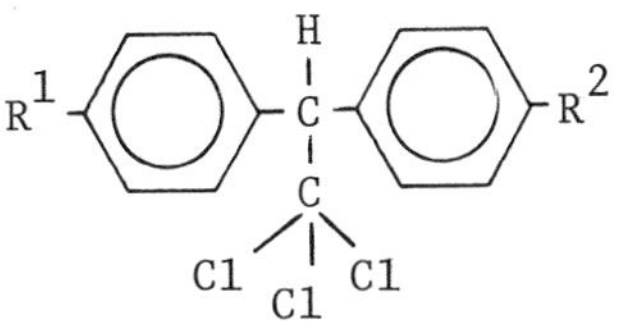

$$R^1\text{-}C_6H_4\text{-}CH(CCl_3)\text{-}C_6H_4\text{-}R^2$$

R^1	R^2	Ecological Magnification (ppm fish/ppm H_2O)
Cl	Cl	84,500
CH_3O	CH_3O	1,545
C_2H_5O	C_2H_5O	1,536
CH_3	CH_3	140
CH_3S	CH_3S	5.5
CH_3O	CH_3S	310
CH_3	C_2H_5O	400
Cl	CH_3	1,400

[a]Kapoor et al., 1973.

Further exploration of the principles of biodegradability is beyond the space limitations of this paper. However, the idea is being applied either deliberately or serendipitously to the design of new detergents, insecticides, and herbicides, and deserves very serious consideration in the design of new organic molecules for any predominantly environmental use (National Academy of Sciences, 1972).

ACKNOWLEDGMENTS

Preparation of this paper was supported by research data obtained by a grant from the National Science Foundation RANN Program, GI 39843X; and from the University of Illinois Water

Resources Center, Grants B-050 and B-070, Illinois, supported by the U. S. Department of Interior. The writer acknowledges the contributions of Drs. Po-Yung Lu, Keturah Reinbold, and James R. Sanborn.

REFERENCES

1. Abelson, P. 1970. Science 170, October 30, editorial.

2. Adamson, R. H., Dixon, R. L., Francis, F. L., and Rawl, D. P. 1965. Proc. Nat. Acad. Sci. 54, 1386.

3. Anon. 1972. Chem. Eng. News, January 10, p. 6.

4. Anon. 1974a. Chem. Eng. News, May 13, p. 14.

5. Anon. 1974b. Pestic. Chem. News, April 3, p. 3.

6. Anon. 1975. Chem. Eng. News, February 24, p. 7.

7. Aulerich, R. J., Ringer, R. K., and Iwamoto, S. 1973. J. Reprod. Fertil. 19, Suppl. 365.

8. Breidenbach, A. W., Gunnerson, C. G., Kawahara, F. K., Lichtenberg, J. J., and Green, R. S. 1967. Publ. Health Rept. 82, 139.

9. Brodie, B. B. and Maickel, R. P. 1961. Proc. First Int. Pharmacol. Meet. 6, 299.

10. Buerger, A. A. 1966. J. Theoret. Biol. 11, 131.

11. Casida, J. E. 1970. J. Agr. Food Chem. 18, 753.

12. Environmental Protection Agency. 1972. An Evaluation of DDT and Dieldrin in Lake Michigan, Ecol. Res. Ser. EPA R3-72-003, Washington, D. C., August.

13. Environmental Protection Agency Draft Analytical Report, New Orleans Area Water Supply. 1974. EPA 966/10-74-002, Dallas, Texas, November.

14. Fujita, T., Iwasa, J., and Hansch, C. 1964. J. Amer. Chem. Soc. 86, 5175.

15. Goldstein, A., Aronow, L., and Kalman, S. M. 1969. Principles of Drug Action, Harper & Row, New York.

16. Gruger, E. H., Karrick, N. L., Davidson, A. I., and Hruby, T. 1975. Environ. Sci. Technol. 9, 121.

17. Grzenda, A. R., Paris, D. F., and Taylor, W. J. 1970. Trans. Amer. Fish. Soc. 99, 385.

18. Grzenda, A. R., Taylor, W. J., and Paris, D. F. 1971. Trans. Amer. Fish. Soc. 100, 215.

19. Hammelink, J. L., Waybrant, R. C., and Ball, R. C. 1971. Trans. Amer. Fish. Soc. 100, 207.

20. Hites, R. A. and Biemann, K. 1972. Science 178, 158.

21. Kapoor, I. P., Metcalf, R. L., Hirwe, A. S., Coats, J. R., and Khalsa, M. A. 1973. J. Agr. Food Chem. 21, 310.

22. Kapoor, I. P., Metcalf, R. L., Hirwe, A. S., Lu, Po-Yung, Coats, J. R., and Nystrom, R. F. 1972. J. Agr. Food Chem. 20, 1.

23. Kapoor, I. P., Metcalf, R. L., Nystrom, R. F., and Sangha, G. K. 1970. J. Agr. Food Chem. 18, 1145.

24. Kerr, S. R. and Vass, W. P. 1973. Pesticide Residues in Aquatic Invertebrates In Environmental Pollution by Pesticides (C. A. Edwards, Ed.), Plenum, London, Chap. 4.

25. Kleopfer, R. D. and Fairless, B. 1972. Environ. Science Technol. 6, 1036.

26. Kuschner, M., Laskin, S., Drew, R. T., Cappiello, V., and Nelson, N. 1975. Arch. Environ. Health 30, 73.

27. Lee, D. H. K. 1964. Amer. J. Pub. Health 54, Suppl. 7.

28. Lu, Po-Yung. 1973. Model Aquatic Ecosystem Studies of the Environmental Fate and Biodegradability of Industrial Compounds, Ph.D. dissertation, University of Illinois, Urbana-Champaign, Illinois.

28a. Lu, Po-Yung and Metcalf, R. 1975. Environ. Health Perspect. 10, 269.

28b. Maltoni, C. and Lefemine, G. 1975. Ann. N.Y. Acad. Sci. 246, 195.

29. Merck, E., Editor. 1889. Merck's Index of Fine Chemicals and Drugs for the Materia Medica and the Arts (1st. Amer. edition), Darmstadt, New York.

30. Metcalf, R. L. 1966. Metabolism and Fate of Pesticides in Plants and Animals. In Scientific Aspects of Pest Control, National Academy of Sciences, Publication 1402, Washington, D. C., p. 280.

31. Metcalf, R. L., Booth, G. M., Schuth, C. K., Hansen, D. J., and Lu, Po-Yung. 1973. Environ. Health Perspect., June, p. 27.

32. Middleton, F. M. 1960. Proc. Conf. on Physiol. Aspects Water Quality, Washington, D. C.

33. National Academy of Sciences. 1972. Degradation of Synthetic Molecules in the Biosphere, Washington, D. C.

34. National Institute for Occupational Safety and Health. 1973. Toxic Substances List, U. S. Department of Health, Education, and Welfare, Rockville, Maryland, June.

35. Neely, W. B., Branson, D. R., and Blair, G. E. 1974. Environ. Sci. Technol. 8, 1113.

36. Quarles, J. 1973. Fed. Reg. 38, 24342.

37. Rainey, R. H. 1967. Science 155, 1242.

38. Reinbold, K. A., Kapoor, I. P., Childers, W. F., Bruce, W. N., and Metcalf, R. L. 1971. Bull. Ill. Natural Hist. Surv. 30, 405.

39. Reinbold, K. A. and Metcalf, R. L. 1976. Pestic. Biochem. Physiol. 6, 401.

40. Ringer, R. K., Aulerich, R. J., and Polin. 1974. Report of Pesticide Research Center, Michigan State University, East Lansing, Michigan.

41. Sanborn, J. R., Childers, W. F., and Metcalf, R. L. 1975. Bull. Environ. Contam. Toxicol. 13, 209.

42. Singh, A. R., Laurena, W. H., and Autian, J. 1973. J. Pharmacol. Sci. 61, 51.

43. Stecker, Paul G., Editor. 1968. Merck Index; An Encyclopedia of Chemicals and Drugs, 8th ed., Merck, Rahway, New Jersey.

44. Terriere, L. C. 1968. The Oxidation of Pesticides: The Comparative Approach. In Enzymatic Oxidation of Toxicants, (E. Hodgson, Ed.,), North Carolina State University, Raleigh, North Carolina.

45. U. S. Tariff Commission Data. 1970. Synthetic Organic Chemical, U. S. Production and Sales 1968, Government Printing Office, Washington, D. C.

46. Warner, R. E. 1967. Bull. World Health Organization 36, 181.

47. Youngs, W. D., Gutermann, W. H., and Lisk, D. J. 1972. Environ. Sci. Technol. 6, 451.

The Biological Fate of Radionuclides in Aquatic Environments

STEVEN M. GERTZ
Porter-Gertz Consultants, Inc.
Ardmore, Pennsylvania

and

IRWIN H. SUFFET
Drexel University
Department of Chemistry
Environmental Studies Institute
Philadelphia, Pennsylvania

I. INTRODUCTION

A. Environmental Fate

Trace amounts of radioactive materials are now present in natural water systems. Future uses of nuclear energy for power production may add more. In order to fully appreciate and quantitatively understand the consequences of radionuclide introduction to the aquatic environment, the environmental fate of these substances must be understood.

"Environmental fate" is commonly used to describe the disposition of pollutants in the environment. Efforts aimed at a formal definition can lead to controversy; however, it is generally agreed that the environmental fate of a chemical pollutant relates to how and where, in what forms, and in what concentration it is finally distributed in the environment. The physical, chemical, and biological influences on the radionuclide must be understood. As defined here, physical influences effect transport or dispersion of radionuclides in the aquatic environment. Chemical influences relate to those physiochemical reactions of the radionuclide that cause it to precipitate, adsorb, coagulate, and so on, and biological influences are the physiochemical and biochemical interractions or metabolisms associated with receptor organisms as radionuclides interact with the aquatic food chain.

Figure 1 depicts the probable paths that most radionuclides follow after introduction to the environment. Radioactivity can

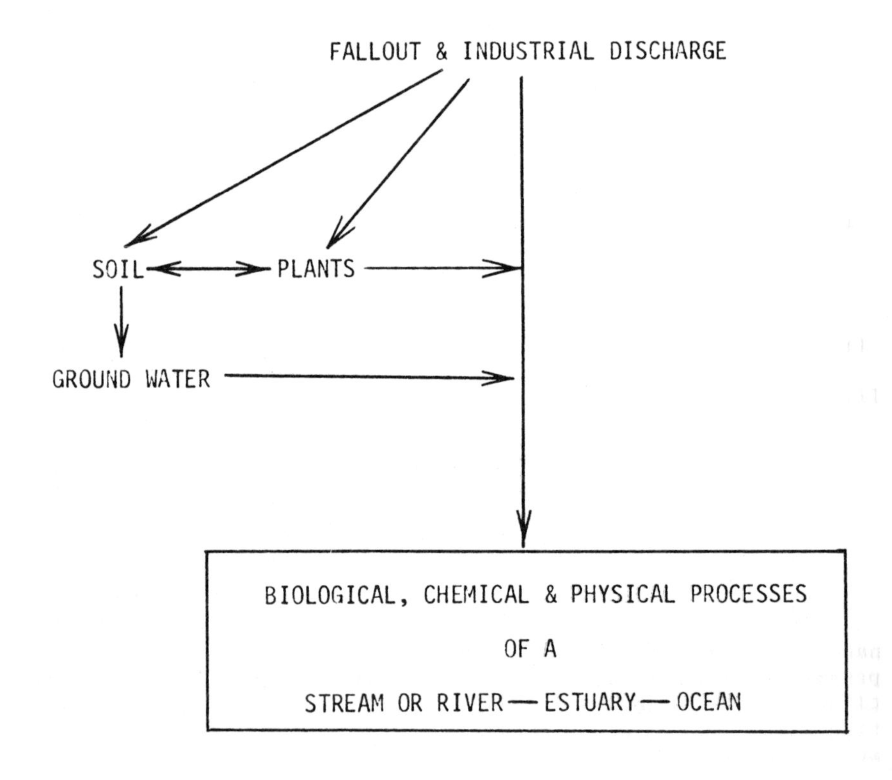

FIGURE 1. The most probable paths of radionuclides in the environment.

enter the air, water, or land environments directly. Those radionuclides injected to the atmosphere tend to return to earth due to dust settlement and rainfall. Thus they may impinge on either the land or water environments.

The radioactivity that falls to the terrestrial environment may be accumulated by soil, surface litter, cover crops, other plants, or soil animals. Plants are able to accumulate radionuclides through their leaves following foliar deposition or from the soil by root uptake; however, rainfall leaches leaf sorbed radionuclides to the soil (Peters and Witherspoon, 1972; Witherspoon and Taylor, 1969). Radionuclide uptake by soil animals has not been thoroughly investigated in terms of its ecological significance.

Radionuclides in soil or litter are subject to infiltration, percolation, diffusion, and erosion. Infiltration and percolation of radionuclides through soil is very limited and most of the radioactivity is found in the upper few centimeters of soil (Ritchie et al., 1970, 1972). Some radioactivity may, in time, infiltrate to rock strata and enter ground water supplies; however, this removal mechanism is minor. The principle mechanism of removal of radioactivity from the soil is erosion to the aquatic environment (Rogowski and Tamura, 1970).

Radionuclides in the aquatic environment have their chemical fates governed by the sorptive capacities of the bottom muds and suspended solids. Some suspended solids will settle out and by adsorption and ion exchange bring much of the suspended activity to the bottom muds. Bottom muds also adsorb and exchange materials directly with the water. The chemistry involved in these processes is similar to that of soils. One difference is the higher water concentrations in aquatic settings that allows for greater diffusion and infiltration.

For example, at the outfall of the Knoll's Atomic Power Plant, levels of 1400, 1000, and 700 pCi of cesium-137/kg wet wt. have been found in bottom muds at 0 to 6, 12 to 18, and 18 to 24-inch depths, respectively. A similar study in a pond having artificial sediment (9.13 percent sand > 50 μm, 7.5 percent silt 2-50 μm, and 1.2 percent clay < 2 μm) noted cesium-137 to be maximized at a 36-inch depth. It was also noted that 99.44 percent of the applied cesium was associated with the sediments at an equilibrium time of 80 days (Brungs, 1965).

Sediment accumulation of radionuclides occurs by sedimentation of suspended solids, surface adsorption, diffusion, and infiltration. These accumulation processes are dependent on the rate of water flow, as higher flows below that flow rate at which scouring occurs produce higher sorption rates and have little effect on the desorption rate (Kudo and Gloyna, 1971). Similar results, although by a different mechanism, sedimentation, were obtained with a slow moving experimental stream using attapulgite and kaolinite clays (Purushothaman, 1970). The rate at which radionuclides are accumulated by river muds has been defined in a model flume by Armstrong and Gloyna (1970).

In this discussion of environmental fate of radionuclides, tritium (^3H) is excluded. While much is known about ^3H, its environmental behavior is unique and must be treated as a separate topic. Most of the specific references in this presentation are to cesium and strontium, groups I and II ions, since the environmental behavior of these radionuclides has been studied more than most others. However the general mechanisms described for environmental fate of radioactivity are applicable to most radionuclides.

Specifically, this chapter discusses the biological fate of radionuclides in aquatic environments. The understanding of this part of environmental fate is important since it relates to man's use of water for drinking, food sources, recreation, and the possible effects on living organisms in the aquatic ecosystem.

B. Biological Fate

The term "biological fate" is commonly used to describe the biological interractions in an aquatic ecosystem that cause a pollutant to become associated with receptor organisms. When this definition is used to describe radionuclide fate, concentration factors are often employed. The concentration factor is defined as the ratio of the concentration of the radionuclide associated with the receptor organism to that of the water milieu. Mathematically, the concentration factor can be defined as

$$\text{C.F.} = \frac{C_o}{C_w} \tag{1}$$

where C_o is the concentration associated with the receptor organism and C_w is the concentration in the water phase.

Concentration factors may be expressed in terms of wet (live) weight, dry weight, or ash weight of the aquatic organism. Calculation of the concentration factor with respect to ash weight defines the extent to which the concentration of a radionuclide in the mineral residue exceeds that of the water. The concentration factor calculated in relation to dry weight defines the capacity of the organic component to accumulate radionuclides from the aqueous medium. And concentration factors calculated in relation to the organism's wet weight reflect the role of the living organism with respect to its ability to accumulate radionuclides. It should be noted that concentration factors in terms of wet weight are required if internal radiation doses are to be calculated. It is the internal radiation dose that is a part of the overall environmental impact caused by radionuclides in the biosphere.

Concentration factors are widely used to assess the impact that any pollutants may have on an ecosystem. The concept of biological magnification is important, but the qualifying assumptions and detailed theory underlying this concept have not been fully developed. This can lead to misuse and misunderstandings of this phenomenon. Two extensive reviews of previous research (Thompson et al., 1972; Polikarpov, 1966) on concentration factors of many radionuclides show that most of this research was done under constant environmental conditions where the only parameter varied was the radionuclide concentration. Yet some research on algal uptake of radionuclides has shown a marked dependence on the algal's chemical and physical environment (Austin et al., 1967; Williams and Swanson, 1958; Jacobson and Cember, 1970). Therefore a true description of biological fate must consider the environmental influences that can alter uptake and transfer of radionuclides.

One method of approach to describe the biological fate of a radionuclide is the development of a source-pathway-receptor model (Reichle et al., 1970). This method should describe trophic transfer—the process by which chemicals from the environment move up a food chain as shown in Figure 2.

Successful food chains in aquatic environments are autotrophic; they begin with green plants, the primary producers. These organisms, which are on the first trophic level, are responsible for converting inorganic materials, CO_2 and H_2O, to organic materials. As a consequence of normal plant metabolism, many elements and compounds are accumulated above their environmental concentrations.

Herbivores, which eat green plants, occupy the second trophic level. Carnivores, which eat the herbivores, occupy the third trophic level, and so on. Food chains may have as many as nine trophic levels. The use of trophic levels for classification purposes is functional and many organisms occupy more than one trophic level. A fish may eat algae (second level), amphipods (third level), and insects (fourth level), and so on.

A by-product of trophic transfer within a food chain is the passage of radionuclides up successive trophic levels. As a consequence of trophic transfer, these radionuclides may be accumulated by higher organisms, or less often, discriminated against (Reichle et al., 1970; Davis and Foster, 1958; Harvey, 1964; King, 1964). Many organisms can accumulate radionuclides

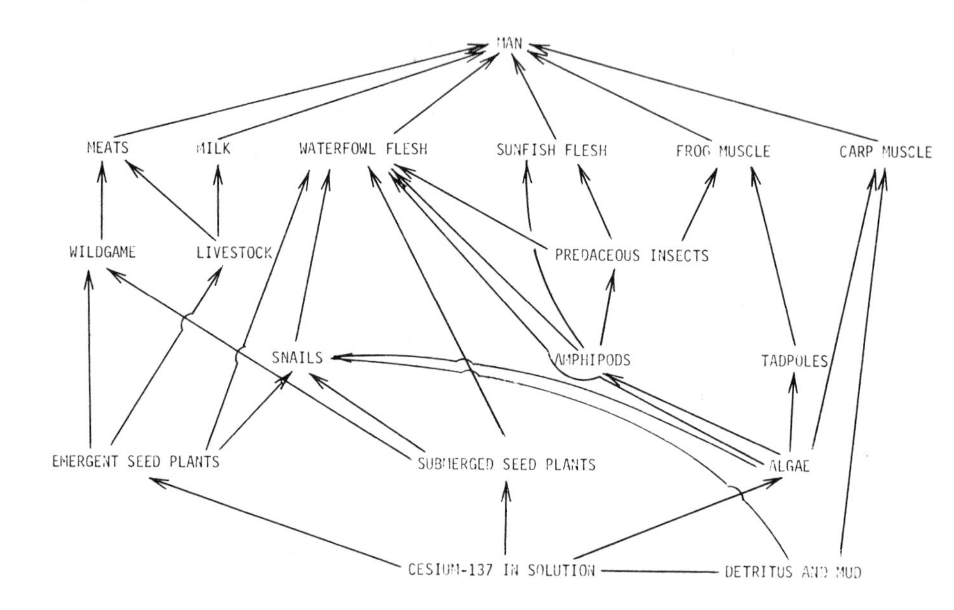

FIGURE 2. A food pathway by which Cesium-137 can move from fresh water to man (after Pendleton and Hanson, 1958). Reprinted with permission from United Nations International Congress on Peaceful Uses of Atomic Energy.

directly from the environment. However, at low levels of radioactivity, food chain uptake of radionuclides is most important (Krumholz and Foster, 1957; Garder and Skilberg, 1966).

To use the source-pathway-receptor model to describe the biological fate or radionuclides requires identification of the food chain and data on the retention of radioactivity by each organism. A description of biological fate within this scope should start with a consideration of algal uptake. The source is the water environment, the pathway is direct, and algae are the receptors.

II. RADIONUCLIDE UPTAKE BY ALGAE

Figure 3 depicts a generalized algal cell and notes those structures which are discussed.

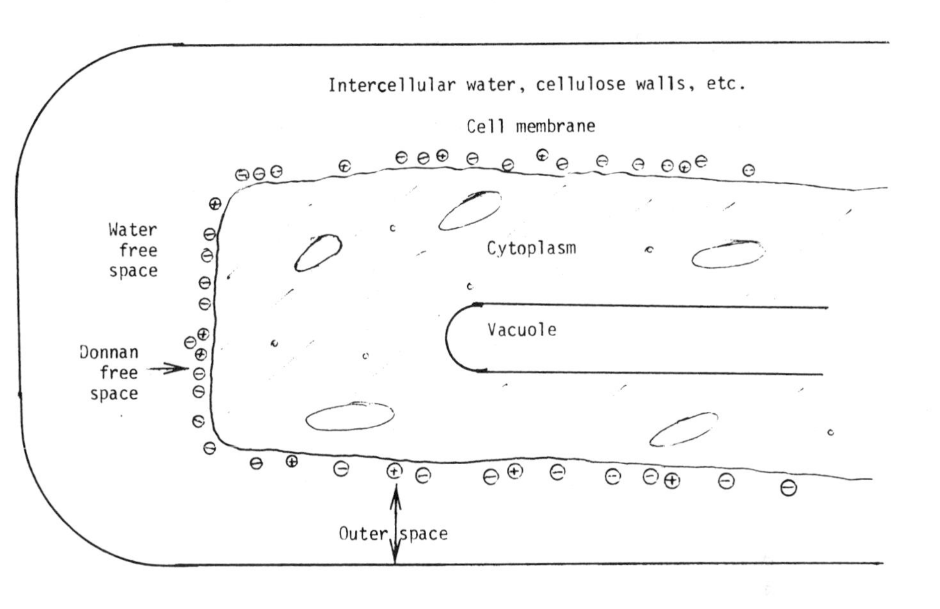

FIGURE 3. A generalized algal cell showing some of the structures associated with radionuclide uptake.

The first cellular structure of an alga that is encountered by an entering radionuclide is the algal cell wall. The cell wall of a green alga is characterized to be an outer layer of pectoses and mucilage, and an inner layer of pectin, cellulose, carbohydrates, polysaccharides, and lipid material (Northcote and Goulding, 1958). These layers contain many sorption sites which can accumulate radionuclides.

The second cellular structure that a radionuclide encounters is the cell membrane. The cell membrane contains protein and lipid material in an undefined spatial arrangement. Small pores may be present in this membrane to allow for diffusion.

To fully appreciate the biological fate of radionuclides in algae (the first trophic level in most aquatic environments), three levels of understanding can be defined for predictive purposes. These levels are described below.

1. Level I. Level I describes the biological fate of radionuclides in terms of the concentration factors. As described previously, the concentration factor is a value arrived at both theoretically and by a number of studies. These values may not necessarily be the same in all aquatic environments.

2. Level II. Level II considers the biological fate of radionuclides in terms of the concentration factor and those environmental stresses that affect the concentration factor for a radionuclide and an organism. The environmental variables that may affect an organism's concentrating ability are temperature, water quality, stream flow rate, and intensity and duration of the photo-period.

3. Level III. Level III considers the biological fate of radionuclides in terms of the biological mechanisms that cause an organism to accumulate radioactivity. This level represents the highest level of understanding, for if the biological uptake mechanisms are known then the concentration factors and the environmental variables which affect the concentration factors must necessarily be known.

Level III considerations lead to two general mechanisms by which algae accumulate radionuclides: active transport (requiring metabolic energy) and passive transport (not requiring metabolic energy).

Active Transport. Active transport is the moving of a substrate across the cell membrane by means of a carrier enzyme at the expense of cellular energy, usually in the form of ATP. It allows for rapid accumulation of substrate, as compared to passive transport, and most commonly occurs against an electrochemical gradient. This process is also used to eliminate some materials from the cell (Price, 1970).

Passive Transport. Passive transport is any uptake mechanism that does not require the expenditure of metabolic energy. Adsorption, adsorption exchange, exchange diffusion, facilitated diffusion, and leakage are all forms of passive transport.

Adsorption. Adsorption is the process by which a substance becomes bound to the cell wall. There are both physical and chemical adsorption processes. Physical adsorption is relatively nonspecific and is due to the operation of weak attractive forces. It is often reversible. Chemical adsorption is, on the other hand, the result of forces approaching in magnitude those of a chemical bond. It is seldom reversible.

Adsorption Exchange. Adsorption exchange, variously called ion exchange or chemical adsorption, is the electrostatic binding of one ion by an ion of the opposite charge. It is the predominant uptake mechanism in the Donnan Free Space (D.F.S.). The D.F.S. seems to be located on and in the cell wall and membrane of plants. It exists only in the kinetic sense. The ions migrate to this area by free diffusion (Laties, 1959).

Exchange Diffusion. Exchange diffusion is the process whereby an ion, by means of a carrier ion complex, is exchanged for a like ion in the cytoplasm of a cell. It does not require metabolic energy. Its main function is that it increases the turnover rate of some substances and allows for the turnover and transfer of others, for example, sodium for potassium in seaweeds (MacRobbie and Dainty, 1958).

Facilitated Diffusion. Facilitated diffusion allows for the movement across the membrane of various substances by means of a carrier system. This is accomplished at the expense of kinetic energy and is solely dependent on the external concentration of the substrate (Lerner, 1971).

Leakage. Leakage is the process whereby materials, without the aid of a carrier system, migrate through pores into the cell. It is dependent on external concentrations of substrate and appears to be nonselective. This process is usually demonstrated in killed cells.

Active transport is probably the most important process for radionuclide accumulation by living algae. Active transport usually occurs against a concentration or electropotential gradient. It is accomplished by the formation of an enzyme complex on one side of the membrane, translocation, and release on the other side. If the ion is incorporated into a product, the Michaelis-Menten equations follow:

$$E + S \underset{k_2}{\overset{k_1}{\rightleftharpoons}} \overline{ES} \xrightarrow{k_3} E + P \tag{2}$$

and

$$v = \frac{V_{max} S}{(K_m + S)} \tag{3}$$

where v is the velocity of the reaction, V_{max} is the maximum attainable velocity for ideal substrate concentrations, S is the substrate concentration, and K_m is the Michaelis constant. The Michaelis constant is mathematically equal to

$$\frac{k_1 + k_3}{k_2} \tag{4}$$

under steady conditions.

If no product is formed in Eq. 2, the ion is simply transported; Eq. 3 becomes

$$v = \frac{V_{max} I}{(K_t + I)} \tag{5}$$

where K_t is the sum of the forward reactions divided by the reverse reaction, which is the same as the Michaelis constant defined in Eq. 4, and I is the external concentration of the ion. The general form of Eq. 5 is

$$C_t = C_i + C_e (1 - e^{K_d t}) + vt \tag{6}$$

in which C_t is the concentration of ions at time t, C_i is the concentration of ions in the cell at t = 0, C_e is the equilibrium concentration of exchangable ions in the outer space, K_d is the diffusion limited constant for movement into the outer space, and v is the rate of movement into the inner space. This equation can also be used to describe the accumulation by passive methods.

Radionuclides may be accumulated in the D.F.S. by adsorption exchange, as this area of the cell presents an unequal concentration of ionic charge. For example, it is thought to be responsible for the accumulation of sodium ions by *Porphyra*, a marine alga (Eppley, 1957).

The radionuclide once in the cell may be sorbed by cellular inclusions, bound strongly or loosely, to various tissues, remain unbound as an ion, or be bound to soluble cell fractions, and thereby give the appearance of being unbound. For example, it was determined that approximately 17 and 39 percent of the accumulated cesium in Euglena and Chlorella, respectively, was unbound; the remainder was associated with cellular particulates (Williams, 1960).

Cesium uptake has repeatedly been shown to be directly proportional to external concentration (Williams and Swanson, 1958; King, 1964; Williams, 1960; Gutknecht, 1965). This accumulated steady-state concentration can be defined by the Freundlich isotherm, but this is not sufficient proof for an adsorptive mechanism (Briggs et al., 1961). Mathematical adherence to an adsorption isotherm can also indicate ion exchange. Active transport has also been suggested as a possible mechanism but was not proved (King, 1964).

Unfortunately, to the best of the authors' knowledge, there has not been a definitive study on algal uptake mechanisms of radionuclides. Much of the present knowledge is based on studies with stable isotopes, such as Na and K. The applicability of this data is questionable. This is true since the environmental levels of radionuclides are on the order of 10^{-10} to 10^{-15} M while the stable element studies used millimolar quantities. Thus there is no reason to believe the same mechanisms are operable at these lower concentrations.

This brings us to our other levels of understanding, concentration factors and environmental stresses. Reported concentration factors for cesium-137 and strontium-90 in algae are variable. Each can range over an order of magnitude for the same algal specie. These reported ranges could be due to differences in growth media, water quality, the physical environment, or the culturing technique employed. For example, it has been shown that potassium depresses cesium-137 uptake by Chlorella (Williams, 1960). This depressant effect was shown to be linear at the concentrations studied.

Temperature appears to have no effect on the equilibrium concentrations of cesium accumulated by algae. The blue-green alga Plectonema boryanum did not show any essential difference in concentration factors for cesium-137 at temperatures of 25 to 40°C (Harvey, 1969). Temperature does affect the rate of cesium uptake in Chlamydomonas moewusii; lowering the temperature lowers the uptake rate (King, 1964). This could indicate that cesium is intracellular, for if surface sorption occurred, the equilibrium concentration factors would probably change with changes in temperature. The amount and intensity of sunlight, metabolic rate, and population division rate would probably not affect the amount of cesium accumulated at equilibrium, but these factors might affect the rate of uptake.

Studies on strontium-90 uptake by Chlorella have shown Mg and Ca to depress the concentration factor (Austin et al., 1967). Similar results were noted with higher aquatic plants (Timofeeva and Kulekov, 1966).

The Freundlich isotherm was proposed as a method to describe the levels of radioactivity accumulated by algae in the environment (Armstrong and Gloyna, 1969). Strontium-85 was the model nuclide, Vallisneria, the target organism.

The isotherm was followed initially but failed to describe the equilibrium concentrations. Cesium levels in seaweeds have been described by the Freundlich isotherm, but the effects of environmental stresses were not considered (Gutkneckt, 1965). Cesium levels in Oedogonium have been ascertained in aquarium studies, but the data could not be applied accurately to the environment (Pendleton and Hanson, 1958).

Some recent work in our laboratory utilized the approach embodied in Levels II and III. Chlamydomonas reinhardii, a fresh water green alga, was grown in a chemostatic flow culture system. The use of this system allowed us to vary the environment while keeping the algae in their log growth phase. In this manner, we were able to best simulate a single species aquatic environment on a laboratory scale.

Cesium-137 was the nuclide studied as it is an important radionuclide present in nuclear power plant effluents. It was shown that algal concentration factors for cesium-137 were governed by the total $[PO_4^{3-}]$, $[Na^+]$, and the algal biomass. Temperature, $[K^+]$, $[Cl^-]$, and algal reproduction rate had no effect on the observed concentration factors which ranged from 115 ± 2 to 586 ± 9. The following equation was developed representing the concentration factors of cesium-137 by C. reinhardii as a function of water quality. It is

$$\text{C.F.} = \frac{598 - (4.56 \times 10^4 [Na^+])}{1 + \frac{448 - (4.56 \times 10^4 [Na^+])}{150} e^{-5.29(P_a - 0.0065)}} \tag{7}$$

where C.F. is the concentration factor, $[Na^+]$ is the aqueous Na concentration, and P_a is the moles of $[PO_4^{3-}]$ per kilogram dry weight per liter of algae ($P_a = [PO_4^{3-}]/\text{kg/liter}$). The derivation of this equation, and other experimental evidence, indicated active transport to be the most probable uptake mechanism and that sodium and cesium have the same carrier site on the cell membrane (Gertz, 1973).

Algae accumulate many different radionuclides. Movements through the food chain will then determine the biological fate and effect of these radionuclides (Reichle et al., 1970). It has been shown that carp receives 80 percent of its accumulated cesium directly from algae and 7 percent from benthic sediments (Kevern, 1958). It was also demonstrated that bluegills obtain 91.6 percent of their accumulated cesium via the food chain (Reichle et al., 1970), and rainbow trout a maximum of 60 percent (Gallagos, 1970). Therefore, any attempt to completely define the biological fate of radionuclides in the aquatic environment must start with algal uptake of radioactivity.

III. SUMMARY

Three levels of understanding were presented to describe algal uptake of radioactivity. Level I described this phenomenon in terms of concentration factors. This level of understanding is basic to the study of radionuclide uptake but it is not complete. Level II considered algal radionuclide uptake in terms of concentration factors and those environmental stresses which may affect an organism's concentrating ability. This level, while a higher level of understanding than Level I, is still not complete, for while the phenomenon is better quantified, the reason for the phenomenon is still not known. Level III, the highest level of understanding, considers the biological mechanisms that permit and cause an alga to accumulate radioactivity. Not only does this level permit quantification of radionuclide uptake, but most important, it answers why.

Research at each level of understanding that was presented fills a definite need, and adds to the body of scientific knowledge. But in order to add the most usable knowledge, our research efforts must not be satisfied with just Levels I or II investigations, we must aim for Level III.

REFERENCES

1. Armstrong, N. E. and Gloyna, E. F. 1969. Proc. Second Nat'l Symp. on Radioecology (D. J. Nelson and F. C. Evans, eds.), Ann Arbor, Michigan, AEC Conf. #670503, May 15-17, 1967.

2. Austin, J. H., Klett, C. A., and Kaufman, W. 1967. Int. J. Oceanol. Limnol. 1, 1.

3. Briggs, G. E., Hope, A. B., and Robertson, R. M. 1961. Electrolytes and Plant Cells, Blackwell, Oxford.

4. Brungs, W. A. 1965. Distribution of Co-60, Zn-65, Sr-85, and Cs-137 In a Fresh Water Pond, U. S. Dept. of Health, Education, and Welfare. U. S. Public Health Service Publ. #999-RH-24.

5. Davis, J. J. and Foster, R. F. 1958. Ecology 39, 530.

6. Eppley, R. W. 1957. Exp. Cell Res. 13, 173.

7. Friend, A. G., Story, A. H., Henderson, C. R., and Busch, K. A. Behaviour of Certain Radionuclides Released into Fresh Water Environments, U. S. Dept. of Health, Education, and Welfare, U. S. Public Health Service Publ. #999-RH-13.

8. Gallegos, A. F., Whicker, F. W. and Hakonson, T. E. 1970. Proc. Health Physics Aspects of Nuclear Facility Siting. Idaho Falls, Idaho, AEC Conf. #701106, Nov. 3-6, 1970.

9. Garder, K. and Skilberg, O. 1966. Arch. Hydrobiol. 62, 151.

10. Gertz, S. M. 1973. Ph.D. dissertation, Drexel University.

11. Gutknecht, J. 1965. Limnol. Oceanogr. 10, 58.

12. Harvey, R. S. 1964. Health Phys. 10, 243.

13. Harvey, R. S. 1969. Proc. Second Nat'l. Symp. on Radioecology, Ann Arbor, Michigan, AEC Conf. #670503, May 15-17.

14. Jacobson, A. F. and Cember, H. 1970. Proc. Health Physics Aspects of Nuclear Facility Siting, Idaho Falls, Idaho, AEC Conf. #701106, Nov. 3-6, 1970.

15. Kevern, N. R. 1958. Trans. Amer. Fish Soc. 95, 363.

16. King, S. F. 1964. Ecology 45, 852.

17. Krumholz, L. A. and Foster, R. F. 1957. Nat. Res. Council, Publ. 551, Washington, D. C.

18. Kudo, A. and Gloyna, E. F. 1971. Water Res. 5, 71.

19. Laties, G. C. 1959. Ann. Rev. Plant Physiol. 10, 87.

20. Lerner, J. 1971. J. Chem. Educ. 49, 391.

21. MacRobbie, E. A. C. and Dainty, J. 1958. J. Plant Physiol. 11, 782.

22. Northcote, D. H. and Goulding, K. T. 1958. Biochem. J. 70, 391.

23. Pendleton, R. C. and Hanson, W. C. 1958. Proc. Second U. N. Intl. Conf. on Peaceful Uses of Atomic Energy, Geneva, Vol. 18:1, New York, New York.

24. Peters, L. N. and Witherspoon, J. P. 1972. Health Phys. 22, 261.

25. Polikarpov, G. G. 1966. Radio-Ecology of Aquatic Organisms, Reinhold, New York.

26. Price, C. A. 1970. Molecular Approaches to Plant Physiology, McGraw-Hill, New York.

27. Purushothaman, K. 1970. Fourth Ann. Conf. on Trace Substances in Environmental Health, Columbia, Missouri, June 23.

28. Reichle, D. E., Dunaway, P. B., and Nelson, D. J. 1970. Nucl. Safety 11, 43.

29. Ritchie, J. C., Clebsch, E. E. C., and Rudolph, W. K. 1970. Health Phys. 18, 479.

30. Ritchie, J. C., McHenry, J. R., and Hill, A. C. 1972. Health Phys. 22, 197.

31. Rogowski, A. S. and Tamura, T. 1970. Health Phys. 18, 467.

32. Thompson, S. E., Burton, C. A., Quinn, D. J., and Ng, Y. C. 1972. Concentration Factors of Chemical Elements in Edible Aquatic Organisms, University of Cal., Lawrence Livermore Laboratory, Publ. #UCRL 50564 (Rev. 1).

33. Timofeeva, N. A. and Kulekov, N. V. 1966. Proc. Int. Symp. on Radioecological Concentration Processes, AEC Conf. #660405, Stockholm, April 25-29.

34. Williams, L. G. 1960. Limnol. Oceanogr. 5, 301.

35. Williams, L. G. and Swanson, H. D. 1958. Science 12, 3291.

36. Witherspoon, J. P. and Taylor, F. G., Jr. 1969. Health Phys. 17, 825.

A Transport Model for Long Term Release of Low Level Radionuclide Solutions into a Stream Ecosystem

YOUSEF A. YOUSEF
College of Engineering
Florida Technological University
Orlando, Florida

and

ERNEST F. GLOYNA
College of Engineering
The University of Texas
Austin, Texas

I. INTRODUCTION

Since the beginning of nuclear power development, safety to man and his environment has been a prime consideration. Highest standards of engineering practice and "fail safe" principals have been employed in reactor design, and effluent standards have been set at conservative levels. As a result, data on the spectrum of hazards involved under various degrees of accidental release and the threshold limits of controlled environmental release of radionuclides is incomplete.

Experimentation on real rivers, while valuable, is not always practical. On the other hand, the results from small scale ecosystems that simulate stream processes can be most helpful in giving a qualitative and quantitative insight into ecological responses to radioactive stress. At the University of Texas, Center for Water Resources, a model river was used over the years as a tool to study interactions between stream ecosystems and various contaminants such as radionuclides. The project was supported by the United States Atomic Energy Commission, Division of Reactor and the Department of Interior, Office of Water Resources Research.

II. EXPERIMENTATION

A stream ecosystem typical of the local area was simulated in a metal flume 200 feet in length, 2.5 feet in width, and 2.0 feet in depth. A center partition divided the flume into two identical channels, each 1.25 feet in width. Flow in the east channel consisted of potable water supplied from municipal water systems. Into the west channel, water rich in phytoplankton was pumped from a supply reservoir.

Bottom sediments typical of the local area, from Lake Austin, were placed in both channels of the flume to a depth about 4 to 6 inches. The sediments were classified according to the U. S. Department of Agriculture, 23 percent by weight coarse and medium sand, 59.5 percent by fine sand, 5 percent by very fine sand, 7.5 percent by silt, and 3 percent by clay (Akira and Gloyna, 1969). The organic content is about 2 percent and the clay fraction consisted of illite, vermiculite, and other degraded clay minerals that could not be classified (Yousef, Kudo, and Gloyna, 1971).

Plant growth was established in the last 50 feet of the west channel. The predominant plants at the beginning of the test were Potamogeton and Vallisneria. One month later when the experiments were initiated, the predominant specie was Chara.

The water flow was maintained constant at a rate of 10 liters/minute and an average velocity of 0.44 feet/minute. Cesium-134 was added at a rate of 2.16 μCi/minute and strontium-85 was at a rate of 0.46 μCi/minute for the period of 35 days in the east channel and 21 days in the west channel.

Before beginning the experimentation, dye studies, water analysis, and diurnal changes in numbers and species of the plankton community were investigated. Temperature, pH, and oxygen were continuously monitored by thermistors, pH meters, and oxygen probes installed along the flume. A detailed description on equipment, experimenting procedure, and analysis techniques were outlined in published reports (Yousef et al., 1970; Yousef et al., 1971, 1975).

III. RESULTS AND DISCUSSION

The most important influences on radionuclide transport into a stream ecosystem are hydrodynamic characteristics and surface retention. Hydrodynamic characteristics involve mixing, dispersion, diffusion, and dilution. Surface retention involves the ability of sorbing surfaces such as sediments and plankton to concentrate radionuclides. The mechanisms involved in retaining and storing radioactivity include sorption, sorption and sedimentation, chemical precipitation, and biological uptake. However, activity stored in a hydrological environment may be released by changes in dynamic, physical, and biochemical conditions. An increase in velocity will scour bottom sediments and return radioactivity to the flowing portion of the stream, create new sorbing surfaces, and relocate adsorbed radionuclides.

IV. DISPERSION

Radionuclides are dispersed downstream from the point of release by turbulence, molecular diffusion, and convection. The dispersion mechanism is responsible for the mixing that takes place before contaminants are gradually extracted from solution by physical, chemical, and biological processes. Dispersion, dilution, and physical movement of radionuclides downstream are greatly influenced by hydraulic processes that involve the velocity of the stream flow and its uniformity in relation to stream cross section and stream reach.

The hydraulic characteristics of the model river were evaluated by releasing organic dyes such as Rhodamine B and Fluorescein prior to radionuclides. The longitudinal dispersion coefficients D_x values were calculated from time-concentration curves under different flow conditions and plant distribution at various sampling locations along the stream. For velocities between 0.33 and 3.3 feet/minute, D_x values can be expressed by an empirical relationship in the form (Yousef et al., 1970; Gloyna et al., 1971):

$$D_x = 3.26\ U^{0.607} \tag{1}$$

where D_x is the longitudinal dispersion coefficients (ft^2/min) and U is the mean velocity (ft/min).

From Eq. 1, dispersion is shown to be directly related to velocity. However, wind and temperature effects may have an exaggerated effect. Temperature effects may stimulate mixing in one case and in another case establish a stratified flow regime (Gloyna et al., 1971). Wind in all cases will stimulate mixing. Thermal effects could not be controlled and some error could occur in the experimental determination of longitudinal coefficient of dispersion and in the prediction of concentration at a location and point in time. Wind effects, however, were eliminated by covering the top of the flume with a clear plastic material.

A. Dispersion of Dye

During the initial two days of radioactivity release, Rhodamine B dye was also simultaneously released through the same dosing apparatus (Yousef et al., 1975) and samples were taken at stations 2, 22, 52, 102, 152, and 197 feet from the inlet at both channels in the flume. The samples were prepared for counting in a multichannel analyzer for radioactivity and for Rhodamine B measurement in a Turner Fluorometer. In the case of radioactivity corrections were made for decay, self-adsorption, and efficiency. In the case of Rhodamine B, corrections were made for temperature and decay by sunlight.

The average time concentration curves of the dye at the sampling stations in the east and west channels are shown in Figures 1 and 2. The plotted points from which the curves were constructed were average values of water samples taken one inch below the water surface and one inch above the bed sediments. It can be seen from the figures that east and west channels did not behave exactly the same. Stratification occurred from time to time and was noticed by the difference between dye concentration from top samples as compared to bottom samples. Figure 3 shows a stratified flow in the west channel at the 2-foot station as

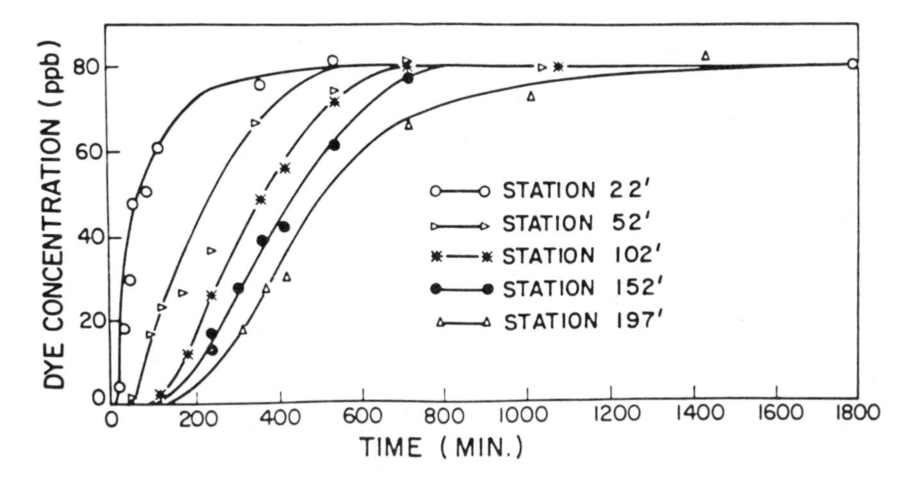

FIGURE 1. Time concentration curves for continuous dye release in the east channel.

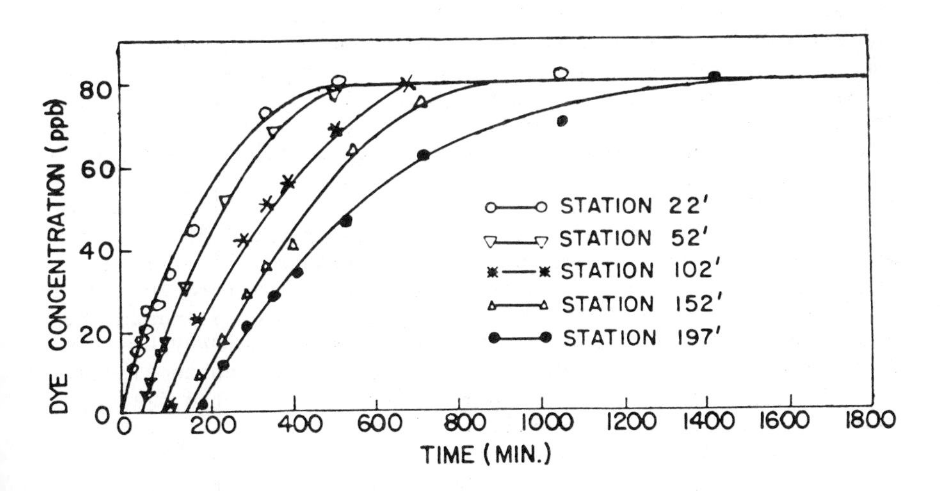

FIGURE 2. Time concentration curves in the west channel for continuous dye release.

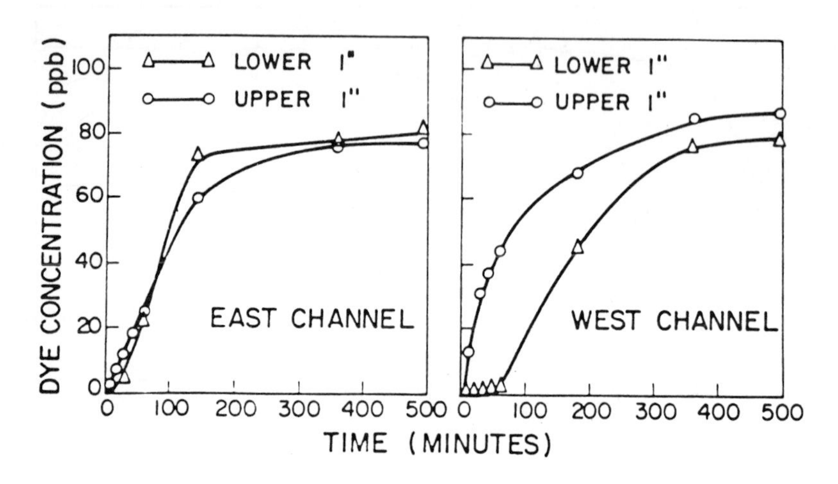

FIGURE 3. Time concentration curves at stations E2 and W2 for continuous release of dye.

compared with the nonstratified flow in the east channel. In general, the upper layers of water had higher dye concentrations during the day than did the lower layers. However, at night the dye concentrations were evenly distributed from top to bottom. Thermal and wind effects encountered in experimentation with the flume are discussed in detail by Padden (1970). Thermal effects in some cases resulted in convection currents that stimulate mixing action and in other cases resulted in stratification with underflow or overflow.

From the dye concentration curves, the flow through time of the dye can be estimated as the time elapsing until 50 percent of the equilibrium dye concentration had reached the effluent end of the flume. From Figures 1 and 2, the flow through time in the east channel would be established at 445 minutes and in the west channel at 460 minutes. This would establish the velocity of flow at approximately 0.44 feet/minute, a reasonable figure, considering depth to bottom sediment varied from approximately 6 to 7.5 inches, width was 1.25 feet, and quantity flow was 10 liter/minute.

B. Dispersion of ^{134}Cs and ^{85}Sr

Cesium-134 and strontium-85 were continuously released at a uniform rate of approximately 0.216 μCi/liter for cesium-134 and 0.046 μCi/liter for strontium-85 for a period of 35 days in the east channel and 21 days in the west channel.

Whenever the radionuclides are continuously injected into a waterway, concentration gradients are created within the water and across the water solid interface. Several diffusive mechanisms act to reduce the concentration gradient until a steady-state condition is achieved. Bed sediments, rooted plants, and phytoplankton concentrated radioactivity at a rate directly proportional to the difference between a saturation level and its concentration at that time. The saturation concentration is estimated by multiplying the maximum retention factor of sorbing surface times the radioactivity in water. In approximately 3 weeks, a quasiequilibrium period was reached, where radionuclides uptake by sorbing surface is equal to the radioactivity lost as a result of desorption and/or migration through sediments.

Under continuous release conditions, at quasiequilibrium, the extraction of radionuclides from the water phase on a mass balance basis has a minor effect on the specific activity in the water as shown in Figures 4 and 5. Most of the radionuclides released were discharged through the flume. A higher fraction of strontium than cesium was discharged through the flume; hence, relatively more cesium was retained in the flume system than strontium. Retention of radioactivity was greater in the west channel as compared to the east channel due to the presence of phyto and zooplankton.

V. RETENTION OF RADIOACTIVITY

When radioactivity is released into streams, part of this radioactivity will be extracted by bed sediments, by suspended solids, and by biota. Some of the retained activity may subsequently be released and some may be stored for indefinite periods of time. The mechanisms involved in retaining and storing radioactivity include sorption, sedimentation, chemical precipitation, and biological uptake. However, the activity stored in a hydrological environment may be released by changes in dynamic, physical, and biochemical conditions. As increase in velocity will scour bottom sediments and return radioactivity to the flowing portion of the stream. Toxic material and/or organic stress introduced into the stream will upset normal biological balances and will alter the numbers and types of aquatic

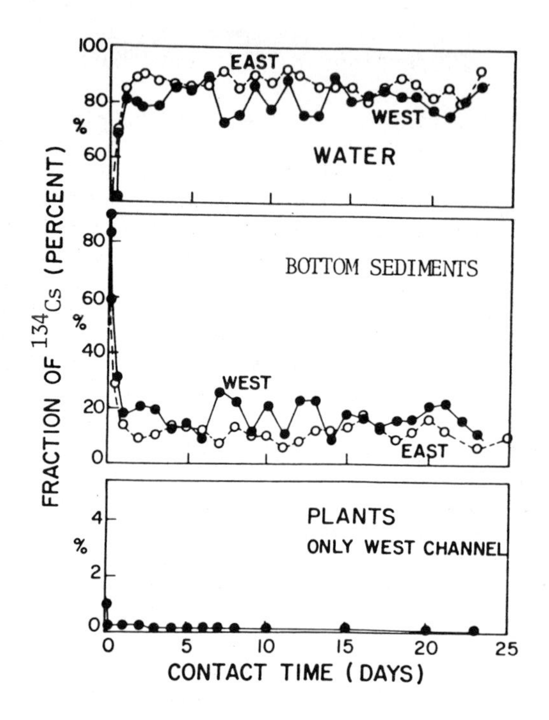

FIGURE 4. Mass balance of cesium-134 released in the flume.

flora and fauna and may cause sudden release of stored radioactivity, which will redistribute in the environment.

A. Retention by Bed Sediments

Six sediments core samples were taken from each channel each day, sliced, and counted. The average sediment retention of radioactivity increased with time to an equilibrium level after a period of approximately 25 days or equivalent to 80 times the hydraulic flow through times. The bed sediments in the west channel exhibit higher concentration than did the sediments in the east channel due to radioactivity associated with deposited phytoplankton that was obvious on the collected samples.

The average surface concentration factor (K_s) by the bed sediments, which is the ratio of the radioactivity retained by the bottom sediments per square centimeter of surface area to the radioactivity contained in one ml of water flowing in the system

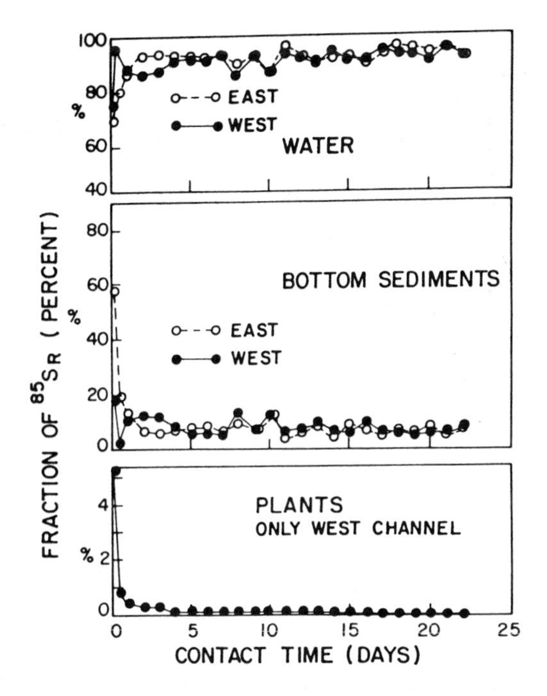

FIGURE 5. Mass balance of strontium-85 released in the flume.

was measured daily. The surface concentration factors (K_s) at equilibrium in the case of ^{134}Cs was approximately 275 and for ^{85}Sr was approximately 85 as shown in Figures 6 and 7. Bottom sediments had a greater affinity for ^{134}Cs than for ^{85}Sr.

At the end of the experiment, after 35 days of continuous release of radioactivity, water in the east channel was drained off. Two sediment cores were taken at each foot along the channel. The top 0.5 inch of each of ten cores was composited and prepared for counting. Each composited sample represented a 5-foot length of the channel. The concentration of ^{134}Cs and ^{85}Sr in each composited sample is shown in Figure 8. The average areal concentration of ^{134}Cs by bed sediments in the east channel was 55.1 μCi/cm^2 and ^{85}Sr averaged 3.34 μCi/cm^2. These concentrations correspond to K_s values of 260 for ^{134}Cs and 74 for ^{85}Sr.

Radionuclide concentration was relatively higher at both ends of the channel with lower concentrations at the midsection

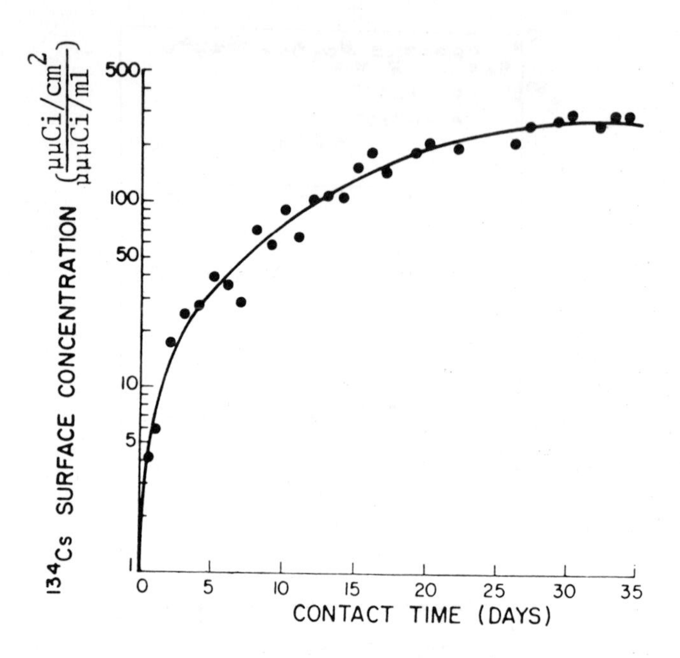

FIGURE 6. Average surface concentration of cesium-134 by bed sediments in the flume related to contact time.

between 70 and 150 feet. Just why this distribution occurred can only be conjectured. It may have been caused by thermal stratification in which a higher velocity underflow resuspended and relocated deposited radioactivity. Thermal stratification and underflow was noted in previous experiments (Gloyna et al., 1971). The area that had the highest concentration was from 150 feet to 180 feet. This could be due to the end weir effect. The end weir established a settling area that was the recipient of all of the particles moving along the bed sediment.

B. Retention by Plants

Fresh water plants accumulate radionuclides in varying amounts depending on the radionuclide, presence of stable element, biological availability of the element, and plant species.

Species of rooted plants typical of the Southwest, that is, *Potamogeton*, *Vallisneria*, *Myriophyllum*, *Utricularia*, and *Chara*, were transplanted to the west channel of the flume between stations 150 and 200 feet. At the beginning of release period

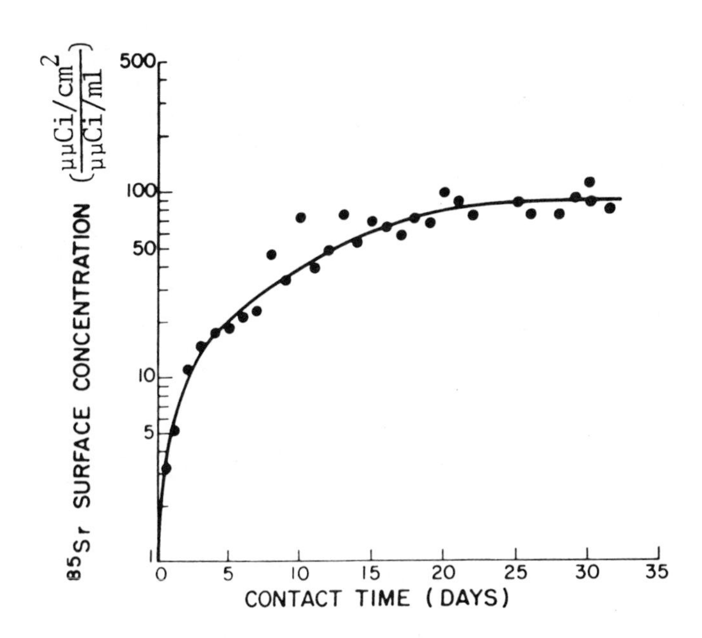

FIGURE 7. Average surface concentration of strontium-85 by bed sediments in the flume related to contact time.

of radioactivity the predominant species were Potamogeton and Vallisneria. One month later, toward the end of the experiment, the macro algae Chara dominated the plant area, and the species Myriophyllum and Utricularia disappeared. The distribution of the plant species was 20.7 g/ft^2 Chara, 11 g/ft^2 Potamogeton, and 5.4 g/ft^2 Vallisneria. These weights were based on air dry weight; however, oven dry weight was found to average 5.7 percent of the air dry weight.

On continuous releases of ^{134}Cs and ^{85}Sr, plants accumulated radionuclides to an equilibrium level as shown in Figure 9. The immediate uptake of ^{134}Cs and ^{85}Sr by plants was relatively high and a quasi-equilibrium state was reached in the first 2 days. The average concentration factors, K_p, for ^{134}Cs and ^{85}Sr, which are the ratio of the specific activity of plants (μCi/g oven dry weight) to the specific activity of water (μCi/ml) are shown in Figure 10. Between the second to the twenty-first day

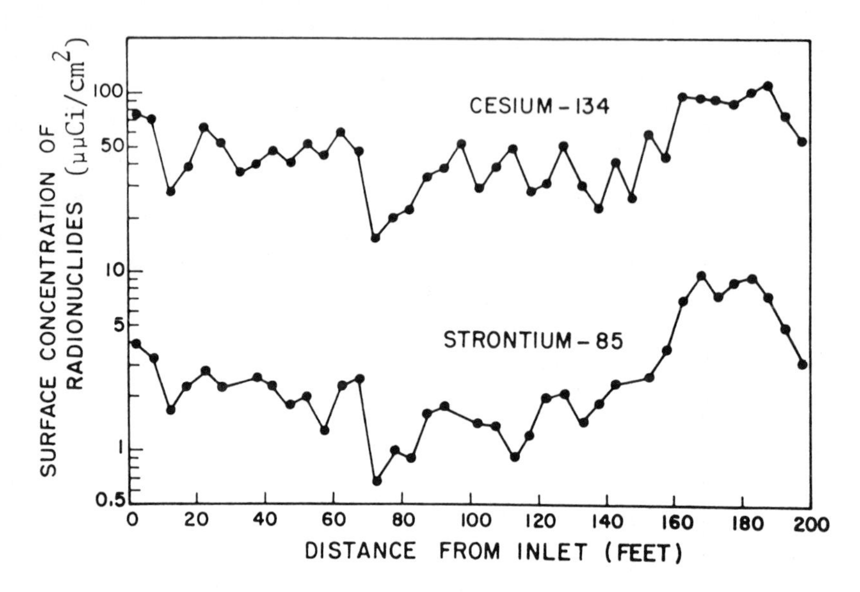

FIGURE 8. Concentration of radionuclides by bed sediments along the length of the flume.

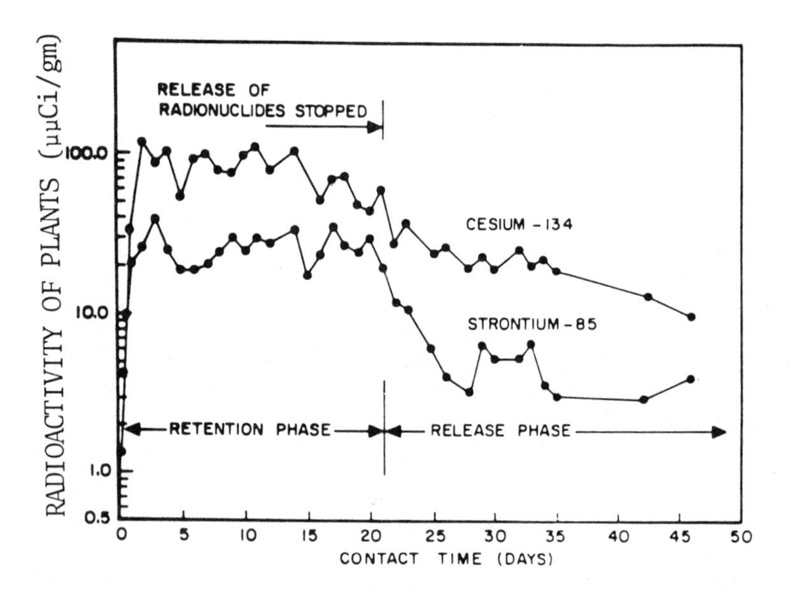

FIGURE 9. Concentration of ^{134}Cs and ^{85}Sr by aquatic plants resulting from continuous release of radionuclides in the flume.

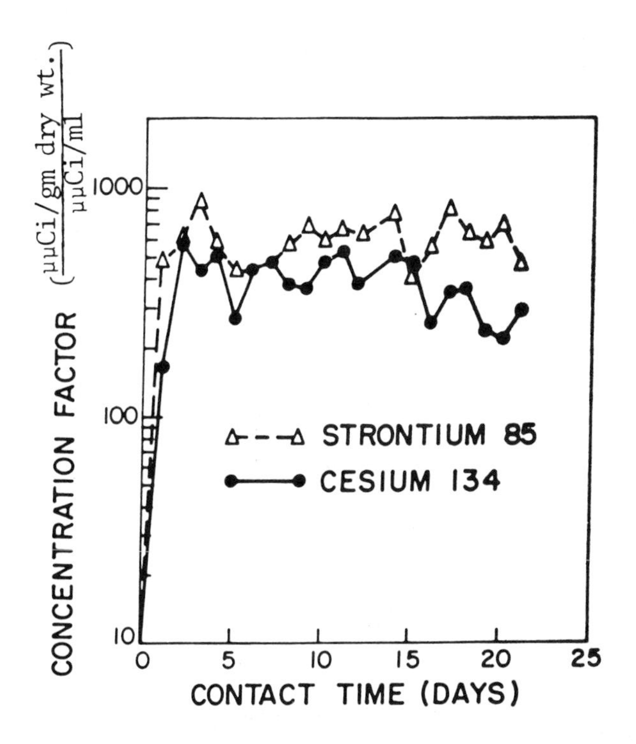

FIGURE 10. Average concentration factors for ^{134}Cs and ^{85}Sr by aquatic plants.

after radionuclides release, K_p values for ^{134}Cs varied between 220 and 580 with an average of 400. During the same time, concentration factors K_p for ^{85}Sr varied from 430 to 860 with an average of 600.

After 3 weeks' release, the flow of ^{134}Cs and ^{85}Sr was discontinued in the west channel, and the radioactivity of the plants continued to be monitored. Initially, the plants released activity at a relatively high rate, which continued for approximately a week. Thereafter, release was at a much slower rate as shown from Figure 9. Release of radioactivity from plants followed an exponential function as shown in Figure 11.

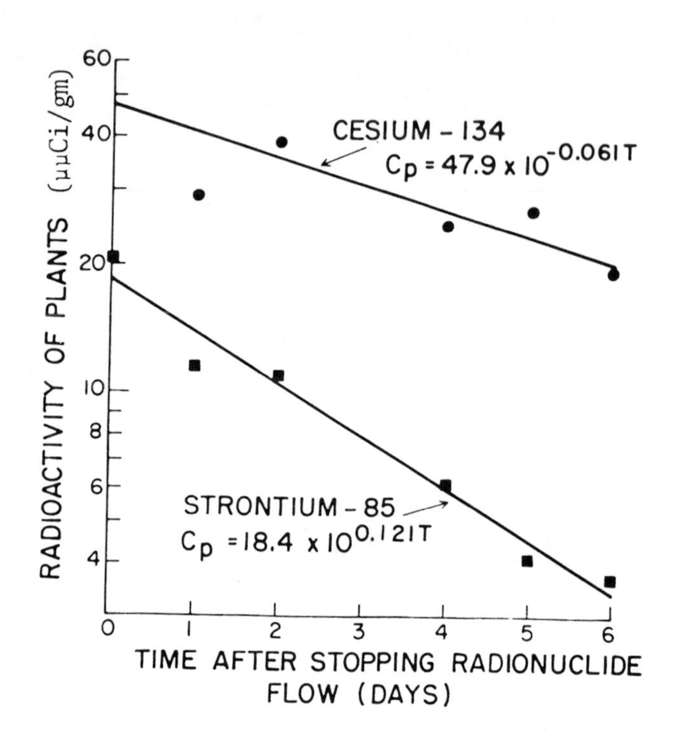

FIGURE 11. Release of radionuclides by plants after discontinuing the source.

C. Retention by Suspended Biota

The overall radioactivity transported through the west channel of the flume is made up of the radioactivity of ions dissolved in the water and the radioactivity associated with suspended solids. The suspended solids in the west channel consisted primarily of phytoplankton retained on millipore filters (0.45 μm pore size), the concentration of which ranged between 15 and 25 mg/liter. Diurnal changes in radionuclides uptake was observed (Yousef et al., 1975). The percent of the activities related to ^{134}Cs in the phytoplankton averaged between 8.9 and 26.7 percent of the total activity in the water samples depending upon the time of the day during which the samples were taken. Under test conditions, ^{85}Sr associated with phytoplankton did not show a pattern similar to that of ^{134}Cs. A detailed study on the relative abundance of plankton, in the west channel, distribution of ^{134}Cs and ^{85}Sr between suspended phytoplankton and solution and diurnal changes in radionuclides uptake, has been recently published by the authors (Yousef et al., 1975). The rate of change in ^{134}Cs concentration associated with suspended algae in 1 liter of water/hour was correlated with the rate of net photosynthetic oxygen production in mg/liter/hour. The uptake of ^{134}Cs by phytoplankton was related to diurnal photosynthetic production of oxygen and a linear relationship exists between the rate of change in ^{134}Cs concentration and the rate of net photosynthetic oxygen production as shown in Figure 12.

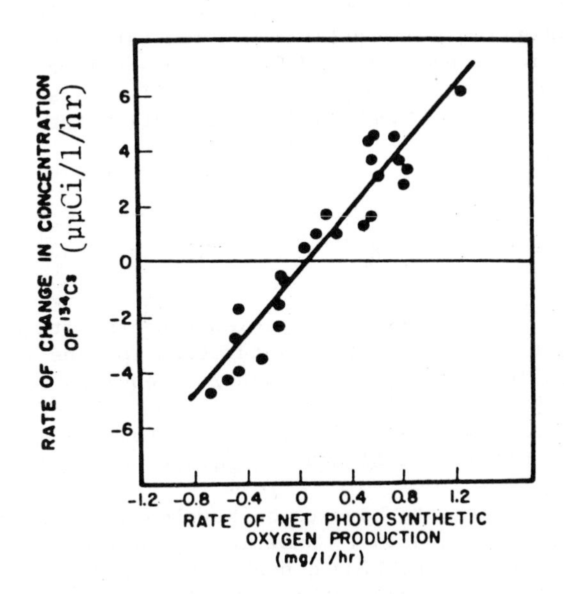

FIGURE 12. Correlation between rate of change in ^{134}Cs concentration and the rate of net photosynthetic oxygen production.

The concentration factors for radiocesium by suspended biota varied between 2880 to 19200 at various locations along the flume and at different times of the day. However, concentration factors for radiostrontium showed less than 1000.

VI. TRANSPORT MODEL

A basic transport equation that combines mixing characteristics, dilution, uptake, and release by various surfaces in the aquatic system was developed as follows:

$$\frac{\partial C}{\partial t} = D_x \frac{\partial^2 C}{\partial x^2} - u \frac{\partial C}{\partial X} - \sum_{i=1}^{i=n} f_i K_i \, [G_i(c) - C_i] \cdots \quad (2)$$

where D_x is the longitudinal dispersion coefficient (ft^2/min), u is the velocity (ft/min), f_i is the mass of ith sorbent affecting a unit volume of the flow zones, K_i is the mass transfer coefficient for phase i, $G_i(c)$ is the transfer function relating the concentration of activity in the water to the equilibrium level in the phase i, and C_i is the specific activity in the ith position of the n-sorption phases. The mass transfer coefficients for bed sediments and plants follow the general relationship:

$$\frac{dC_i}{dt} = K_i \, [G_i \, C_w - C_i] \quad (3)$$

where K_i is the transfer rate coefficient for the ith sorbent, G_i is the concentration factor at saturation for the ith sorbent, C_w is the specific activity of the water at equilibrium conditions, and C_i is the specific activity of the ith sorbent at time t. The mass transfer coefficient for plants and sediments are shown in Figure 13 and Figure 14 and are summarized in Table 1.

Accurate coefficients have to be developed in order to verify the transport equation. Numerical solution of the model was used in predicting the time concentration curves and equilibrium levels for continuous release of Rhodamine B in both the channels of the flume. The input values were as follows: D_x = 3.8 ft^2/min, $\overline{u}$ = 0.44 ft/min, dosing rate of dye - 1.78 mg/min, and flow rate - 10 liters/min.

The predicted and observed time concentration curves are shown in Figures 15 and 16. Discrepancies may be observed between predicted and observed concentration of dye; however, the equilibrium concentration seems to occur at the same time as predicted.

Also, the time concentration curves for ^{134}Cs and ^{85}Sr along the east and west channels of the flume were predicted.

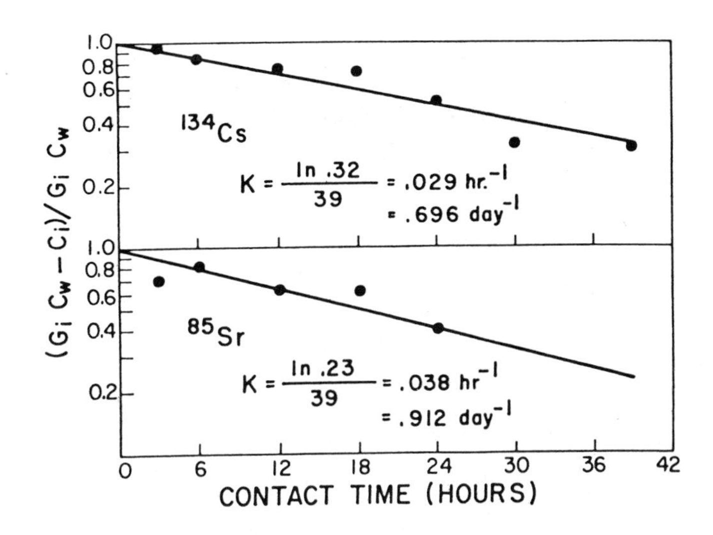

FIGURE 13. Transfer rate coefficients for concentration of radionuclides by plants in the flume.

TABLE 1. Mass Transfer Coefficients of Sediments and Plants

	Mass Transfer Coefficients (Day^{-1})			
	Sediments		Plants	
Location	^{134}Cs	^{85}Sr	^{134}Cs	^{85}Sr
East channel	0.033	0.030	---	---
West channel	0.043	0.036	0.696	0.912

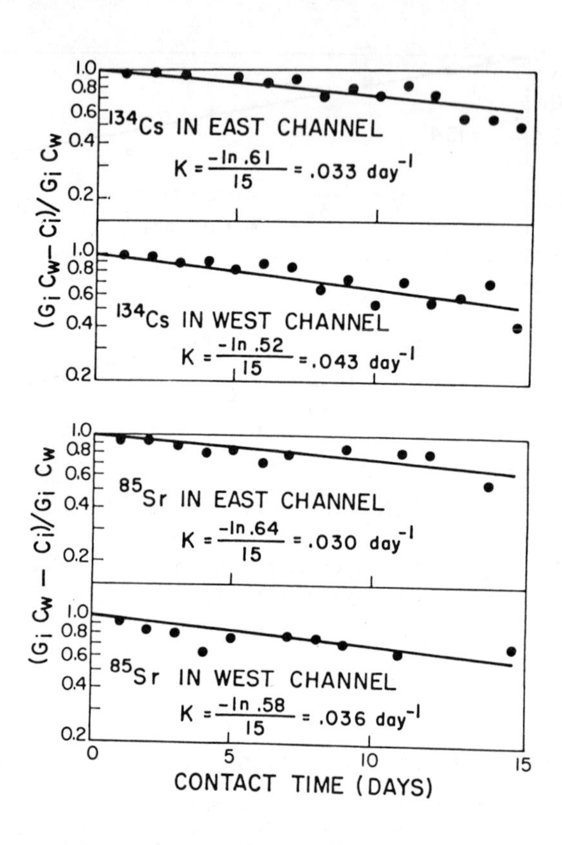

FIGURE 14. Transfer rate coefficients for concentration of radionuclides by bed sediments in the flume.

In the east channel, the draw-down of radionuclides by sediments was considered, while plants and sediments were considered in the west channel. The data obtained from the transport model predicts that equilibrium levels could be reached within three weeks of continuous discharge of radionuclides. Equilibrium is reached when the specific activity of the inflow water is equal to the specific activity of outflow water through the flume. This period of 3 weeks was also stated earlier from the observed data. However, the observed data oscillates around an equilibrium level due to changes in physicochemical and/or biological conditions, which are not predictable by the transport model.

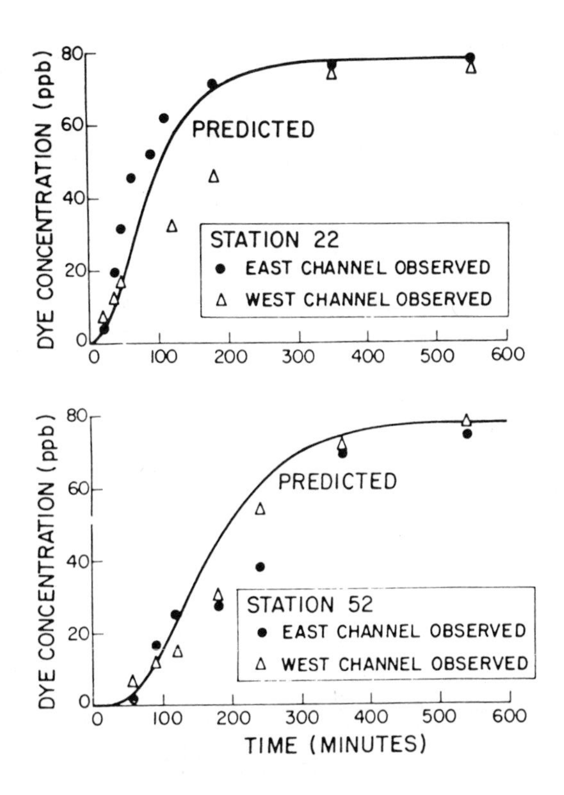

FIGURE 15. Predicted and observed time concentration values for continuous release of Rhodamine B.

VII. CONCLUSIONS

1. Stream processes in flowing water bodies were simulated in small scale ecosystems. Problems encountered with the simulated system such as stratification, deposition, erosion, and relocation indicate the complexity of defining a real river system.

2. Under continuous release of radionuclides, bottom sediments and plants concentrate radioactivity until such time as an equilibrium state is reached. Plants reach equilibrium much faster than sediments. The rate of uptake is directly proportional to the difference between the saturation

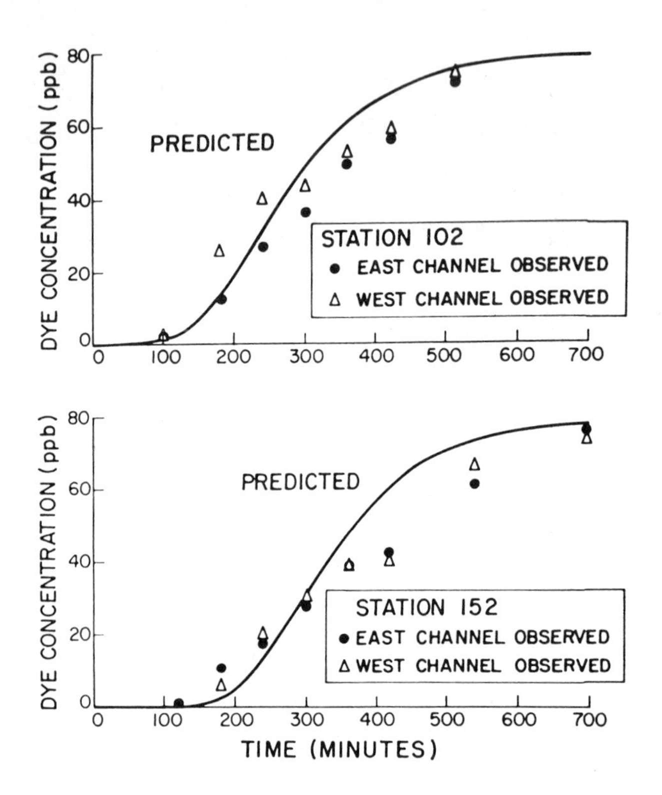

FIGURE 16. Predicted and observed time concentration values for continuous release of Rhodamine B.

concentration and the concentration in sediments or plants at a given point in time. Release of radioactivity back to solution was observed after cessation of the source.

3. Under test conditions, cyclical diurnal uptake of ^{134}Cs by suspended phytoplankton in the flume was noted. The uptake followed a pattern similar to the phytosynthetic oxygen production and a linear correlation was found to exist (Figure 12). Also, suspended phytoplankton will tend to deposit on the plant leaves, stems, and bottom sediments. Therefore, the surface concentration by sorbing surfaces increases.

REFERENCES

1. Akira, K. and Gloyna, E. F. 1969. Radioactivity Transport in Water—Interaction between Flowing Water and Bed Sediments, Technical Report submitted to the U. S. Atomic Energy Commission, Contract AT (11-1)-490, Center for Research in Water Resources, University of Texas at Austin EHE 69-03, CRWR-36.

2. Gloyna, E. F., Yousef, Y. A., and Padden, T. J. 1971. Transport of Organic and Inorganic Materials in Small Scale Ecosystems. In Adv. Chem. Ser. 106, Nonequilibrium Systems in Natural Water Chemistry, Amer. Chem. Soc., Washington, D. C.

3. Padden, T. J. 1970. Simulation of Stream Processes in a Model River, Ph.D. dissertation, University of Texas at Austin, December.

4. Yousef, Y. A., Kudo, A., and Gloyna, E. F. 1970. Radioactivity Transport in Water—Summary Report, Technical Report 70 submitted to the U. S. Atomic Energy Commission, Contract AT (11-1)-490, Center for Research in Water Resources, University of Texas at Austin, EHE-70-05, CRWR-53, ORO-20, February.

5. Yousef, Y. A., Padden, T. J., and Gloyna, E. F. 1971. Radioactivity Transport in Water—Continuous Release of Radionuclides in a Small Scale Ecosystem, Technical Report 21 submitted to the U. S. Atomic Energy Commission, Contract AT (11-1)-490, Center for Research in Water Resources, University of Texas at Austin, EHE-71-1, CRWR-75, ORO-21, September 30.

6. Yousef, Y. A., Padden, T. J., and Gloyna, E. F. 1975. Diurnal Changes in Radionuclides Uptake by Phytoplankton in Small Scale Ecosystems, Int. J. Water Res. 9(2):181-187.

Mechanisms of Bioaccumulation of Mercury and Chlorinated Hydrocarbon Pesticides by Fish in Lentic Ecosystems

JERRY L. HAMELINK
Lilly Research Laboratories
Division of Eli Lilly and Co.
Box 708
Greenfield, Indiana

RONALD C. WAYBRANT
Michigan Department of Natural Resources
Stevens T. Mason Building
Lansing, Michigan

and

PHILIP R. YANT
Department of Zoology
University of Michigan
Ann Arbor, Michigan

I. INTRODUCTION

The hypothesis that the biological magnification of pesticides in aquatic environments is controlled by mass transfers of residues through the food chain (Hunt, 1966; Rudd, 1964; Woodwell, Wurster, and Isaacson, 1967; Woodwell, 1967; Harrison, et al., 1970) was first questioned about 6 yrs ago (Hamelink, Waybrant, and Ball, 1971; Woodwell, Craig, and Johnson, 1971). We offered an alternative hypothesis (Hamelink, Waybrant, and Ball, 1971) that stated that biological magnification was controlled by absorption and partition from water. This basic hypothesis has been accepted or restated by many investigators (Kapoor, et al., 1973; Neely, Branson, and Blau, 1974; Kenaga, 1975; Veith and Konasewich, 1975; Derr and Zabik, 1974; Edwards, 1975; National Academy of Science, 1973). Briefly, we proposed that algae, invertebrates, and fish continuously exchange stable pesticide residues with their environment, namely water, and given sufficient time, this exchange should reach a finite equilibrium point wherein the organisms lose residues at the same rate that they are acquired. Hence, we proposed that the organisms essentially reach a chemical equilibrium with their environment such that a series of constants, analogous to partition coefficients, might be used to describe each component of an aquatic ecosystem that would then "define" the degree of biomagnification possible for a given compound. Consequently, we use the collective term "exchange equilibria" to characterize the process of biomagnification of trace contaminants in lentic ecosystems.

The bioconcentration of residues by fish has been shown to be directly related to the n-octanol-water partition coefficient (Neely, Branson, and Blau, 1974) and inversely correlated with water solubility (Kapoor et al., 1973) for a variety of compounds. Recently, trifluralin residues in fish from the Wabash River were shown to be determined by the trifluralin concentration in the water (Spacie, 1975). In addition, two separate studies on the uptake of dieldrin from food and water under laboratory conditions both concluded the two sources were not additive (Reinert, 1972; Chadwick and Brocksen, 1969). Nonetheless, the relative significance of food and water sources in determining the concentration of residues found in fish from natural environments (Harrison et al., 1970; Woodwell, Craig, and Johnson, 1971; Macek and Korn, 1970) and the time required to reach equilibrium (Hassett and Lee, 1975) have yet to be elucidated. Consequently, we investigated the bioaccumulation of lindane, DDE, and mercury by rainbow trout (Salmo gairdneri) in a large scale model lake in order to better understand these parameters.

II. EXPERIMENTAL DESIGN AND METHODS

We have been studying the fate and effects of trace contaminants in a model ultraoligotrophic lake since 1971 (Hamelink and Waybrant, 1976; Waybrant, 1973; Yant, 1974). The model ecosystem is a flooded limestone quarry (Figure 1) located in southern Indiana near the town of Oolitic. It is a large rectangular, limestone box, 300-ft. long, 135-ft. wide, and 50-ft. deep.

FIGURE 1. A view of Quarry T from the northeast corner in June 1972.

The water is very clear, with Secchi disc readings ranging from 7 to 41 ft. with a median of 17 ft. The system supports a highly developed trophic structure, including spawning populations of bluegill (Lepomis macrochirus) and largemouth bass (Micropterus salmoides). Rainbow trout were stocked in the fall of 1972 and again in the spring of 1974. During the summer of 1972, the water was treated with a single 0.05 μg/L (ppb) dose of both lindane and DDE (Waybrant, 1973), then 5.0 ppb mercury as mercuric nitrate 21 days later (Yant, 1974). Residue levels and routine limnological parameters have been periodically measured on most components in the system and from an adjacent control quarry for over 3 yrs.

A. Water

Water for pesticide analysis was collected in acetone-washed, evacuated 1-gal. glass jugs. The jugs were sealed with a glass tube inserted through a rubber stopper. They were individually lowered to a desired depth, and the sealed tube was broken to let in water.

Water for mercury analysis was collected with a 2.2-liter Wildco Kemmerer Plus (TM) water sampler. Each sample was immediately transferred to a 500-ml Nalgene bottle containing 32-ml concentrated HNO_3. This resulted in a 1 N acid solution, which prevented loss of mercury by adsorption onto the container walls (Robertson, 1968). Six depths (0, 10, 20, 30, 40, and 50 ft.) were sampled in three randomly selected locations each period for both mercury and pesticides.

Water for pesticide analysis was filtered through a glass wool plug to remove zooplankton and measured to 3 liters. The sampling jug was rinsed twice with 50-ml aliquots of hexane, which were also used to extract the water. The sample was extracted for 15 min. with 200 ml of hexane on a magnetic mixer. The water was drawn off and discarded, then the hexane was dried with anhydrous Na_2SO_4 and evaporated to 3 to 5 ml. The extract was cleaned by passage through a microflorisil column with 30% benzene in hexane, which removed phthalates derived from glassware contamination. Recovery efficiencies for water and other samples are given in Table 1.

Water was analyzed for total mercury by digesting the 500-ml sample with 5-ml 1:1 H_2SO_4 and 2.5-ml 2% $KMnO_4$ for 24 hrs. at room temperature. Aliquots of 100 ml were measured into a BOD bottle, reduced with 5-ml 1:1 H_2SO_4 and 5-ml 10% $SnCl_2$ in 1 M HCl, then immediately analyzed on a Coleman MAS-50 mercury analyzer.

B. Zooplankton

Zooplankton were captured with modified light traps (Baylor and Smith, 1953) set at the surface and 40 ft. deep. Both *Daphnia sp.* and at least two species of rotifers in the genus Keratella were found in the quarry, but only the *Daphnia* were readily captured in the traps. The zooplankton were filtered onto a #1 Whatman filter and frozen until analyzed.

For pesticide analysis, the zooplankton were ground with anhydrous Na_2SO_4 and sand in a mortar and pestle. The mixture was extracted three times with 10-ml hexane, evaporated to 2 ml, passed through a microflorisil column with 30% benzene and analyzed.

TABLE 1. Percent Mean (Standard Deviation) Recovery of Standards from Spiked Samples.

Sample Type	Compound		
	Lindane	DDE	Total mercury
Water	78 (13)	93 (6)	104 (15)
Bottom mud	77 (16)	76 (11)	91 (12)
Zooplankton	100	100	100[a]
Fish	83 (13)	76 (8)	94 (8)

[a]There was no significant difference between recoveries of standards with zooplankton present or absent from the methods and essentially 100% of the standards added were recovered.

Zooplankton and fish were digested for mercury analysis following a modified Adrian (1971) method. About a gram of sample was placed in a 2-oz Nalgene bottle, 3-ml $HClO_4$ (70% solution), and 7-ml HNO_3 were added, then the bottles were tightly sealed and allowed to stand overnight at room temperature. The bottles were placed in a hot water bath (70 to 80°C) for 2 to 3 hr, allowed to cool, 2- to 3-ml D-HOH added, then while still open, they were reheated in the water bath until nitric acid fumes were dispelled. After cooling, the samples were transferred to a BOD bottle, diluted to 100 ml, reduced, and analyzed as before.

C. Fish

Fish were caught by hook and line, weighed, measured, and immediately frozen until analyzed. The frozen trout were ground to a fine powder with dry ice in a Hobart Food chopper followed by a Sorval Omni-Mixer. All grinding operations were carried out in a chest-type deep freezer to prevent the tissue from adhering onto the exposed metal cups and blades.

A 5-g subsample was ground with anhydrous Na_2SO_4 and extracted with petroleum ether in the Omni-Mixer for pesticide analysis. The residues were cleaned by elution through a standard florisil column (Mills, Onley, and Gaither, 1963).

III. RESULTS

A. Water

Most of the DDE and mercury added to the epilimnion was translocated out of the water by sedimenting particles and deposited on the bottom. By comparison, the less strongly abosrbed lindane remained in the water phase and slowly disappeared, apparently due to degradation of the compound (Figure 2). Thus within 120 days after treatment, the concentration of DDE in the water stabilized near 1 pptr (parts per trillion). Lindane concentrations in the whole water column slowly declined from 25 pptr on post-treatment day 123 to 13 pptr on day 358. Mercury concentrations in the water fell below our detectable limits of 0.1 ppb 154 days after treatment. Hence, only 1 or 2% of the DDE and mercury persisted in the water for any appreciable length of time, while lindane had a half life of about 120 to 150 days. However, this does not mean the system was decontaminated 6 months after treatment. On the contrary, both DDE and mercury were found in high concentrations in the bottom mud (24.21 ± 10.58 and 2234 ± 571 ppb dry wt. in top 1.5-cm layer on day 123, respectively) and significant levels of both were found in the biota for over 2 yr after treatment.

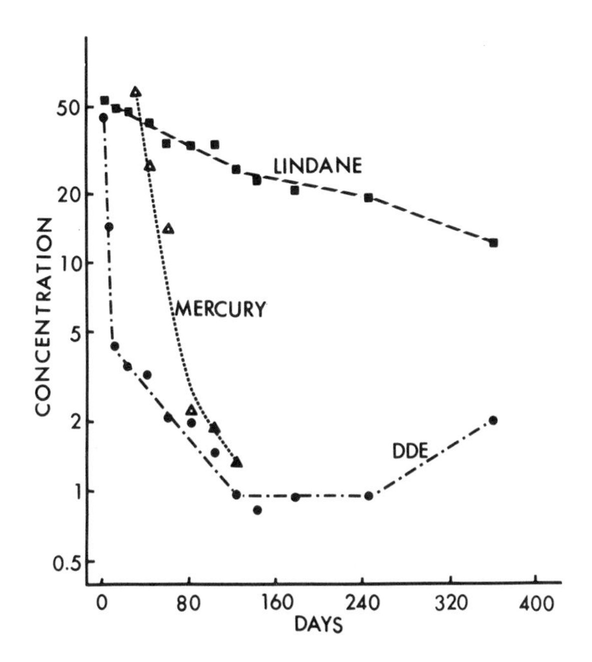

FIGURE 2. Concentrations of mercury (units in 0.1 μg/liter), DDE and lindane (units in 0.001 μg/liter) in the whole water column versus time.

B. Zooplankton

The concentration of DDE residues found in zooplankton from the quarry was directly related to the concentration in the water [$\bar{Y}$ (ppb) = 27.3 + 53.75X (pptr) (R = 0.99; N = 12)] (Hamelink and Waybrant, 1976). In fact, the regression observed was essentially an extension into the parts per trillion range of the correlation we previously observed with DDT (Hamelink, Waybrant, and Ball, 1971) [$\bar{Y}$ (ppm) = 0.424 + 13.521$\underline{X}$ (ppb)] and the correlations reported for DDE in midge larvae [$\bar{Y}$ (ppm) = -2.5 + 28.9X (ppb) and Y (ppm) = -3.2 + 28.6X (ppb)] (Derr and Zabik, 1972). Hence, these small animals apparently concentrate and retain body burdens of DDE residues in direct proportion to the concentration found in the water. Because of this they may be useful "detectors" for the presence of minute quantities of these compounds in natural waters.

The zooplankton also took up and retained lindane, but at levels much lower than those for DDE (Table 2). Furthermore, the amounts measured were quite variable. The concentrations

TABLE 2. The Concentrations and the Concentration Factors Relative to Water of the Same Time Period for the DDE and Lindane in the Invertebrates. (PPB Wet Wt.)

Day	DDE $\bar{X}$ ± SD	Concentration Factor	Lindane $\bar{X}$ ± SD	Concentration Factor	DDE-C.F. / Lindane-C.F.
Prtr	5.03 ± 0.72	- -	N.D.[b]	-	-
1	58.30 ± 1.53	1.31×10^3	TR.[c]	-	-
5	775.08 ± 86.86	5.16×10^4	26.17 ± 2.07	414	124
21	223.47 ± 83.04	6.35×10^4	10.00	170	373
42	120.07 ± 48.95	3.64×10^4	22.60 ± 17.58	448	81
60	60.15 ± 29.61	2.86×10^4	9.38 ± 0.50	314	91

[a]SD, standard deviation

[b]N.D., none detected with level of sensitivity of ca. 0.1 ppb.

[c]TR., trace, ca. 0.2 ppb.

in the zooplankton and the whole water column gave a correlation coefficient of 0.556, which accounted for only 31% of the variation. Thus while lindane uptake did occur, it was not very consistent. Presumably this was due to the low concentrations in the animals, the analytical procedure employed, and the stratified distribution of lindane in the quarry water during the summer months.

The concentration factors (C.F.) observed for the zooplankton were relatively constant after the first day, ranging from 2.86×10^4 to 6.35×10^4 for DDE and from 170 to 448 for lindane (Table 2). Furthermore, the relative concentration factors (i.e., C.F. for DDE/C.F. for lindane) were usually around 100-fold, except for the unusually high value observed on day 21 when the low weight of invertebrates captured made quantification difficult. Lindane is between 1000 to 10,000 times more soluble in water than DDE (Gunther, Westlake, and Jaglan, 1968), so the relative concentration factors, and the direct correlation of DDE in zooplankton with DDE in water continue to support the exchange equilibria hypothesis (Hamelink, Waybrant, Ball, 1971).

We were not able to make a similar comparison for the concentrations of mercury in zooplankton and in water due to an insufficient number of paired water and zooplankton samples. This arose because mercury concentrations in the water rapidly fell below detectable limits, relative to the sampling schedule employed, while considerable problems were experienced with the light traps during the same time period. However 1 year after treatment, significantly higher concentrations of mercury were observed in zooplankton captured near the bottom compared to those captured at the same time in the epilimnion (Yant, 1974). The exact cause for this type of inverted concentration gradient, beyond the presence of high concentrations of mercury in the mud, is not known. Nonetheless, zooplankton appeared to provide a relative measure of the degree of contamination available to the biota as time progressed.

C. Fish

All the fish collected showed significant accumulations of each compound during the course of the study. Unfortunately, as a result of sampling problems due to low fish population density, which were aggravated by poaching, the data for fish is not complete. Bluegills were readily captured through the day 81 sampling period and then none were caught until day 358, almost a year later. Largemouth bass were always difficult to capture. Therefore, the data gathered on contaminant uptake by these species of fish is limited. However, 180 rainbow trout were stocked in the fall and captured with some regularity for almost 1 year after introduction. The trout were added 134 and 113 days after the DDE and mercury treatments, respectively, so they were not exposed to greatly elevated concentrations of either compound in the water. Consequently, the somewhat more complex interplay of those factors which are believed responsible for controlling residue body burdens in fish will be illustrated by the data obtained from the trout.

For 346 days after being planted in the quarry, the trout grew exponentially following ($R = 0.83$; $N = 33$; $F = 53.69$) the equation:

$$W = 61.16\ (1.0030)^{t} \tag{1}$$

where W is the mean weight in grams of a fish, and t is the number of days after being introduced with an estimated mean weight of 61.16 g. Their body burdens of both DDE and mercury, determined as the product of individual body weight and contaminant concentration, increased even faster than the fish grew as evidenced by the base of the exponent in the following equations.

The body burden (B) in nanograms of DDE followed ($R = 0.84$; $N = 31$; $F = 73.67$) the equation:

$$\text{B-DDE} = 3122\ (1.0083)^{t} \tag{2}$$

as shown in Figure 3. The trout accumulated mercury at a somewhat faster rate than DDE following ($R = 0.88$; $N = 31$; $F = 103.4$) the equation:

$$\text{B-Hg} = 1867\ (1.0103)^{t} \tag{3}$$

as shown in Figure 4.

There was little correlation between body weight and contaminant concentration with DDE displaying a linear correlation coefficient of 0.50 and mercury a value of 0.69. Body weight versus concentration was not strongly correlated because the body weight of the fish did not necessarily relate to the exposure time, or the starting weight of the fish. Thus although heavy fish often contained high concentrations of both products and vice versa, there were many exceptions because both growth and exposure time interacted to permit a given body burden to be acquired.

Since the accumulation rates for DDE and mercury were very similar, a strong ($R = 0.87$; $N = 31$; $F = 94.68$) correlation between the body burdens of the compounds in individual fish was also observed:

$$\text{Log B-Hg} = -0.2044 + 1.0351\ (\log \text{B-DDE}) \tag{4}$$

as shown in Figure 5.

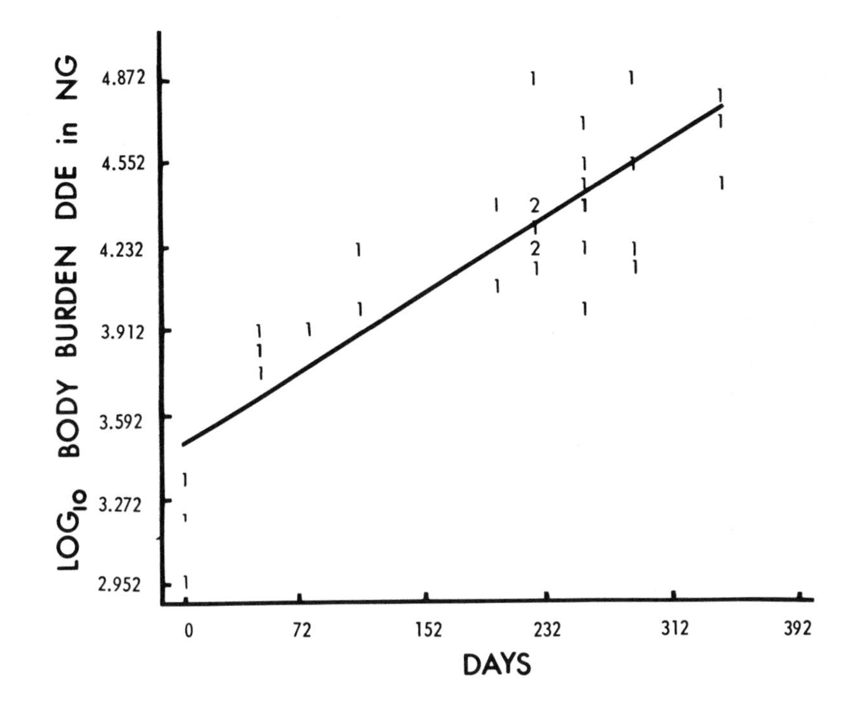

FIGURE 3. Body burden in nanograms of DDE in rainbow trout from Quarry T versus days after introduction. Values within graphical field denote the number of individuals plotted at a given locus by computer.

This does not mean the body burden of either contaminant was dependent on the other. Rather, the slope value and strong correlation observed implies that the body burden of both contaminants may have been determined by the same factors operating in their environment.

In contrast to the mercury and DDE, the concentration of lindane in the trout did not increase with time. The average concentration of apparent lindane in the trout when stocked was 2.2 ± 2.1 ppb (N = 3). Forty-five days after stocking, the concentration observed was 27.3 ± 11.9 ppb (N = 3), which stabilized at 13.3 ± 4.2 ppb (N = 7) for 300 days thereafter.

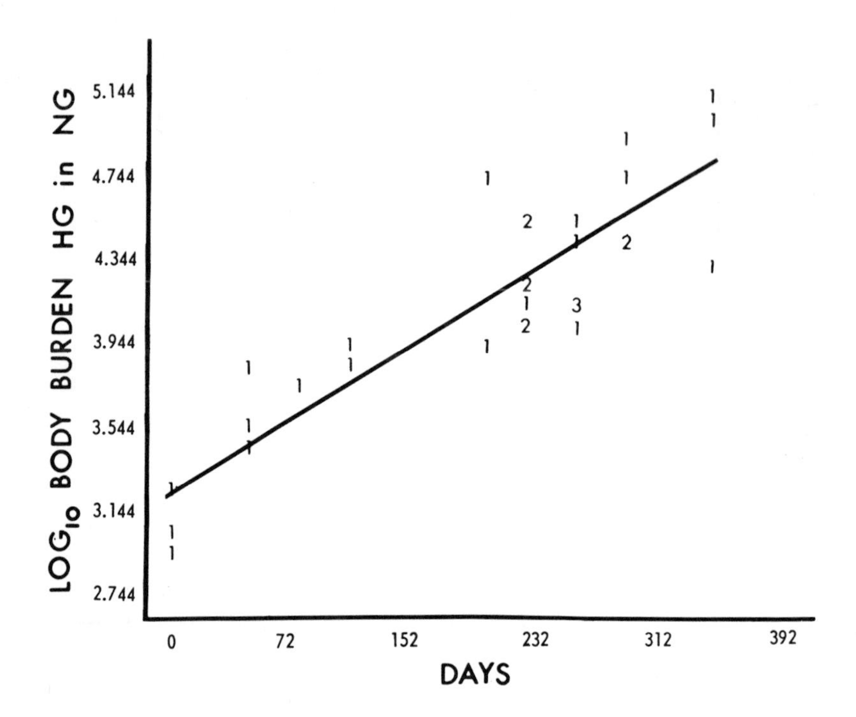

FIGURE 4. Body burden in nanograms of mercury in rainbow trout from Quarry T versus days after introduction.

IV. DISCUSSION

The trout were presumably able to acquire the three compounds both directly from the water and from their food. The quantity of residues available directly from the water depends upon many factors. The three primary factors are the concentration in the water, the volume of water passed over the gills, and the efficiency with which fish extract the compound from the water. The volume of water ingested by fresh water fish is inconsequential (Prosser and Brown, 1961).

The volume of water passed over the gills can be estimated from oxygen consumption. Trout consume about 0.35-ml O_2 at STP/g wet wt/hr at 15°C (Prosser and Brown, 1961), which is equivalent to 0.50-mg oxygen/g/hr or 12 mg/g/day. Water saturated with oxygen contains about 10 mg oxygen/liter at 15°C. The extraction efficiency of trout for oxygen is about 80% (Randall, 1970). Hence, a trout must extract at least 1500 ml of water/g/day

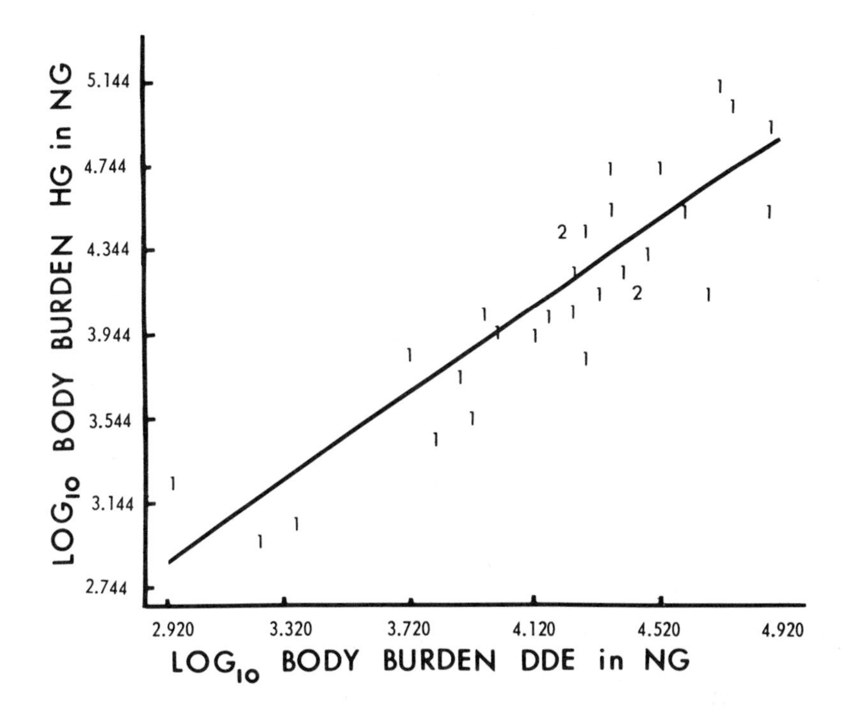

FIGURE 5. Body burden mercury versus body burden of DDE in individual trout from Quarry T (units in nanograms).

at a temperature of 15°C. Given the growth equation, the total volume of water respired by an "average" trout after 346 days in the quarry would be about 56 X 10^6 ml. This means that each fish passed a volume of water equivalent to 0.1% of that in the whole quarry. Thus the ratio of fish biomass to water volume may be an important factor to consider when conducting studies of this nature in model systems. The concentration of inorganic and organic mercury in the water was generally less than 0.1 ng/ml after the trout were introduced. We have assumed the concentration of organic mercury, presumably as methyl mercury, in the water was 1% of the total (de Freitas, Qadri, and Case, 1974) or 0.001 ng/ml. The concentration of DDE was 0.0010 ± 0.0004 ng/ml (N = 54) during the entire period of exposure. The mean concentration of lindane in the water was 0.0215 ± 0.00149 ng/ml (N = 54) and was declining throughout this period. Therefore this value constitutes a slight overestimate for the total period of exposure.

The uptake efficiency displayed by fish for organic mercury is much greater than that observed for ionized inorganic forms. Direct uptake from the water for inorganic form is about 0.2% efficient, while the same for methyl mercury may approach 20% (Freitas, Qadri, and Case, 1974; Burrows and Krenkel, 1973; Olson, Bergman, and Fromm, 1973). Lindane and DDE are presumably acquired with an efficiency no greater than DDT which may also range up to 20% (Reinert, Stone, and Willford, 1974).

The trout grew from an average of 61 to 173 g, or by 112 g in 346 days. Fish may eat up to ten times the weight gained as food. Thus, the average trout may have consumed as much as 1120 g of food in 346 days. This is equivalent to 3% of their body weight per day, which may be somewhat high for fish in a natural environment (Lagler, Bardach, and Miller, 1962).

Cursory examination of stomach contents revealed that zooplankton, primarily large *Daphnia sp.*, were the major autochthonous food consumed by the trout during the fall, winter, and spring. This agrees with the observations of Galbraith (1967). Terrestrial insects and invertebrates, such as earthworms, were a significant but unquantified component of their diet during the warmer months. Thus the overall level of contamination present in their diet was probably somewhat less than that observed in the zooplankton.

The zooplankton contained 356 ± 158 ppb (N = 16) of mercury and 42.6 ± 47.2 ppb (N = 3) of DDE during the summer between day 387 and 481. Twenty percent of the mercury found in zooplankton 1 yr after treatment was toulene-extractable organic mercury (Yant, 1974). Lindane levels in the zooplankton during this period were below our detectable limits, due to the low weight captured. However, given the bioconcentration factor of 336 observed during the first 60 days after treatment and a concentration in the water of 21 pptr, lindane concentrations of about 7.0 ppb should have been present in the zooplankton. Finally, the three organic compounds were assumed to be taken up with an efficiency of 85% from a natural food source, while inorganic mercury uptake was assumed to be 15% efficient (Macek and Korn, 1970; Freitas, Qadri, and Case, 1974; Giblin and Massaro, 1973; Macek et al., 1970; Grzenda, Paris, and Taylor, 1970).

The quantity of each compound available for uptake by the trout from each source was estimated from the parameters above, and compared to the amounts actually retained (Table 3). Whenever a parameter was estimated, we selected the value resulting in the highest calculated uptake. Therefore, the quantities reported as being available in Table 3 are believed to be maximum values.

The exchange equilibria hypothesis (Hamelink, Waybrant, and Ball, 1971) would predict that trace contaminants are taken up until an equilibrium is reached. After reaching equilibrium the concentration would not rise, although the body burden would increase proportional to growth. Thus, after equilibrium there should be an increasing difference between the amount available for uptake and the amount retained by the fish.

This was observed for lindane (Table 3). In fact, 1255 times more lindane appeared to be available for uptake than was retained. This implies that the body burden could have been turned over 3.6 times daily. Unfortunately, little work has apparently been done on the metabolism of lindane by fish. Hence, the significance of degradation and excretion of lindane by fish is an unknown variable. However, Gakstatter and Wiess (1967) observed that fish exposed to lindane in water apparently reached equilibrium within a few hours. Furthermore, nearly all of the lindane was eliminated from the treated fish in less than 2 days after being transferred to clean water (Gakstatter and Weiss, 1967). Thus lindane may continuously undergo a rapid (i.e., half life of 3 to 6 hr) exchange in fish, similar to the very rapid (i.e., half life of 0.5 to 1.5 hr) exchange observed with tritiated water (Morgan, Landolt, and Hamelink, 1971; Motais, et al., 1969).

The difference between the amount of mercury available and the amount retained is much smaller than that observed for lindane. Mercury is bound to protein within fish, making it less available for exchange with the water (Olson, Bergman, and Fromm, 1973; Giblin and Massaro, 1973). Thus we expected most of the mercury to be retained. Some of the available mercury was not retained because it can be excreted. This is supported by the observation that a larger percentage of the inorganic mercury than organic mercury apparently was not retained (de Freitas, Qadri, and Case, 1974; Burrows and Krenkel, 1973; Giblin and Massaro, 1973).

The difference between the amount of DDE retained versus the amount estimated to be available is negligible. The concentration factor for DDE in trout should be about 1×10^6 (Hamelink, Waybrant, and Ball, 1971). Given 1 pptr in the water the equilibrium level for DDE in trout would be about 1 ppm or a body burden of 171 μg DDE. By day 346, the amount of DDE estimated to be available was only 62.1 μg. Thus the trout were not able to obtain enough DDE to reach equilibrium, and essentially all of the DDE acquired appeared to be retained by the trout.

TABLE 3. A Comparison of the Sources of Contamination Versus the Quantities of Contaminants Retained by Rainbow Trout During 346 Days of Exposure (Units in Micrograms).

Compound	Source (adj. for uptake efficiency)		Total "available"	Total "retained"[b]
	Water	Food		
Lindane	288	7.8	295.8	0.23
DDE	14.4	47.7	60.1	55.7
Org.-Hg[a]	14.4	79.7	94.1	39.2
Inorg.-Hg	14.4	47.8	62.2	26.1

[a]Organic mercury was assumed to be 60% of the total mercury found in the trout.

[b]Total retained estimated from Eq. 2 and 3 for Hg and DDE, respectively, while lindane was the product of estimated weight from Eq. 1 times 13.3 ppb.

V. CONCLUSIONS

Our estimates indicate that a large share of the DDE and mercury residues in the trout must have been acquired from their food. Hence, we agree that the food chain is probably a major source of DDT-R (Macek and Korn, 1971) and mercury (Jernelov and Lann, 1971) contamination in fish from natural environments. However, as previously stated, the source of supply is not the only factor that determines contamination levels in fish. Rather, many factors interact to determine the residue concentration found in any given fish. Nonetheless, we believe residues dissolved in the water serve as the origin for any contamination in the fish. Thus residue concentrations in the water must be regulated in order to regulate contamination levels in fish stocks and the exchange equilibria concept provides a rationale for setting tolerance limits for residue concentrations in water supplies.

Unfortunately, aquatic organisms do not behave like ideal solvents in a separatory funnel, so setting tolerance limits in particular situations requires some judgment. For example, the biomagnification potential of many compounds can be readily assessed based on the bioconcentration factor derived from the uptake and clearance rates observed when fish are exposed to compounds dissolved in the water (Neely, Branson, and Blau, 1974). We have also demonstrated that the rate at which DDE and mercury are acquired from both food and water by fish versus their growth rate may largely determine residue concentrations in fish. This agrees with the observation that concentrations of DDE (Anderson and Fenderson, 1970) and mercury (Scott, 1974; Cross et al., 1973) are positively correlated with weight or age of fish from natural habitats. Yet for compounds like lindane or dieldrin (Scott, 1974) residue concentrations were not observed to increase with weight or age.

These differences between compounds arise because fish can reach and maintain a state of equilibrium with lindane (Gakstalter and Weiss, 1967) or dieldrin (Reinert, 1972; Chadwick and Brocksen, 1969) in the water in a relatively short period of time, but not with DDE or mercury (Reinert, Stone, and Willford, 1974). The bioconcentration factor for lindane is about 1×10^2 (Gakstalter and Weiss, 1967) and for dieldrin about 1×10^4 (Reinert, 1972), while that for DDE would undoubtedly exceed 1×10^5. Hence as the bioconcentration factor increases above 1×10^5, a considerable amount of time would be required to acquire and equilibrate relatively large amounts of a compound from a highly dilute source. Thus from a practical standpoint, direct application of the exchange equilibria principle would appear to be limited to those compounds displaying a bioconcentration factor of something less than 1×10^5.

Although it is extremely difficult for a fish exposed to static concentrations of those compounds which may be highly biomagnified to achieve equilibrium (Hassett and Lee, 1975), an effective state of equilibrium could be rapidly reached by fish if they were periodically exposed to elevated levels in the water (Grzenda, Paris, and Taylor, 1970), or their food (Macek and Korn, 1970; Macek et al., 1970). For example conditions of this nature might be encountered in the vicinity of contaminated effluent or tributary in a large lake. The concentration in the water may fluctuate seasonally (Godsil and Johnson, 1968; Vanderford, 1974) or change with depth as implied by the stratified distribution of mercury we observed in zooplankton (Yant, 1974). Thus wherever and whenever some elevated levels were encountered by a fish, the net effect would be to hasten the time required to bring the concentration in the fish into equilibrium with lower concentrations elsewhere in a lake. These facts inevitably lead to the corollary that static concentrations in nature are seldom static. Consequently, defining an equilibrium state for fish in natural environments becomes a more arbitrary process as the bioconcentration factor increases but the concept still provides a useful rationale for setting water quality criteria and conducting environmental hazard evaluations.

REFERENCES

1. Adrian, W. J. 1971. A A Newsletter 10, 96.

2. Anderson, R. B., and Fenderson, O. C. 1970. J. Fish. Res. Board Can. 27, 1.

3. Baylor, E. R., and Smith, F. E. 1953. Ecology 34, 223.

4. Burrows, W. D., and Krenkel, P. A. 1973. Environ. Sci. Tech. 7, 1127.

5. Chadwick, G. G., and Brocksen, R. W. 1969. J. Wildlife Manag. 33, 693.

6. Cross, F. A., Hardy, L. H., Jones, N. Y., and Barber, R. T. 1973. J. Fish. Res. Board Can. 30, 1287.

7. Derr, S. K., and Zabik, M. J. 1972. Trans. Amer. Fish. Soc. 101, 323.

8. Derr, S. K., and Zabik, M. J. 1974. Arch. Environ. Contam. Toxicol. 2, 152.

9. Edwards, J. G. 1975. Science 189, 1974.

10. de Freitas, A. S. W., Qadri, S. U., and Case, B. E. 1974. Origins and Fate of Mercury Compounds in Fish. In Proc. Int. Conf. on Transport of Persistent Chemicals in Aquatic Ecosystems, National Research Council of Canada, Ottawa, Canada, p. III, 31-36.

11. Gakstatter, J. H., and Weiss, C. M. 1967. Amer. Fish. Soc. 96, 301.

12. Galbraith, M. G. 1967. Trans. Amer. Fish. Soc. 96, 1.

13. Giblin, F. J., and Massaro, E. J. 1973. Toxicol. Appl. Pharmacol. 24, 81.

14. Godsil, P. J., and Johnson, W. C. 1968. Pestic. Monit. J. 1, 21.

15. Grzenda, A. R., Paris, D. F., and Taylor, W. J. 1970. Trans. Amer. Fish. Soc. 99, 385.

16. Gunther, F. A., Westlake, W. E., and Jaglan, P. S. 1968. Residue Rev. 20, 1.

17. Hamelink, J. L., and Waybrant, R. C. 1976. Trans. Amer. Fish. Soc. 105, 124.

18. Hamelink, J. L., Waybrant, R. C., and Ball, R. C. 1971. Trans. Amer. Fish. Soc. 100, 207.

19. Harrison, H. L., Loucks, O. L., Mitchell, J. W., Parkhurst, D. F., Tracy, C. R., Watts, D. G., and Yannacone, V. J., Jr. 1970. Science 170, 503.

20. Hassett, J. P., and Lee, G. F. 1975. Modeling of Pesticides in the Aqueous Environment. In Environmental Dynamics of Pesticides. (R. Hague and V. H. Freed, Eds.). Plenum, New York, pp. 173-184.

21. Hunt, E. G. 1966. Biological Magnification of Pesticides. Symp. Sci. Aspects of Pest Control, Publ. #1402. Nat. Acad. Sci., Washington, D.C., pp. 251-262.

22. Jernelov, A., and Lann, H. 1971. Oikos 22, 403.

23. Kapoor, I. P., Metcalf, R. L., Hirwe, A. S., Coats, J. R., and Khalsa, M. S. 1973. J. Ag. Food Chem. 21, 310.

24. Kenaga, E. E. 1975. Partioning and Uptake of Pesticides in Biological Systems. In Environmental Dynamics of Pesticides. (R. Hague and V. H. Freed, Eds.), Plenum, New York, pp. 217-273.

25. Lagler, K. F., Bardach, J. E., and Miller, R. R. 1962. Ichthyology. Wiley, New York, p. 545.

26. Macek, K. J., and Korn, S. 1970. J. Fish Res. Board Can. 27, 1496.

27. Macek, K. J., Rodgers, C. R., Stalling, D. L., and Korn, S. 1970. Trans. Amer. Fish. Soc. 99, 689.

28. Mills, P. A., Onley, J. H., and Gaither, R. A. 1963. J.A.O.A.C. 46, 186.

29. Morgan, T. J., Landolt, R. R., and Hamelink, J. L. 1971. Behavior of Tritium in Fish Following Chronic Exposure. In Tritium. (A. A. Moghissi and M. W. Carter, Eds.). Messenger Graphics, Las Vegas, pp. 378-381.

30. Motais, R., Isaia, J., Rankin, J., and Maetz, J. 1969. J. Exp. Biol. 51, 529.

31. National Academy of Science, Committee on Water Quality Criteria. 1973. Toxic Substances. In Water Quality Criteria 1972, EPA. R3.73.033, March, p. 183.

32. Neely, W. B., Branson, D. R., and Blau, G. E. 1974. Environ. Sci. Tech. 8, 1113.

33. Olson, K. R., Bergman, H. L., and Fromm, P. O. 1973. J. Fish Res. Board Can. 30, 1293.

34. Prosser, C. L., and Brown, F. A., Jr. 1961. Comparative Animal Physiology, 2nd ed. Saunders, Philadelphia, p. 158.

35. Randall, D. J. 1970. Gas Exchange in Fish. In Fish Physiology, Vol. IV. (W.S. Hoar and D.J. Randall, Eds.). Academic Press, New York.

36. Reinert, R. E. 1972. J. Fish. Res. Board Can. 29, 1413.

37. Reinert, R. E., Stone, L. J., and Willford, W. A. 1974. J. Fish. Res. Board. Can. 31, 1649.

38. Robertson, D. E. 1968. Anal. Chem. Acta 42, 533.

39. Rudd, R. L. 1964. Pesticides and the Living Landscape. Univ. Wis. Press, Madison, Wisconsin, p. 320.

40. Scott, D. P. 1974. J. Fish. Res. Board Can. 31, 1723.

41. Spacie, A. 1975. The Bioconcentration of Trifluralin from a Manufacturing Effluent by Fish in the Wabash, River, Ph.D. thesis, August. Purdue Univ., unpublished.

42. Vanderford, M. J. 1974. Factors Affecting Pesticide and Mercury Levels in Sport Fish from Indiana Lakes and Reservoirs, M.S. thesis, August. Purdue Univ., p. 109, unpublished.

43. Veith, G. D., and Konasewich, D. E., Eds. 1975. Structure-Activity Correlations in Studies of Toxicity and Bioconcentration with Aquatic Organisms. Proc. Symp. March 11-13, 1975. Sponsored by Int. Joint Commission Res. Advisory Board, Windsor, Ontario, p. 347.

44. Waybrant, R. C. 1973. Factors Controlling the Distribution and Persistence of Lindane and DDE in Lentic Environments, Ph.D. thesis, December. Purdue Univ., p. 175, unpublished.

45. Woodwell, G. M. 1967. Sci. Amer. 216, 24.

46. Woodwell, G. M., Craig, P. P., and Johnson, H. A. 1971. Science 174, 1101.

47. Woodwell, G. M., Wurster, C. F., Jr., and Isaacson, P. A. 1967. Science 156, 821.

48. Yant, P. R. 1974. Movement and Distribution of Mercury in a Model Lake, M.S. thesis, December. Purdue Univ., p. 65, unpublished.

ADDENDUM

Since this report was prepared, several studies on the accumulation of pollutant chemicals by fish have been published. One study by Norstrom, McKinnon and deFreitas (1976) is particularly noteworthy. They also reasoned that the uptake rate of pollutants by fish would depend upon those factors which controlled metabolism and growth of the fish as we have previously described. However, they developed a bioenergetics based model to quantify pollutant accumulation by fish that provides a more rigorous method for analyzing the process than the approximation we employed. It is encouraging to note that we are in general agreement with their conclusions. We did select lower respiratory volumes and extraction efficiencies from the water than they derived. Applying these differences to our results would yield about a three fold increase in the amount of DDE and organomercury potentially available from the water. Hence these parameters need further study and their model deserves more validation. Nonetheless, we recognize the superiority of their analysis and encourage our readers to pursue the techniques they have developed.

REFERENCE

Norstrom, R. J., McKinnon, A. E., and deFreitas, A. S. W. 1976. J. Fish. Res. Board Can. 33, 248.

Dynamics of Phthalic Acid Esters in Aquatic Organisms

B. THOMAS JOHNSON, DAVID L. STALLING, JAMES W. HOGAN, and RICHARD A. SCHOETTGER
Fish-Pesticide Research Laboratory
U.S. Department of Interior
Fish and Wildlife Service
Columbia, Missouri

I. INTRODUCTION

In 1972, Mayer, Stalling, and Johnson (1972) reported residues of phthalic acid esters (PAE's) in wild and cultured stocks of fish. They found that fish caught from various parts of North America for the National Pesticide Monitoring Program contained several unknown compounds that subsequently were identified by gas chromatographic-mass spectrometric techniques as di-*n*-butyl phthalate (DBP) and di-2-ethylhexyl phthalate (DEHP). For example, channel catfish collected in agricultural and industrial areas of

Mississippi and Arkansas contained DEHP residues as high as 7500 μg/kg. (All residues mentioned herein are based on weight/weight (wet weight) unless otherwise indicated). Whole body residues of the butyl ester were 200 μg/kg in channel catfish obtained from the Fairport (Iowa) National Fish Hatchery, where water for the hatchery is obtained from the Mississippi River downstream from Rock Island and Moline, Illinois, and Davenport, Iowa.

Identification of PAE's taken up by fish from the Nation's waterways poses a number of important questions: Are PAE's continuous contaminants that are rapidly degraded in aquatic ecosystems, but where the residues in water and biota are biologically inert and reflect a steady state between input and degradation; or are they biologically important pollutants that threaten fishery resources? Answers to these questions are of paramount importance to natural resource managers and to industry.

The occurrence of PAE residues in the North American environment was reviewed at a conference on PAE's sponsored by the National Institute of Environmental Health Science (1973). Additionally, Autian (1973) and Mathur (1974) published substantial literature reviews. Metcalf, et al. (1973) have studied the fate of di-2-ethylhexyl phthalate in a model ecosystem. In the present paper, we summarize the research completed on the phthalates DBP and DEHP at the Fish-Pesticide Research Laboratory and address ourself to four questions: What are PAE's? How do they form residues in fish? Do they persist in aquatic organisms? Do they or can they affect fishery resources?

II. PAE's

PAE's are the esters of benzene ortho dicarboxylic acid; in general, they are colorless liquids, have high boiling points and low volatility, and are lipophilic (Graham, 1973). They are synthesized commercially from phthalic anhydride and the appropriate alcohol (Graham, 1973). The U.S. Tariff Commission estimated that one-half billion kilograms of 20 different phthalates were manufactured in the United States in 1972 (U.S. Tariff Commission, 1974). A tenfold increase in production of PAE's is predicted in the next three decades (Anon., 1973). Of the phthalates produced in 1972, DEHP was by far the most important; it accounted for more than 39% of the market and DBP for only about 2% (Table 1).

Although PAE's have a wide variety of uses ranging from antifoaming agents in the paper industry to perfume vehicles in cosmetic production, they are used chiefly as plasticizers with vinyl chloride polymers (PVC) (U.S. Tariff Commission, 1974).

TABLE 1. Production and Sale of Phthalate Plasticizers in the United States in 1972 (U.S. Tariff Commission, 1974)

Phthalate[a]	Percent of total market
_n_-Octyl _n_-decyl	13.7
Di-2-ethylhexyl (DEHP)	39.4
Diisodecyl	13.8
Di-_n_-butyl (DBP)	2.3

[a]Total production of 4.4×10^5 kg represented 65% of total plasticizer market.

Vinyl chloride polymers plastic formulations may contain 30 to 60% of a phthalate ester plasticizer (Graham, 1973). The plasticizers impart flexibility, workability, and extensibility to the PVC product (U.S. Tariff Commission, 1974).

III. SOURCES AND TRANSPORT OF PAE's

Since PAE plasticizers have such a diversified market in automotive, construction, clothing, home furnishings, medical, and packaging industries (Table 2), their appearance in the aquatic habitats and biota is not surprising. Within aquatic ecosystems, both DBP and DEHP residues have been detected in fish, water, and sediment (Mayer, Stalling, and Johnson, 1972; Corcoran, 1973; Hites, 1973; Lake Michigan Toxic Substances Committee, 1974). The sources of these phthalates are probably municipal and industrial effluents (Lake Michigan Toxic Substances Committee, 1974). Monitoring surveys by several Great Lake States (Lake Michigan Toxic Substances Committee, 1974) showed that the concentrations of PAE's in effluents of industrial and municipal waste treatment facilities ranged from less than 1 μg/liter to as high as 1200 μg/liter. Phthalate residues in tributaries to Lake Michigan were 1 μg/liter or less. Although the fate of PAE's discharged into these tributaries is not well defined, analysis of settleable solids showed residues ranging from 1 to 75 μg/g (dry wt.). These results suggest that PAE's may be adsorbed to particulate materials in streams and ultimately deposited in bottom sediments. Another possible means of transporting PAE's in the aquatic environment has been suggested by Ogner and Schnitzer (1970), who reported that DBP and

TABLE 2. Major Plasticizer Markets[a]

Use patterns	Amounts ($kg \times 10^6$)
Building and construction	387
Home furnishings	203
Transportation	114
Apparel	72
Food and medical packaging	46
Total	822

[a]Source: U.S. Tariff Commission (1974).

DEHP formed complexes with fulvic acid, a chemical fraction found in humic substance in soil and water. Because the phthalate-fulvic acid complex is water soluble, it is an ideal vehicle for transfer of relatively insoluble PAE's in aquatic ecosystems.

IV. TOXICITY

The acute 96-hr LC_{50} values for DBP with fathead minnow (*Pimephales promelas*), channel catfish (*Ictalurus punctatus*), rainbow trout (*Salmo gairdneri*), scud (*Gammarus pseudolimnaeus*), and crayfish (*Orconectes nais*) fell between 730 and 10,000 μg/liter (Mayer and Sanders, 1973). Although the toxicity of DEHP is more difficult to determine in static tests because it is less soluble in water, 96-hr LC_{50} values were estimated to be above 10,00 μg/liter; flow-through tests for scud (*G. fasciatus*) yielded a 9-week LC_{50} value of 210 μg/liter (McKim, 1974). The 96-hr LC_{50} values for the alcohol moieties, *n*-butanol and 2-ethylhexanol with bluegill (*Lepomis macrochirus*) were greater than 100 and 10 mg/liter, respectively (Julin, 1975). The acute toxicities of both DBP and DEHP are considerably below those of most organochlorine insecticides, which are usually toxic at concentrations between 0.1 and 50 μg/liter.

Johnson (1975) showed, with one exception, that growth of mixed bacterial cultures isolated from pond hydrosoil was not inhibited at concentrations as high as 1000 mg/liter of DBP, DEHP *n*-butanol, or 2-ethylhexanol. He cultured the bacteria on nutrient-yeast agar at 22°C and tested them against the phthalates and alcohols by the gradient plate, sensitivity disc, and broth-serial dilution methods. Selective inhibition of the growth of some colonies by 2-ethylhexanol was noted at 100 mg/liter; however, this effect was lost at the 10 mg/liter concentration. We suspect that at environmental concentrations of 1200 μg/liter (Lake Michigan Toxic Substances Committee, 1974) 2-ethylhexanol would have no effect on indigenous hydrosoil microflora and would not influence the degradation of DEHP.

V. DEGRADATION IN SLUDGE

Whether PAE's, such as DEHP and DBP, are biologically degraded in waste treatment plants or in sediment of a natural ecosystem has been partly investigated. Graham (1973) reported that laboratory scale, activated sludge processes degraded 91% of DEHP introduced within 38 hr. Similarly, Saeger and Tucker (1973) demonstrated that all PAE's tested underwent complete aerobic degradation in activated sludge and river water. However, analysis of sewage sludge from 54 municipal sewage treatment plants showed DEHP residues of 17 to 884 μg/kg (dry wt.) (Lake Michigan Toxic Substances Committee, 1974). Thus either activated sludge processes are not efficient in degrading PAE's, or raw sewage contains large quantities of them. We tend to suspect the latter.

VI. DEGRADATION IN HYDROSOIL

Laboratory incubation of ^{14}C-labeled carboxyl DBP and DEHP with pond hydrosoil suggests that natural microflora do, in time, hydrolyze the ester linkage and decarboxylate the phthalic acid moiety (Johnson and Lulves, 1975). Johnson and Lulves (1975) found a significant difference between the conditions and rates of degradation of DBP and DEHP in hydrosoil (Table 3). Within 24 hr, 46% of the [^{14}C]DBP was degraded to mono-*n*-butyl phthalate in samples incubated aerobically; after 5 days nearly 98% of all radioactivity disappeared from the soil. In contrast, 14 days were necessary for 50% of the [^{14}C]DEHP to disappear. Anaerobiosis significantly slowed or stopped the degradation of both phthalates. DBP was degraded only one-sixth as fast in hydrosoil overlaid with nitrogen, whereas no degradation of DEHP was detected after 30 days of anaerobic incubation.

Thin-layer chromatographic-autoradiographic analyses of degradation products of DBP and DEHP from hydrosoil extracts revealed the monoester and phthalic acid (Table 4). In the DBP-treated hydrosoil extracts Johnson and Lulves (1975) found several

TABLE 3. Biodegradation of [^{14}C] Di-$\underline{n}$-butyl Phthalate and [^{14}C] Di-2-ethylhexyl Phthalate in Freshwater Hydrosoil

Phthalate and incubation period (days)	Percent recovery of radioactivity from hydrosoil[a]	
	Aerobic	Anaerobic
[^{14}C] Di-n-butyl		
1	95	100
5	3*	69*
7	5*	59*
14	8*	29*
30	3*	2*
Heat-killed control		
30	100	100
^{14}C Di-2-ethylhexyl[b]		
7	100	100
14	53*	100
30	41*	100
Heat-killed control		
30	100	100

[a]Precision = 85 ± 5%; asterisk indicates significant difference from the control ($P < 0.05$, Fisher's t test).

[b]Degradation products were 2% of total radioactivity recovered.

TABLE 4. Effect of Incubation on the Recovery of Di-n-butyl Phthalate from Freshwater Hydrosoil (Johnson and Lulves, 1975)

Radioactive compound	R_f value[b]	Percent radioactivity recovered after day(s) incubation[a]				
		1	5	7	14	30
Aerobic Incubation						
Di-n-butyl phthalate	0.82	46.3 ± 5.7	85.3 ± 2.8	70.0 ± 2.1	74.0 ± 2.1	76.0 ± 6.5
Mono-n-butyl phthalate	0.33	46.3 ± 5.0	7.0 ± 2.1	26.0 ± 2.0	22.0 ± 1.5	18.6 ± 6.1
Unknown I	0.25	2.3 ± 0.3	Trace[d]	0	0	0
Unknown II	0.15	0.1 ± 0.01	0	0	0	0
Phthalic acid	0.09	0.9 ± 0.06	2.3 ± 1.2	3.0 ± 0.5	2.3 ± 0.6	3.6 ± 1.3
Origin	0.00	2.3 ± 0.05	3.3 ± 0.3	1.0 ± 0.05	1.0 ± 0.01	1.0 ± 0.01
Recovery DBP compared to control[c]		95.0	2.9	5.4	7.9	2.7
Anaerobic Incubation						
Di-n-butyl phthalate	0.82	68.3 ± 10.0	29.6 ± 8.9	29.0 ± 8.1	5.0 ± 1.5	37.6 ± 12.1
Mono-n-butyl phthalate	0.33	30.7 ± 5.0	45.3 ± 4.5	63.0 ± 9.0	85.6 ± 5.0	16.9 ± 6.5
Unknown I	0.25	T[d]	4.3 ± 0.3	1.3 ± 0.3	1.0 ± 0.03	0
Phthalic acid	0.09	0.5 ± 0.1	16.3 ± 4.6	4.6 ± 3.2	6.3 ± 2.7	35.0 ± 11.7
Origin	0.00	T	4.6 ± 1.4	T	T	9.0 ± 2.5
Recovery DBP compared to control[c]		100.0	69.1	58.5	39.1	2.0

[a] Data represent mean value of triplicate samples ($\bar{\chi}$ = ±S.E.).
[b] Solvent system petroleum ether:diethyl ether:acetic acid (77:20:3 v/v/v).
[c] Autoclave-killed control: aerobic 158,109 ± 3165 dpm and anaerobic 166.018 ± 14,758 dpm ($\bar{\chi}$ ± S.E.).

unknowns that they suspected to be hydroxy butyl phthalates; they extracted no such unknowns from hydrosoil containing DEHP. Oxygen tension did not appear to influence the degradation products of the phthalates. However, ester hydrolysis of DBP required twice as long and decarboxylation about six times longer under anaerobic conditions in the hydrosoil. After 30 days, they found no significant loss or change in DEHP in hydrosoil overlaid with nitrogen.

Johnson and Lulves (1975) attributed the rapid loss of radioactivity from the hydrosoil to decarboxylation. They believed that the ^{14}C-labeled portion of the phthalate molecule, the carbonyl group, is exposed to further degradation after initial hydrolysis of the ester-linkage. Subsequently, the labeled carbon is removed from the monoester (or phthalic acid) moiety in the form of $[^{14}C]O_2$ by decarboxylation.

Johnson and Lulves (1975) proposed a pathway for the degradation of DBP and DEHP in freshwater hydrosoil (Figure 1) based on evidence obtained by TLC-autoradiography and gas chromatography-mass spectrometry. Under either aerobic or anaerobic conditions, the diesters are initially hydrolyzed to form the phthalic acid half ester and corresponding alcohol. They believed that the monoester is later degraded to phthalic acid by continuous esterase activity in the hydrosoil, and postulated that phthalic acid is rapidly decarboxylated to form either 1,2-dihydroxy-benzene (cathechol) or benzene, but that the hydroxylated compound is the most likely product. Meikle (1972) indicated that when a carboxyl group is attached directly to the aromatic ring, the most common route of microbial attack on the molecule is by the oxidative decarboxylation pathway. Investigations of $[^{14}C]$n-butyl phthalate (monoester) and ^{14}C-labeled phthalic acid under both aerobic and anaerobic conditions tend to suggest this pathway. Comparisons of TLC-autoradiograms and radiorespirometer data of $[^{14}C]$butyl monoester extracts recovered from incubated hydrosoil indicate that the phthalate's rapid disappearance is concomitant with the appearance of phthalic acid, the loss of radioactivity from hydrosoil, and the recovery of $[^{14}C]O_2$ in respirometer samples.

There is extensive literature on the microbial destruction of aromatic nuclei and the assimilation of the carbon-fragments by microorganisms. Johnson and Lulves (1975) believe that the phthalic acid moiety undergoes ring cleavage, but they were unable to obtain substantial support for this hypothesis because the phthalic acid moiety was only ^{14}C-labeled carbonyl.

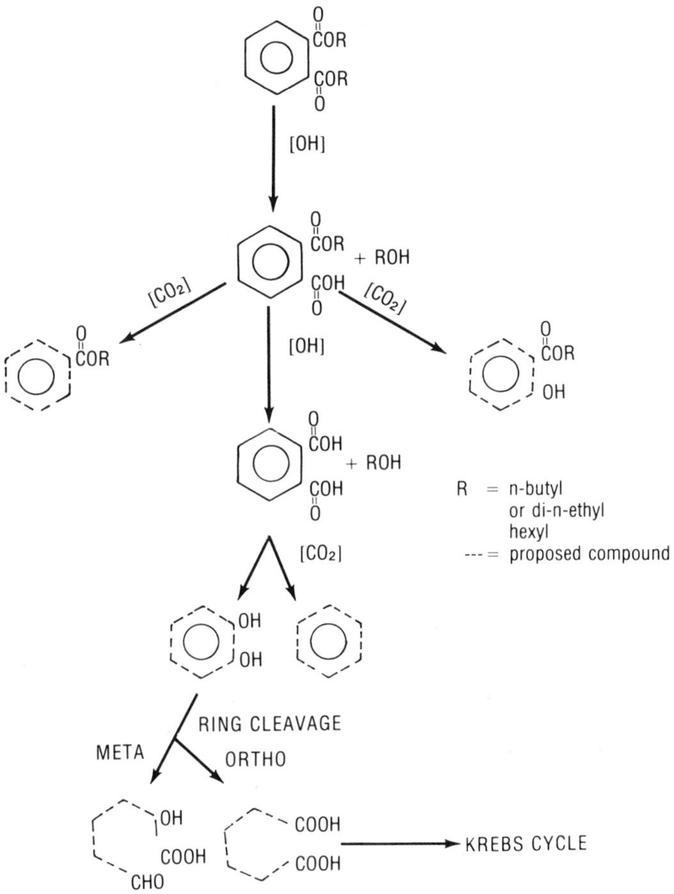

FIGURE 1. Postulated pathway for microbial degradation of DBP and DEHP in hydrosoil (Johnson and Lulves, 1975).

VII. METABOLISM IN FISH

Fathead minnows were exposed to concentrations of [^{14}C]DEHP ranging from 1.9 to 62.0 μg/liter for 56 days in a flow-through system (Mayer and Sanders, 1973); channel catfish were treated with 1 μg/liter [^{14}C]DEHP for only 24 hr under static conditions (Stalling, Hogan, and Johnson, 1973). PAE residues in tissue obtained from these fish separated and identified by TLC-autoradiographic methods, were 2-ethylhexyl phthalate, the corresponding monoester glucuronide (unknown aglycone), phthalic acid, and a phthalic acid glucuronide (Tables 5 and 6). The monoester was the predominant residue recovered from both the fathead minnows and catfish.

Channel catfish exposed to [^{14}C]DBP for 2, 4, and 48 hr rapidly degraded the parent phthalate to its monoester (Hogan, 1975). Within 4 hr, 75% of the residue extracted from the catfish was in the form of monobutyl phthalate. Employing gas chromatography-radioactivity monitoring techniques, Stalling and Hogan (1975) found that nearly 75% of the radioactivity recovered from the water exposed to fish was the monoester. DBP was rapidly hydrolyzed, and the resulting monoester was eliminated from fish. No phthalic acid or conjugate was recovered in either fish or water. After 48 hr, less than 2% of the DBP was recovered. Control samples indicated that DBP was stable in water over the 48 hr test.

In vitro, hepatic microsomes taken from male channel catfish degraded DBP 16 times more rapidly than DEHP (Table 7). At least four degradation products were recovered: the monoester, phthalic acid, and two unknown products (Stalling and Hogan, 1975). No evidence of conjugates was found.

Inhibition and cofactor experiments suggested that at least two separate microsomal enzyme systems degrade DEHP and DBP. One enzyme system is responsible for the production of at least

TABLE 5. Composition of Residues in Fathead Minnows Continuously Exposed to [^{14}C]Di-2-ethylhexyl Phthalate (DEHP) (Mayer and Sanders, 1973).

Exposure (days)	Percent radioactivity as metabolite[a,b]					
	Phthalic acid	Phthalic acid conjugate	DEHP	MEHP[c]	MEHP conjugate	Other[d]
28	5.2	3.0	49.6	37.1	0.7	4.4
56	4.9	3.7	60.0	28.7	1.4	1.3

[a]Mean of residues in groups of four fish; groups were exposed to concentrations ranging from 1.9 to 62 μg/liter DEHP.

[b]Radiolabeled components separated by TLC with Chrom AR sheets, recovered, and counted with liquid scintillation.

[c]Mono-2-ethylhexyl phthalate.

[d]Material remaining at origin after enzymatic hydrolysis.

TABLE 6. Composition of Radioactive Phthalate Metabolites in Channel Catfish Exposed to 1 μg/liter of Di-2-ethylhexyl Phthalate for 24 hr[a] (Stalling, Hogan, and Johnson, 1973).

Radioactive fractions	Percent composition
Di-2-ethylhexyl phthalate	14.0
Mono-2-ethylhexyl phthalate	66.0
Phthalic acid	4.0
Conjugates	
Phthalic acid	0.3
Mono-2-ethylhexyl phthalate	13.7
Other	2.0

[a]Total residue equivalent to 2.6 μg/g di-2-ethylhexyl phthalate.

TABLE 7. Metabolism of Phthalic Acid Esters in a Hepatic Microsomal Preparation from a Male Channel Catfish (Stalling, Hogan, and Johnson, 1973).

Radioactive fraction	Percent total radioactivity recovered	
	Di-*n*-butyl phthalate	Di-2-ethylhexyl phthalate
Monoester	55	1
0, 1, and 2[a]	42	5
Parent compound recovered	3	94

[a]Phthalic acid and unknown 1 and 2, respectively.

two unknowns, is carbon monoxide-sensitive, and requires NADPH (Tables 8 and 9). The other enzyme system responsible for production of the monoester is inhibited by diisopropyl fluorophosphate (DFP). This latter system is not inhibited by carbon monoxide and does not require NADPH. From these data, Stalling, Hogan, and Johnson (1973) concluded that the unknowns are formed by a mixed function oxidase, whereas the monoester and probably phthalic acid are produced by esterase activity.

TABLE 8. Effects of NADPH and O_2 on Metabolism of Phthalic Acid Esters by Hepatic Microsomal Enzymes from Male Channel Catfish (Stalling, Hogan, and Johnson, 1973).

Metabolites	Di-n-butyl phthalate (dpm)	Di-2-ethylhexyl phthalate (dpm)
Monoester		
+NADPH, + O_2	36,234	5676
-NADPH, + O_2	31,376	4776
-NADPH, - O_2	27,662	5498
+NADPH, - O_2	34,569	4259
0, 1, and 2[a]		
+NADPH, + O_2	21,125	6130
-NADPH, + O_2	1992	864
-NADPH, - O_2	2955	478
+NADPH, - O_2	26,980	2768

[a]Phthalic acid and unknowns 1 and 2.

TABLE 9. Effect of Air, Nitrogen and Carbon Monoxide on Metabolism of Phthalic Acid Esters by Hepatic Microsomal Enzymes from Male Channel Catfish (Stalling, Hogan, and Johnson, 1973).

	Di-n-butyl phthalate (dpm)			Di-2-ethylhroxyl phthalate (dpm)		
Metabolites	Air	N_2	CO	Air	N_2	CO
Monoester	36,234	34,569	29,423	5676	4234	5724
0,1 and 2[a]	21,125	26,980	6436	6130	2768	639

[a]Phthalic acid and unknowns 1 and 2.

VIII. ACCUMULATION AND ELIMINATION

Residue concentrations of DBP in fish from several areas of North America have ranged from nondetectable to 0.5 μg/g, and those of DEHP have ranged as high as 3.2 μg/g (Mayer, Stalling, and Johnson, 1972; Stalling, Hogan, and Johnson, 1973). PAE residues in fish from the Great Lakes region range from nondetectable to 1.3 μg/g (Lake Michigan Toxic Substances Committee, 1974). However, an additional 30 to 50% as much residue may also be present in fish in the form of the monoester or conjugates of the monoester and phthalic acid (Mayer, Stalling, and Johnson, 1972; Mayer and Sanders, 1973). Mayer and Sanders (1973), who exposed fathead minnows to 1.9 μg/liter of DEHP for 56 days, found that residues reached an equilibrium concentration of 2.6 μg/g within 28 days. Thus the accumulation factor was nearly 1400, which agrees well with data for DEHP in bluegills exposed to 0.1 μg/liter (Johnson, 1975). Mayer (1976), however, found that the accumulation factor for DEHP in fathead minnows was reduced to 160 when the fish were exposed to the higher concentration of 60 μg/liter. Accumulation factors for DEHP and DBP in aquatic crustacea and insects were generally between 350 and 3900 after exposures to concentrations of the phthalates from 0.08 to 0.3 μg/liter (Mayer and Sanders, 1973). Accumulation factors are calculated according to the Federal Register Proposed Rules of 1975 (Anon., 1975).

In model food chains tested at the Fish-Pesticide Research Laboratory in either the 96-hr model (fish-daphnid) or 30-day model (fish-daphnid-microorganism), the water and food accumulation factor was less than 25 for DEHP (DDT = 100) (Johnson, 1974).

When fish and invertebrates containing PAE residues are placed in untreated water, they eliminate 50% of the residue within 3 to 7 days. Recently, Johnson (1975) found that daphnids (*Daphnia magna*) rapidly eliminated both [^{14}C]DEHP and [^{14}C]DBP after exposure to 100, 200, and 1000 ng/liter (Johnson, 1975). The daphnids contained less than 50% of the original phthalate residues in 1 hr and less than 10% within 24 hr. Residues in fish and invertebrates have not been correlated with adverse biological effects.

IX. CHRONIC TOXICITY

The chronic toxicities of DEHP and DBP have not been as well defined as is desired. Studies thus far completed suggest, however, that both are biologically active well below acutely toxic concentrations. McKim (1974) reported that growth of brook trout (Salvelinus fontinalis) was reduced significantly when the fish were held at a DBP concentration of 300 μg/liter, but not at 90 μg/liter. However, aquatic invertebrates appear to be more sensitive than fish. Reproduction in daphnids was impaired by DBP and DEHP concentrations of 20 and 3 to 5 μg/liter, respectively (Mayer and Sanders, 1973; McKim, 1974). The emergence of adult midges (Chironomus tetans) was reduced significantly at a DEHP concentration of 14 μg/liter (McKim, 1974). However, exposure of a different species of midge, C. plumosus, through a complete life cycle to DEHP at concentrations greater than 100 mg/liter did not significantly reduce adult emergence or egg hatchability (Streufert, Jones and Sanders, 1977).

X. SUMMARY

1. Residues of di-n-butyl phthlate (DBP) and di-2-ethylhexyl phthalate (DEHP) are frequently found in fish from various parts of North America.
2. Important sources of phthalates in water are undoubtedly municipal and industrial effluents.
3. PAE's are probably adsorbed by suspended solids and deposited in bottom sediments. Both DEHP and DBP can be transported in water via a water-soluble fulvic acid-phthalate complex.
4. In general, brief exposure of fish and aquatic invertebrates to DEHP and DBP concentrations below 500 μg/liter should not pose an acute toxicity hazard.
5. DEHP and DBP appear to be more readily degraded aerobically than anaerobically in pond hydrosoils. The monoesters and phthalic acid are the major degradation products. Phthalic acid is rapidly decarboxylated in hydrosoil.
6. In fish, the monoester is the predominant metabolite of DEHP and DBP. In vivo, enzymatic production of the monoesters is inhibited by DFP. Phthalic acid and the corresponding monoester are also found as glucuronide conjugates.
7. An accumulation factor of 1400 was found for bluegill exposed to 0.1 μg/liter DEHP for 30 days.
8. In a 96-hr or 30-day compartmentalized food chain, DEHP had an accumulation factor of only 2 to 5 (DDT = 100).
9. Residues of DEHP are at least partly eliminated by fish and invertebrates that are placed in untreated water. Daphnids rapidly eliminate DEHP and DBP.
10. Chronic exposures of fish to DBP concentrations above 90 μg/liter may be detrimental.
11. Reproduction in daphnids and midges is probably affected by continuous exposure to DEHP concentrations of 3 to 5 and 14 μg/liter, respectively.

REFERENCES

1. Anonymous. 1973. Plastics Heading for a Boom. Chem. Ind. 13, 597.

2. Anonymous. 1975. Proposed Rules. Fed. Reg. 40, 26906.

3. Autian, J. 1973. Toxicity and Health Threats of Phthalate Esters: Review of the Literature. Environ. Health Perspect. 4, 3-26.

4. Corcoran, E. F. 1973. Gas Chromatographic Detection of Phthalic Acid Esters. Environ. Health Perspect. 3, 13-15.

5. Graham, P. R. 1973. Phthalate Ester Plasticizers - Why and How They Are Used. Environ. Health Perspect. 3, 3-12.

6. Hites, R. 1973. Phthalates in the Charles and Merrimack Rivers. Environ. Health Perspect. 3, 17-21.

7. Hogan, J. W. 1975. Unpublished data, Fish-Pesticide Research Laboratory, Columbia, Missouri.

8. Johnson, B. T. 1974. Aquatic Food Chain Models for Estimating Bioaccumulation and Bio-degradation of Xenobiotics. Proc. Inter. Confer. Transp. Persistent Chem. Aquat. Ecosystems, Ottawa, Canada IV, Nat. Res. Council Can., p. 17-22.

9. Johnson, B. T. 1975. Unpublished data, Fish-Pesticide Research Laboratory, Columbia, Missouri.

10. Johnson, B. T. and Lulves, W. 1975. Biodegradation of di-n-butyl Phthalate and di-2-ethylhexyl Phthalate in Freshwater Hydrosoil. J. Fish. Res. Board Can. 32(3), 333-339.

11. Julin, A. 1975. Personal communication, Fish-Pesticide Research Laboratory, Columbia, Missouri.

12. Lake Michigan Toxic Substances Committee. 1974. Phthalates. In Report of the Lake Michigan Toxic Substances Committee, Environmental Protection Agency, Regional Office, Chicago, Illinois.

13. Mathur, S. P. 1974. Phthalate Esters in the Environment: Pollutants or Natural Products? J. Environ. Qual. 3(3), 189-193.

14. Mayer, F. L. 1976. Residue Dynamics of Di-2-ethylhexyl Phthalate in Fathead Minnows (*Pimephales promelas*). *J. Fish. Res. Board Can.* *33(11)*, 2610-2613.

15. Mayer, F. L., Jr. and Sanders, H. O. 1973. Toxicology of Phthalic Acid Esters in Aquatic Organisms. *Environ. Health Perspect.* *3*, 153-158.

16. Mayer, F. L., Jr., Stalling, D. L. and Johnson, J. L. 1972. Phthalate Esters as Environmental Contaminants. *Nature* *238* (5346), 411-413.

17. McKim, J. 1974. Personal communication, Environmental Protection Agency National Water Quality Laboratory, Duluth, Minnesota.

18. Meikle, R. W. 1972. Decomposition: Qualitative Relationship. In Organic Chemicals in Soil Environment (C.A.I. Goring and J.W. Hamaker, eds.), Dekker, New York, pp. 145-251.

19. Metcalf, R. L., Booth, G. M., Schuth, C. K., Hansen, D. J. and Lu, P. Y. 1973. Uptake and Fate of di-2-ethylhexyl Phthalate in Aquatic Organisms and in a Model Ecosystem. *Environ. Health Perspect.* *4*, 27-34.

20. National Institute of Environmental Health Sciences. 1973. Perspectives on PAE's. U.S. Department Health, Education and Welfare. *Environ. Health Perspect.*, Exp. Issue No. 3, p. 182a.

21. Ogner, G. and Schnitzer, M. 1970. Humic Substances: Fulvic Acid-dialkyl Phthalate Complexes and Their Role in Pollution. *Science* *170*, 317-318.

22. Saeger, V. W. and Tucker, E. S. 1973. Phthalate Esters Undergo Ready Biodegradation. *Plast. Eng.* *29*, 45-49.

23. Stalling, D. L. and Hogan, J. W. 1975. RAM-GC-MS-Computer (RGMC) in Environmental Metabolism Studies. Presented at 169th National Meeting Amer. Chem. Soc., Philadelphia, Pennsylvania, April 11, 1975.

24. Stalling, D. L., Hogan, J. W. and Johnson, J. L. 1973. Phthalate Ester Residues - Their Metabolism and Analysis in Fish. Environ. Health Perspect 3, 159-173.

25. U.S. Tariff Commission. 1974. Synthetic Organic Chemicals: United States Production and Sales, 1972. U.S. Tariff Commission, 681.

ADDENDUM

Recently, Fish, Jones and Johnson (1977) demonstrated that DEHP undergoes ring cleavage during biodegradation in lake hydrosoil. In laboratory studies they incubated ^{14}C-ring labeled DEHP with hydrosoil in a radiorespirometer. After 28 days of aerobic incubation they recovered 20% of the introduced radioactivity as $^{14}CO_2$. This would suggest ultimate biodegradation and probable use of DEHP as a carbon source by indigenous hydrosoil microflora.

Mutz, Jones and Johnson (1977) have demonstrated in a laboratory hydrosoil microcosm that 1.0 mg/liter DEHP or 2-ethylhexanol did not significantly affect ($P > 0.05$) the total microbial heterotroph population. At these concentrations they detected no inhibition of nitrification, sulfur oxidation or ammonification in the lake hydrosoil.

REFERENCES

1. Mutz, R. C., Jones, J. R., and Johnson, B. T. 1977. The Effects of Phthalate Esters on Nutrient Cycling in Freshwater Hydrosoil. Presented at 77th Annual Meeting of the American Society of Microbiology, New Orleans, La. May 10, 1977.

2. Fish, T. D., Jones, J. R. and Johnson, B. T. 1977. Characteristics of Various Hydrosoils and Their Relative Importance in the Biodegradation of Di-2-ethylhexyl Phthalate (DEHP). Presented at 77th Annual Meeting of the American Society for Microbiology, New Orleans, La. May 10, 1977.

3. Streufert, J. M., Jones, J. R., and Sanders, H. O. 1977. Chronic Effects of Two Phthalic Acid Esters on Midge (*Chironomus plumosus*). Presented at 25th Annual Meeting of the North American Benthological Society, Roanoke, Va. April 7, 1977.

Bone Development and Growth of Fish as Affected by Toxaphene

PAUL M. MEHRLE AND FOSTER L. MAYER
Fish-Pesticide Research Laboratory
Fish and Wildlife Service
U.S. Department of the Interior
Columbia, Missouri 65201

I. INTRODUCTION

Toxaphene is an organochlorine insecticide used extensively alone and in combination with other insecticides in agricultural areas of the United States. An estimated 30 to 40 million pounds are applied annually on crops and livestock (Hercules, Inc., 1970). Its primary use is as an insecticide on cotton, but it is also registered for use on certain grains, alfalfa, fruit, and vegetables. Toxaphene is the technical grade of chlorinated camphene containing 67 to 69% chlorine and consists of about 175 isomers (Casida et al., 1974). It is a recognized contaminant in the aquatic environment; residues in fish ranging from 0.5 to 48 μg/g have been reported by the National Pesticide Monitoring Program (personal communication, B. L. Berger, Division of Population Regulation, U.S. Fish and Wildlife Service, Washington,

D.C.). Only fish from the southern United States have been found to contain toxaphene; the residues most likely originated from spraying and agricultural runoff.

Although toxaphene is a very effective insecticide on target pests, its effects on nontarget organisms, such as fish, have not been thoroughly elucidated. Since toxaphene can be a persistent contaminant in the aquatic environment, we attempted to evaluate the effects of low concentrations of toxaphene on growth and backbone development of brook trout (*Salvelinus fontinalis*), fathead minnows (*Pimephales promelas*), and channel catfish (*Ictalurus punctatus*). These fish represent three important species of fish. Brook trout are an important sport fish as is the channel catfish which is also of considerable economic importance to fish farmers. The fathead minnow is an important forage and commercial bait fish.

II. MATERIALS AND METHODS

Brook trout, fathead minnows, and channel catfish were continuously exposed to toxaphene in water via proportional diluter systems modeled after Mount and Brungs (1967), as modified by McAllister, Mauck, and Mayer (1972). The diluter systems delivered five concentrations of toxaphene and a control with a dilution factor of 0.5 between the concentrations. We used flow-splitting chambers designed by Benoit and Puglisi (1973) to thoroughly mix and divide each toxaphene concentration for delivery to duplicate exposure tanks. Artificial daylight was provided by the method of Drummond and Dawson (1970) and the water temperatures (Table 1) were maintained at $\pm 0.2^{o}$C. The experimental conditions are summarized in Table 1.

Acetone, which was used as the carrier solvent, did not exceed 0.28, 0.28, or 0.11 mℓ/liter in the brook trout, fathead minnow, and channel catfish studies, respectively. An experimental use sample of toxaphene (X-16189-49), furnished by Hercules, Inc., was used throughout the study. Toxaphene concentrations in water were monitored weekly and residues in fish were determined at the intervals indicated in Table 1. The analytical methodology used was that of Stalling and Huckins (1975).

Bone development was assessed by determining the collagen, calcium, and phosphorus concentrations in the backbone (Mayer, et al., 1975; Mehrle and Mayer, 1975 a,b). The backbone was removed from each fish, dried at 110°C for 2 hr in a forced air oven, split into two fractions, and weighed. Collagen was isolated from one fraction and calcium and phosphorus were determined from the other. In addition, fathead minnows were

TABLE 1. Experimental Conditions and Sampling Periods During Continuous Exposures of Brook Trout, Fathead Minnows, and Channel Catfish to Toxaphene.

Experimental condition	Species		
	Brook trout[a]	Fathead minnows[a]	Channel catfish
Water temperature (°C)	9	25	Adults 16-26, fry 26
Age at exposure initiation	Eyed eggs	10 days	Adults (2.5 yr)
Exposure duration	22 days before hatching; fry, 90 days	150 days	Adults 100 days, before spawning; fry, 90 days
Toxaphene concentration range (ng/l)	39-502	94-1420	49-630
Residue analyses (days)	15, 30, 60, 90	30, 60, 90, 150	15, 30, 60, 90
Growth determinations (days)	30, 60, 90	30, 60, 90, 150	5, 30, 60, 90
Bone development determinations (days)	7, 15, 30, 60, 90	150	90
X-ray analyses (days)	None	150[b]	90[c]

[a] Detailed experimental conditions for brook trout and fathead minnows were given by Mayer *et al.* (1975) and Mehrle and Mayer (1975a,b).

[b] Fish were subjected to sublethal electrical shock before X-ray.

[c] Fish were X-rayed before and after sublethal electrical shock.

subjected to a sublethal electrical shock (three 1-sec, 60-V stimuli at 20-sec intervals) and then X-rayed. Channel catfish fry were X-rayed before and after the sublethal electrical shock. The radiographs were examined for alterations in vertebral structures.

Randomized block designs were used and the data were analyzed by analysis of variance (Snedecor, 1965). A multiple means comparison test (least significant difference) was used to compare significant differences among toxaphene concentrations. The level of significance selected was $P < 0.05$.

III. RESULTS

A. Brook Trout

Exposure to toxaphene had no effect on hatchability of brook trout eggs or on the growth of fry during the 15 days after hatching. However, collagen content, as estimated by direct measurement of hydroxyproline in the protein fraction, was significantly decreased after both 7 and 15 days in all fish except those in the 39 ng/liter exposure.

All of the fry exposed to 502 ng/liter of toxaphene died within 30 days after hatching, and those exposed to 288 ng/liter, within 60 days. Growth of fry exposed to three lower concentrations (39-139 mg/liter) declined significantly within 90 days (Figure 1). Collagen in brook trout backbones was also significantly reduced after 30, 60, and 90 days of exposure in all toxaphene concentrations (Table 2). Effects were more pronounced after 90 days than after 30 or 60 days. The results were similar to those from the 7- and 15-day whole body samplings; that is, collagen content was decreased by toxaphene. Phosphorus decreased significantly in the bone of fish from the 139-ng/liter exposure after 60 days, but calcium content was not changed. The reason for the decrease in phosphorus concentration at 60 days and the later increase at 90 days is not known.

In addition to the decrease in collagen content, both the calcium and phosphorus concentrations were about doubled in fish exposed for 90 days to the 68- and 139-ng/liter concentrations. A trend was also observed toward an increase in calcium and phosphorus in fish exposed to 39 ng/liter (Table 2). The ratio of minerals to organic content, that is, calcium + phosphorus to collagen (Ca + P/collagen) in the backbone of the control group was 1.1, whereas it was 1.9, 2,5, and 2.5 in the 39-, 68-, and 139-ng/liter treatment groups, respectively.

Uptake of toxaphene by fry in all concentrations was highest after 15 days of exposure, declined through 60 days, and then

TABLE 2. Mean Percentage of Dried Backbone Composed of Collagen, Calcium, and Phosphorus in Three Species of Fish Exposed Continuously to Different Concentrations of Toxaphene for 90 to 150 days. (Each value is an average for 6 to 10 fish).

Species and (in parenthesis) days of exposure, and toxaphene concentration (ng/liter)	Backbone constitutent		
	Collagen	Calcium	Phosphorus
Brook Trout (90)			
0	19.5	10.2	10.6
39	16.4[a]	15.3	15.6
68	16.2[a]	20.6[a]	20.3[a]
139	16.4[a]	21.1[a]	19.9[a]
Fathead minnows (150)			
0	32.3	11.8	4.1
94	26.9[a]	15.1[a]	3.6
205	22.9[a]	23.3[a]	3.7
399	19.9[a]	20.8[a]	3.8
727	22.4[a]	24.3[a]	3.5
Channel catfish (90)			
0	26.6	6.5	6.4
49	26.0	8.1[a]	6.6
72	24.1[a]	10.9[a]	6.3
129	24.1[a]	9.5[a]	6.0[a]
299	23.5[a]	8.5[a]	5.7[a]
630	23.2[a]	8.1[a]	5.4[a]

[a]Significantly different from controls, $P < 0.05$.

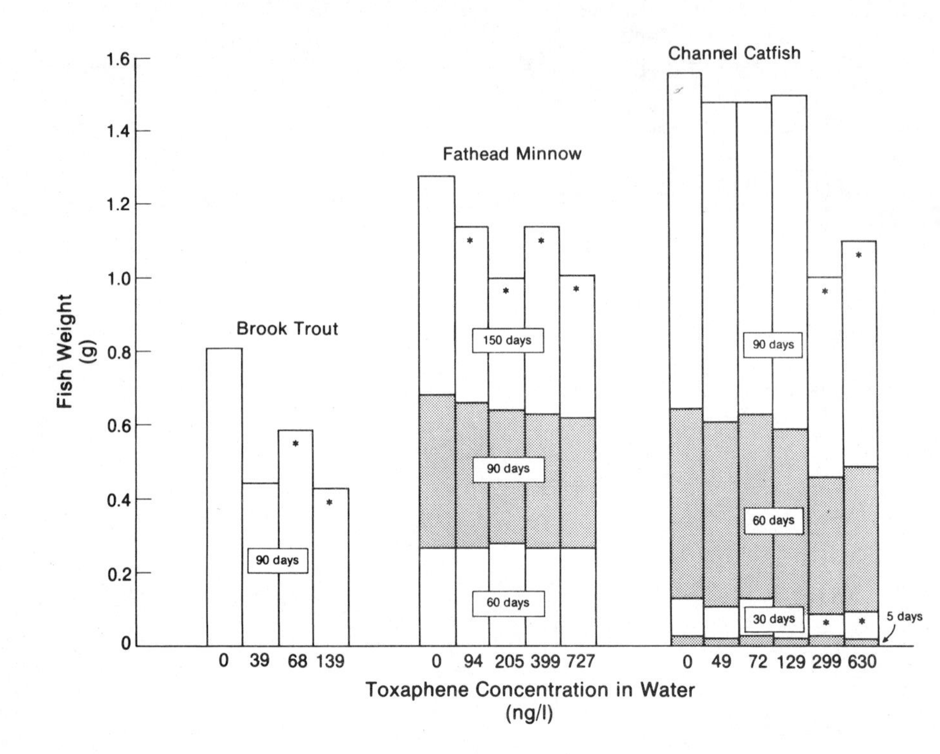

FIGURE 1. Weights of fry of brook trout, fathead minnows, and channel catfish after being exposed to toxaphene (* indicates significant difference from controls, $P < 0.05$).

increased again between 60 and 90 days. The amount of toxaphene in fry at 90 days was 15,000 to 20,000 times that to which they were exposed (Figure 2). Toxaphene residues in eggs were not determined.

B. Fathead Minnows

All concentrations of toxaphene significantly decreased fish growth within 150 days of exposure, but not within 60 or 90 days (Figure 1). Biochemical analyses of the backbone and the length-weight measurements substantiated the same effects of toxaphene after 150 days of exposure, that is, growth and bone development

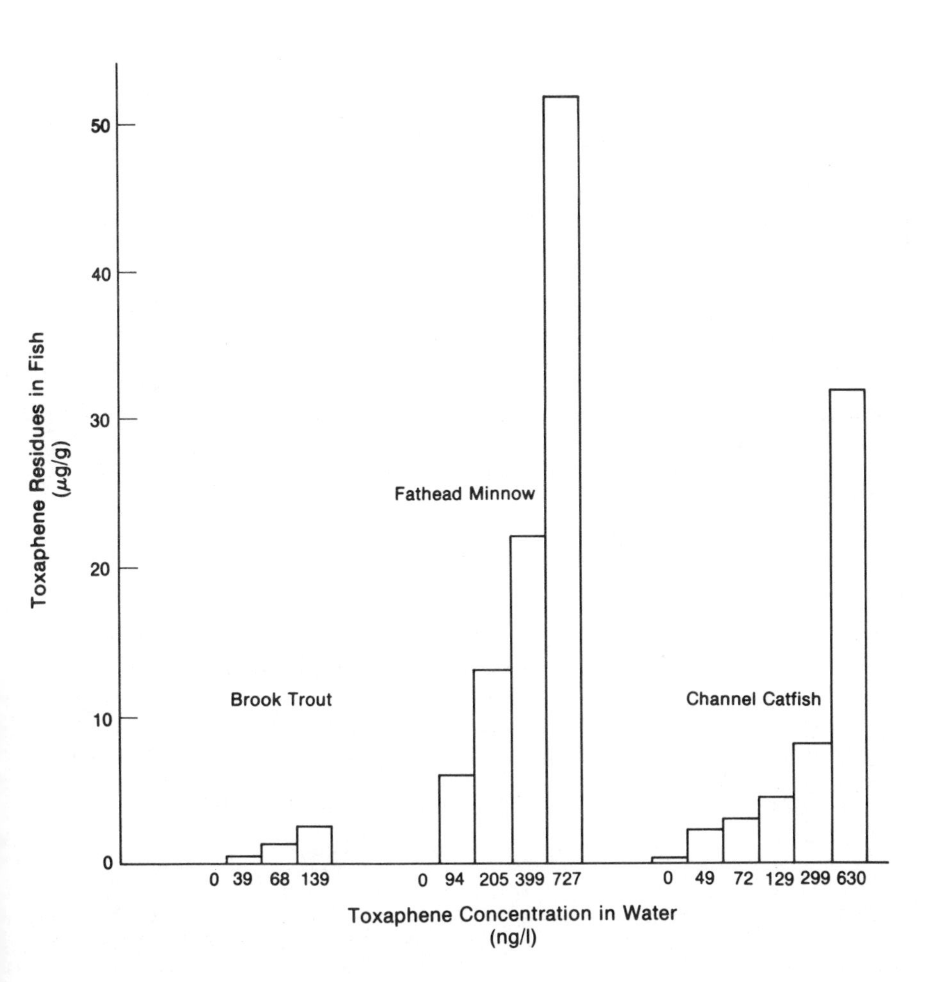

FIGURE 2. Toxaphene residues in fry of brook trout, fathead minnows, and channel catfish after exposures of 90, 150, and 90 days, respectively, to different concentrations of toxaphene.

were significantly altered (Table 2). Collagen content of the backbone was significantly decreased by toxaphene, whereas the amount of calcium was significantly increased. Although the concentration of phosphorus in bone was not changed by toxaphene exposure, a decreasing trend was evident. The Ca + P/collagen ratio was 0.50 in the backbones of control group, whereas it was 0.69, 1.2, 1.2, and 1.2 in the groups receiving 94, 205, 399, and 727 ng/liter of toxaphene, respectively.

Because toxaphene caused a decrease in the collagen content of backbone and a concomitant increase in calcium, we hypothesized that the backbone might be more fragile, and therefore less tolerant to stress. The results of subjecting six fish of each group to an electrical shock supported our hypothesis that toxaphene also altered the quality of the backbone (Figure 3). The backbones of four fish from each group exposed to toxaphene were broken in several locations, whereas those of control fish were not affected. These observations confirmed that toxaphene caused the backbone to be fragile and more susceptible to stress.

Toxaphene residues in fathead minnows after 150 days of exposure to 94, 205, 399 and 727 ng/liter of toxaphene were 5.9, 13, 22, and 52 μg/g, respectively (Figure 2). The residues in fish represent accumulation factors of 70,000 to 107,000.

C. Channel Catfish

Toxaphene did not affect growth and spawning activity of adult channel catfish, or egg hatchability. However, within 30 days after hatching, toxaphene concentrations of 224 and 535 ng/liter had caused a significant increase in fry mortality as well as a significant decrease in growth (Figure 1). The same effects on mortality and growth were observed 60 and 90 days after hatching.

Toxaphene concentrations of 72, 129, 299, and 630 ng/liter significantly decreased collagen and increased calcium in the backbone of catfish fry after 90 days of exposure (Table 1). The results are similar to those we obtained with fathead minnows and brook trout; the decrease in organic matrix and increase in mineral content caused the backbone to be fragile and perhaps susceptible to being broken.

X-ray analyses showed that the backbone was altered by exposure to toxaphene (Figure 4) in at least five of the eight fish X-rayed in the groups exposed to 72, 129, 299, and 630 ng/liter. These changes in backbone structure were not dependent upon toxaphene concentration nor upon the intensity of the electrical shock; toxaphene alone caused the backbone changes, and no dose-dependent response was apparent.

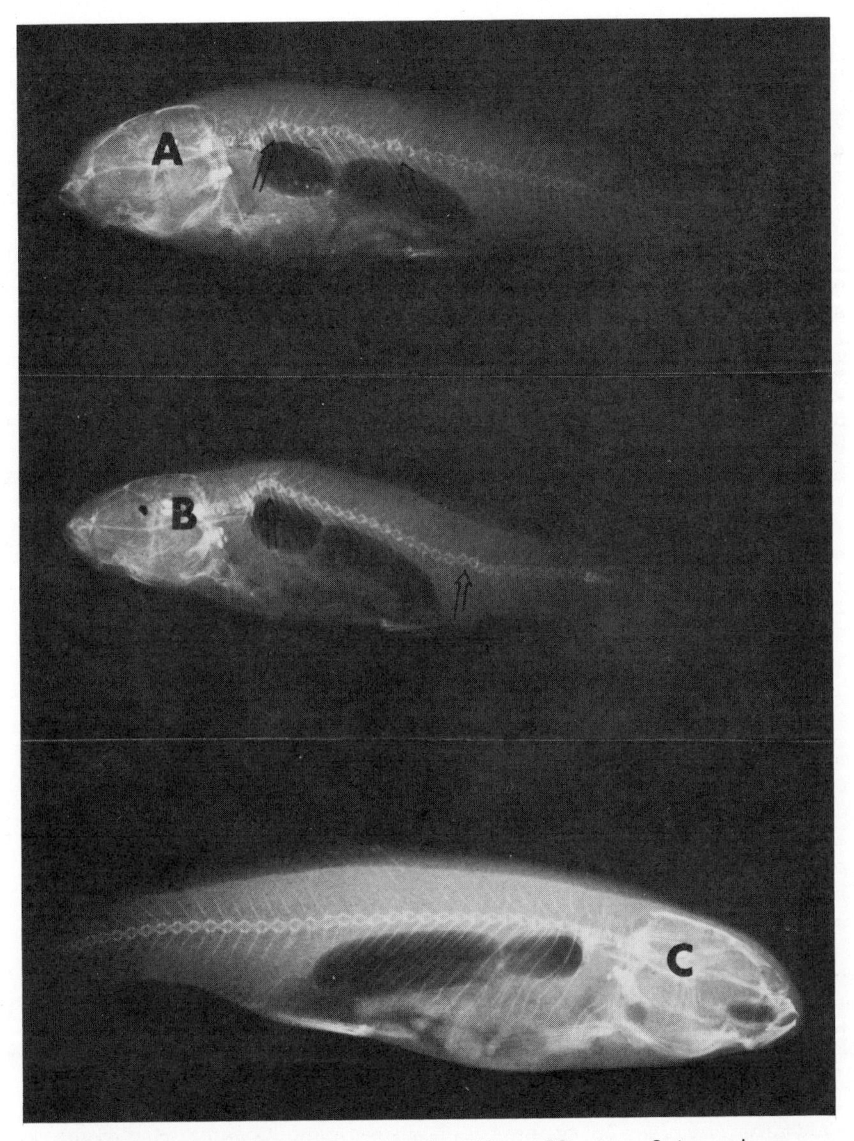

FIGURE 3. Radiographs showing the effects of toxaphene on backbone structure of fathead minnows. A and B represent fish exposed to 94 ng/liter of toxaphene and C a control fish. (Arrows indicate areas of backbone affected. (After Mehrle and Mayer, 1975b. Reprinted with permission from the Journal of the Fisheries Research Board of Canada.

FIGURE 4. Radiographs showing the effects of toxaphene on backbone structure of channel catfish. A represents a fish exposed to 630 ng/liter, B a fish exposed to 72 ng/liter and C a control fish.

Channel catfish fry accumulated toxaphene more rapidly than did adults. After 100 days of exposure, adult catfish accumulated about 54,000 times the lowest toxaphene concentration (49 ng/liter) and 20,000 times the highest concentration (630 ng/liter). The concentration of toxaphene by fry was 40,000 to 91,000 times that to which they were exposed (Figure 2).

IV. DISCUSSION

The backbone consists of an organic matrix, minerals, and water; about 90% of the organic matrix is the protein, collagen, and the remaining organics are mucopolysaccharides, mucoproteins, and lipids (Nusagens, Chantraine, and Lapiere, 1972). Calcification and mineralization occur around and within the collagen fibrils during development and maturation of bone. Most of the mineral fraction consists of calcium, phosphate, and hydroxide crystals called hydroxyapatite $[Ca_{10}(OH)_2(PO_4)_6]$. Determining the effects of toxaphene on collagen, calcium, and phosphorus concentrations in the backbone thus provided an indication of its effects on bone development and bone quality. Interpretation of our data strongly suggests that toxaphene exposure inhibited collagen synthesis. This inhibition depressed backbone collagen in fry and, concomitantly, increased mineralization of the backbones. We do not know whether toxaphene directly altered mineral metabolism.

Incidences of backbone anomalies have been reported in pond-cultured channel catfish by Sneed (1972). Three to five percent of the fish he examined exhibited symptoms such as fractured backbones and the inability to coordinate swimming efforts and maintain normal body position. The causes of this condition were not established, but the possibilities that it was induced by a nutritional deficiency or a pesticide were considered. Several investigators have shown that both nutritional deficiencies and pesticides can induce "broken-back" condition in fish. McCann and Jasper (1972) observed broken backs in bluegills (*Lepomis macrochirus*) within 4 hr after exposure to high concentrations of organophosphate insecticides. Backbone anomalies were observed at concentrations above and below the LC50 for six different organophosphate compounds; however, the broken backs in their study were caused, most likely, by acute muscular contractions rather than by an alteration of bone constituents.

Dietary sources of phosphate, calcium, and the vitamins D and C are critical for proper bone development (Somogyl and Kodicek, 1969). Barnes et al. (1970) reported reduced collagen formation in the skin of scorbutic guinea pigs, and Peterkofsky (1972) showed that vitamin C was required for collagen formation in chick embryo fibroblasts. The synthesized collagen in the fibroblasts under a low vitamin C condition was deficient in hydroxylated amino acids. Wilson and Poe (1973) reported a decrease in collagen content of backbones and a broken-back condition in channel catfish fed a vitamin C deficient diet. Hydroxyproline in the backbone was also decreased. The investigators suggested that the decrease in collagen caused the backbone to be more fragile. Halver, Ashley, and Smith (1969) observed a similar condition in coho salmon (Oncorhynchus kisutch) and rainbow trout (Salmo gairdneri) fed vitamin C deficient diets. In addition, wound repair mechanisms in the skin of these fish were delayed or inhibited because collagen synthesis was impaired. Mehrle and Mayer (1975b) found that the concentration of hydroxyproline, as well as that of seven other amino acids, was significantly decreased in collagen from the backbone of the fathead minnows exposed to toxaphene. On the basis of the amino acid content of collagen in the fish exposed to toxaphene as compared with the amino acid profile of collagen given by Wilson and Poe (1973), toxaphene could be altering several intermediary metabolic pathways associated with collagen and bone metabolism.

We suggest from our results that interference by toxaphene with vitamin C metabolism resulted in impairment of collagen synthesis. Competition between structures or functions for vitamin C could have resulted in a functional deficiency of vitamin C. Wagstaff and Street (1971) showed that vitamin C was necessary for induction of microsomal hydroxylative enzymes that are responsible for detoxification of pesticides and other xenobiotics. Since the use of vitamin C for detoxification of toxaphene was perhaps necessary for survival, a functional deficiency of vitamin C for development and maintenance of bones was induced. If this hypothesis is correct, other xenobiotics and environmental stresses that are related to these detoxification pathways can alter bone metabolism, as toxaphene did. However, further studies of the relationship among bone development, detoxification mechanisms, vitamin C, and xenobiotic metabolism in all three species are needed before definite conclusions can be made.

The whole-body residues of toxaphene in brook trout, channel catfish, and fathead minnows were in the range of those reported in fish by the National Pesticide Monitoring Program. Only the fish in the group exposed to the highest toxaphene concentration in the fathead minnow study were outside that range (0.5 to 48 µg/g). Brook trout, which were exposed at 9°C, accumulated less toxaphene than the other two species which were exposed at 25 to 26°C. Although brook trout accumulated less toxaphene, they had a higher percent of mortality during the exposure than fathead minnows or channel catfish. However, the sublethal effects on bone composition and growth occurred at all concentrations in the brook trout and fathead minnow studies and the "no effect" concentration of toxaphene on growth and bone development was less than 39 and 94 ng/liter for brook trout and fathead minnows. The "no effect" concentration for growth of channel catfish was between 129 and 299 ng/liter, and between 49 and 72 ng/liter for bone composition. Thus, the effect of toxaphene on growth appeared to be relatively less in channel catfish than in the other two species; yet all three species exhibited altered bone composition when exposed to similar concentration ranges of toxaphene. Additional studies are needed to delineate the "no effect" concentrations for brook trout and fathead minnows so that more adequate comparisons among species can be made.

Since the whole-body residues of toxaphene in fish from our studies were similar to those observed in fish from natural environments, the results are probably realistic indications of the adverse influences of toxaphene on essential biological processes of growth and development. Our studies further indicate that descriptions of the environmental fate of residues should include an evaluation of the impact of the pollutant on such critical biological processes as growth and development, if the full significance of the residues is to be ascertained.

REFERENCES

1. Barnes, M. J., Constable, B. J., Morton, L. F., and Kpopicek. 1970. Biochem. J. 119, 575.

2. Benoit, D. A. and Puglisi, F. A. 1973. Water Res. 7, 1915.

3. Casida, J. E., Holmstead, R. L., Khalifa, S., Know, J. R., Ohsawa, T., Palmer, K. J. and Weng, R. Y. 1974. Science 183, 520.

4. Drummond, R. A. and Dawson, W. F. 1970. Trans. Amer. Fish. Soc. 99, 343.

5. Halver, J. E., Ashley, L. M. and Smith, R. R. 1969. Trans. Amer. Fish. Soc. 98, 762.

6. Hercules, Inc. 1970. Toxaphene: Use Patterns and Environmental Aspects, Hercules, Inc., Wilmington, Delaware, November.

7. Mayer, F. L., Mehrle, P. M. and Dwyer, W. P. 1975. Toxaphene Effects on Reproduction, Growth, and Mortality of Brook Trout, Ecological Research Series, U. S. Environmental Protection Agency, EPA-600/3-75-013, Duluth, Minn.

8. McAllister, W. A., Mauck, W. L. and Mayer, F. L. 1972. Trans. Amer. Fish. Soc. 101, 555.

9. McCann, J. A. and Jasper, R. L. 1972. Trans. Amer. Fish. Soc. 101, 317.

10. Mehrle, P. M. and Mayer, F. L. 1975a. J. Fish. Res. Board Can. 32, 609.

11. Mehrle, P. M. and Mayer, F. L. 1975b. J. Fish. Res. Board Can. 32, 593.

12. Mount, D. I. and Brungs, W. A. 1967. Water Res. 1, 21.

13. Nusagens, B., Chantraine, A., and Lapiere, C. M. 1972. Clin. Opthanol. Rel. Res. 88, 252.

14. Peterkofsky, B. 1972. Biochem. Biophys. Res. Commun. 152, 318.

15. Snedecor, G. W. 1965. Statistical Methods, Iowa State Univ. Press, Ames, Iowa.

16. Sneed, K. E. 1970. In Progress in Sport Fisheries and Wildlife, Resource Pub. 106, U. S. Department of the Interior, Washington, D.C., pp. 189-315.

17. Somogyi, J. C. and Kodicek, E. 1969. Nutritional Aspects of the Development of Bone and Connective Tissue, Proceedings of the 9th Symposium of the Group of European Nutritionists, Karger, Basel, Switzerland.

18. Stalling, D. L. and Huckins, J. N. 1975. Analysis and Gas Chromatography-Mass Spectroscopy Characterization of Toxaphene in Fish and Water, Ecological Research Series, U. S. Environmental Protection Agency, EPA-600/3-76-076, Duluth, Minn.

19. Wagstaff, D. J. and Street, J. C. 1971. Toxicol. Appl. Pharmacol. 19, 10.

20. Wilson, R. P. and Poe, W. E. 1973. J. Nutri. 103, 1389.

SECTION V
CHEMICAL FATE OF POLLUTANTS IN THE AQUATIC ENVIRONMENT

Chemical Mechanisms Affecting the Fate of Organic Pollutants in Natural Aquatic Environments

SAMUEL D. FAUST
Department of Environmental Science
Rutgers, The State University
New Brunswick, New Jersey

I. INTRODUCTION

Much concern has been expressed about the introduction and subsequent distribution of organic compounds throughout man's environment. Some of this concern is real and genuine, but much is based upon emotional hysteria and scientific demagogy. If restrictions should be imposed upon organic chemicals, then the decision should be based upon scientific evidence and not upon emotional speculation. This chapter examines chemical mechanisms affecting the fate of organic pollutants in one of man's environments, that is, the aquatic environment. Hopefully, some insight may be derived into the occurrence and distribution of organic compounds throughout various natural aqueous systems so that an evaluation of environmental hazards or damage may be made.

The synthetic organic compound and/or a derivative probably entered natural aquatic environments concurrently with development of its initial manufacturing process. No doubt there was a waste disposal problem that led to discharge into a river or some other body of water. There is evidence in the scientific literature that potable, recreational, irrigational, fish, and shellfish waters are contaminated with foreign (industrially produced) organic compounds. Much of the early evidence was largely circumstantial as observed from physiological responses of aquatic organisms. More recently, the advent of chromatographic separation procedures and of the confirmatory procedures, nuclear magnetic resonance, and mass spectrometry has led to the positive identification of organic compounds in aquatic environments and in the attendant solid phases of bottom sediments.

The presence of a contaminant in any environment poses several questions. How was the contaminant introduced into the environment? How is the contaminant distributed and transported throughout the environment? What is the stability of the contaminant toward the natural chemical and biological forces that would provide the opportunity for degradation? If degradation occurs, then is it complete or partial? If partial, then what are the degradation products and what is their effect on the environment?

This treatise was given the title "Fate of Pollutants in the Air and Water Environments." Admittedly, the word "fate" suggests the large and difficult task of following the organic contaminant and its metabolite into every nook and cranny of man's various environments. This is not the intent of the treatise nor of this chapter. It is, however, the intent to gather some scientific evidence on the distribution and stability of organic compounds in aquatic environments.

There are numerous publications that document, to some extent, the occurrence of foreign organic compounds in aquatic environments. This treatise presents details of the naturally occurring organics of color bodies (Kahn, Section V, Part 2), of polynuclear aromatic hydrocarbons (PAH) (Borneff, Section V, Part 2), and such synthetic compounds as pesticides, detergents, polychlorinated biphenyls, (this paper and others), petrochemicals, (Lysyj and Russell, Section I, Part 1), halogenated hydrocarbons (Zitko and Arsenault, Section V, Part 2), phthalate esters (Johnson, et al., Section IV, Part 2), and so on. The sources and origins of these compounds are many, diverse, and reasonably well known.

The next two sections of this paper attempt to place into perspective our knowledge of the occurrence of organics found in water. Specifically, Section II discusses the naturally occurring organic color and the synthetic organics in ground and surface waters (e.g., soluble hazardous substances, organic pesticides, and PCBs). Section III examines the fact or myth of our concerns for the ubiquitous distribution of potentially toxic organics in the environment (e.g., PAHs, pesticides, and PCBs).

II. OCCURRENCE OF COMPOUNDS OF CONCERN IN THE AQUATIC ENVIRONMENT

A. Naturally-Occurring Organic Color

Some natural waters exhibit a yellow-to-brown color, which is quite common. These waters have various descriptors: "swamp water," "humus water," or "colored water." The latter description should not be confused with colored waters arising from the discharge of industrial waste waters.

Very little information is available about the broad classes of types of organics that occur naturally in surface waters.

Organic matter in soils, aquatic vegetation, and aquatic organisms would be the major sources. Christman and Ghassemi (1966) have reported the most significant information about the chemical nature of organic color and water that arises from: (a) the aqueous extraction of living woody substances, (b) the solution of degradation products in decaying wood, (c) the solution of soil organic matter, or (d) a combination of these processes. Seven products were identified: vanillin, vanillic acid, syringic acid, cathechol, resorcinol, protocatechvic acid, and 3,5-dihydroxybenzoic acid. The most significant point is the presence of phenolic nuclei in the color macromolecule. There is some concern about environmental contamination with phenolic compounds.

B. Polynuclear Aromatic Hydrocarbons (PAH)

These compounds are considered naturally-occurring because their origin is usually a petroleum source. They are released into the environment through a pyrolytic industrial operation of some sort. There is a legitimate concern about the occurrence of these compounds in man's environment because of the carcinogenic properties exhibited to laboratory animals by such PAH as 3,4-benzpyrene (BP).

The ubiquitous distribution of 3,4-benzpyrene throughout man's environment is claimed. Typical concentrations of BP have been reported to range from as low as 0.0001 μg/liter in the Volga River to as high as 13 μg/liter in a Moscow Reservoir (Andelman and Snodgrass, 1974). This should be questioned in view of the analytical uncertainties surrounding detection of organic molecules in environmental samples. Undoubtedly, BP is widespread, but the ubiquitous claim should be questioned until more precise analytical data are available. Andelman and Snodgrass (1974) have published a review of the incidence and significance of PAH in the water environment. Borneff's paper describes his own work in detail.

C. Synthetic Organics in Ground and Surface Waters

Middleton and Lichtenberg (1960) expressed concern about the appearance of synthetic organic chemicals in the rivers of the United States. The isolation method was the chloroform extraction of an activated carbon filter. Rather crude analytical separation procedures were employed to identify these contaminants with infrared analysis. Some of the identified organics were: (a) DDT in the Mississippi, Missouri, Columbia, and Detroit Rivers. (b) Aldrin was recovered from the Snake River at

Pullman, Washington. (c) *o*-Nitrochlorobenzene, also, was found in the Mississippi River at New Orleans in the concentration range of 1 to 2 μg/liter, and (d) phenyl ether was located in the St. Clair, Kanawha, and Ohio Rivers. This was, perhaps, the first report in the United States that identified organic contaminants in natural waters. At that time, the major concern was the organoleptic properties of the water, namely, the taste and odor qualities.

On occasion, some anxiety is expressed about the toxicity and carcinogenesis of organic contaminants in the drinking waters of the United States. Dunham et al. (1967) reported a study whereby carcinogenic effects of chloroform extractables from activated carbon were examined in newborn mice. Apparently the incidence rate for bladder cancer in New Orleans is three times higher than in Birmingham, Alabama, and Atlanta, Georgia (Dorn and Cutler, 1959). The Mississippi River is the source of the water supply for New Orleans. This water was sampled by the activated carbon technique that was extracted with chloroform and alcohol. Results were:

Extract	Range (ppb)	Median (ppb)
Chloroform		
New Orleans	32-76	45
Birmingham Shades	50-150	78
Birmingham Putnam	51-84	62.5
Alcohol		
New Orleans	75-172	98
Birmingham Shades	76-199	122
Birmingham Putnam	79-234	111.5

These data suggest that the New Orleans water was slightly "cleaner" than the supplies in Birmingham. Non-inbred albino, general purpose mice were injected with chloroform and alcohol extractables at the age of 4 to 18 hr. The experiment was terminated when the mice reached the age of 1.5 yr. It was concluded: "Tumors attributable to the pollutants were not induced in the tissues examined, including the bladder, during the experimental period of 78 weeks."

Much of the early concern about organics in surface waters centered around organoleptic properties. Rosen (1962, 1963, 1969) was the forerunner in the separation and isolation of organic compounds from polluted surface waters. The Kanawha River at Nitro, West Virginia, is now a classical study in which many organic compounds of industrial origin were identified.

In recent years, Burnham et al. (1972, 1973) have isolated trace organic contaminants from potable water of Ames, Iowa. The unique situation is that this was a ground water contamination with organics of industrial origin of many years duration. Again, the organoleptic water quality was the primary concern since the supply of Ames, Iowa, had an undesirable taste and odor. The organics had an industrial origin and the probable source of contamination was a pit used for the disposal of coal-tar residues from a coal-gas plant that operated in the 1920s!

In the course of an examination of the toxicity of organic compounds in drinking water, Kopfler and co-workers (1976) report within this book (Part 2, Chapter 18) a list of organics that have been identified from finished or treated water. The list is replete with compounds of industrial origin and of considerable persistence in aquatic environments. Whether these compounds are toxic or not to humans at the concentrations occurring in natural waters is an unanswered question.

D. The Occurrence of Organic Pesticides in Natural Waters

A National Pesticide Monitoring Program was launched in 1967 to establish a network to survey the major drainage rivers in the United States (Green and Love, 1967). Initially, 39 rivers were selected for monthly sampling points near their mouths. A revision of this program was announced in 1971 where water samples would be collected quarterly and bed material samples would be collected semiannually at 161 sites in the conterminous United States, Alaska, Hawaii, and Puerto Rico (Feltz et al., 1971). These pesticides are sought:

Insecticides		Herbicides
Aldrin	Heptachlor	2,4-D
Chlordane	Lindane	2,4,5-T
DDD	Malathion	Silvex
DDE	Methoxychlor	
DDT	Methyl Parathion	
Dieldrin	Parathion	
Endrin	Toxaphene	

The 161 sampling sites are listed by Feltz et al. (1971).

Some of the early river surveys prior to 1967 were conducted by the Federal Water Quality Administration in the U.S. Department of the Interior. Breidenbach et al. (1967), and

Weaver et al. (1965) reported upon the chlorinated hydrocarbons in our major rivers for the years 1957-1965. Samples were collected from 99 stations in the month of September when stream flows are usually the lowest. In order of decreasing concentrations, dieldrin (0.1 μg/liter), endrin, DDT, DDE, DDD, heptachlor, heptachlor epoxide, and BHC were found. Aldrin was found infrequently, but when it was the concentration was low: 0.006 μg/liter.

A five year summary (1964-1968) of the pesticidal content of the major rivers of the United States (100 sites) was reported by Lichtenberg et al. (1970). Several chlorinated hydrocarbons were found (in order of decreasing frequency): dieldrin, endrin, DDT, DDE, DDD, aldrin, heptachlor, heptachlor epoxide, lindane, BHC, and chlordane. The highest concentration was 0.407 μg/liter of dieldrin in the Tombighee River at Columbia, Mississippi. Six organophorus pesticides were sought but only two were found in one sample in the Snake River at Wawawai, Washington: parathion, 0.050 μg/liter and ethion, 0.380 μg/liter. Table 1 shows a comparison of the maximum pesticide concentrations found in this 5-yr survey with the permissible water supply criteria proposed in 1972 (Anon., 1972a). All concentrations are far below these permissible criteria.

There is sufficient evidence to suggest that organic pesticides are in residence in our aquatic environments. Moreover, most of these pesticides have been detected and confirmed as the chlorinated hydrocarbons. The organophosphates, phenoxyacids, carbamates, and ureas are observed infrequently. Concern about the concentrations of pesticides in the water phase leads to the question of safe levels. To date, almost all of the chlorinated hydrocarbon contents have been less than the current water quality criteria for potable water and for fish and other wildlife inhabitants of natural surface waters. Occasionally, an accidental or deliberate discharge will result in a fish kill and other types of wildlife damage. However, none of this has been permanent and no species of aquatic wildlife has been eliminated from the face of this good earth. Also, there is evidence to suggest that some pesticides may have been transported around the world through dust storms and rainfall. This may account for some of the recent (after 1965) world-wide distribution of pesticides.

E. The Polychlorinated Biphenyls

Some anxiety has been expressed about the occurrence and widespread distribution of the somewhat persistent polychlorinated biphenyl (PCB) compounds in aquatic environments. These contaminants may be more ubiquitous than the chlorinated

TABLE 1. Maximum Pesticide Concentration Found Versus Permissible Water Supply Criteria for 1972[a]

Pesticide	Permissible Criteria (μg/liter)	Maximum concentration found (μg/liter)
Dieldrin	1.0	0.407
Endrin	0.5	0.133
DDT	50.0	0.316
DDE		0.050
DDD		0.840
Heptachlor	0.1	0.048
Heptachlor epoxide	0.1	0.067
Aldrin	1.0	0.085
Lindane (BHC)	5.0	0.112
Chlordane	3.0	0.169
Methoxychlor	1000.0	b
Toxaphene	5.0	c
Organophosphates plus carbamates	100.0	0.380
Herbicides:		b
2,4-D	20.0	
2,4,5-TP	30.0	
2,4,5-T	2.0	
Phenols	1.0	b

[a]After Anon (1972a). Reprinted with permission of the National Academy of Sciences, Washington, D.C.

[b]Not determined.

[c]Not detected.

hydrocarbon pesticides. In fact, many of the PCBs are positive interferences with gas-liquid chromatographic procedures for the analytical separation of chlorinated hydrocarbons (Reynolds, 1971). Consequently, many of the early reports on the global distribution of pesticides may have been, in fact, PCBs.

Surprisingly few reports are in the scientific literature, to date, about the occurrence of PCBs in aquatic environments. A report by Duke et al. (1970) indicated the distribution of Aroclor 1254 in the water, sediment, and biota of Escambia Bay, Florida. Apparently, most of the PCBs resided in some sort of a solid phase: that is, fish, crustacean, or sediment since the contents in water were less than 0.1 μg/liter. It was not detected at some sampling stations in the water phase. On the other hand, sediment samples taken near a waste-water outfall reached a high content of 486 ppm. In an effort to develop analytical methodology for PCBs in water, Ahling and Jensen (1970) reported contents in untreated and treated tap water as: 0.50 and 0.33 ppt (parts per trillion), respectively.

F. Soluble Hazardous Substances

The U.S. Environmental Protection Agency (EPA) has directed a considerable portion of its resources and manpower toward the problem of water-polluting spills of oil and hazardous substances. It is estimated that over 10,000 spills occur annually in the United States. Spills are defined as noncontinuous discharges or dumping (nonpoint sources) that occur as a result of accidents, malfunctions of equipment, human error, and deliberate design such as discharge of bilge or ballast water from tankers or convenience dumping of hazardous materials and oil into sewers, streams, estuaries, and coastal waters and upon land areas.

Almost 80% of the reported spills involve oil, including crude and petroleum products ranging from grease to gasoline and waste lubricating oil. The remaining 20% involves hazardous polluting substances other than oils. Because water-soluble chemicals present the greatest threat to a water ecosystem, a priority system estimating theoretical, inherent-hazard chemicals was prepared as part of an EPA study. The ranking system in this study was based on: (a) the lowest concentration range at which a material impairs any of the beneficial uses of water, (b) the quantity shipped annually by each mode of transport, and (c) the probability of an accidental spill to surface waters. Table 2 shows the 20 most hazardous substances based on these criteria of which several are organic compounds. Hazardous was considered to be a danger or a toxicity to aquatic and marine life.

TABLE 2. Priority Ranking of Soluble Hazardous Substances

Rank	
1	Phenol
2	Methyl alcohol
3	Cyclic rodenticides
4	Acrylonitrile
5	Chlorosulfonic acid
6	Benzene
7	Ammonia
8	Miscellaneous cyclic insecticides
9	Phosphorus pentasulfide
10	Styrene
11	Acetone cyanolhydrin
12	Chlorine
13	Nonyl phenol
14	DDT
15	Isoprene
16	Xylenes
17	Nitrophenol
18	Aldrin-toxaphene group
19	Ammonium nitrate
20	Aluminum sulfate

III. UBIQUITOUS DISTRIBUTION IN THE ENVIRONMENT - FACT OR MYTH

There is the indictment that many of the above-mentioned organic contaminants are distributed throughout our environment. This is especially true of the organic pesticide, the PCB, and the PAH. This, of course, implies a widespread and constant threat to living systems. But is this true? Are these contaminants truly ubiquitous? Let us critically examine some of the research reports mainly from the pesticide literature concerning this point.

A. Analytical Difficulties

In reviewing reports on the distribution, occurrence, and persistence of the chlorinated hydrocarbons, especially DDT, dieldrin, and aldrin, in the atmosphere and various bodies of water, careful attention was given to analytical techniques. It is extremely important to confirm the presence of these pesticides in an environmental sample.

Most of the early residue investigations (1960-1965) by analysts used an electron capture detector with gas-liquid chromatography. This detector is nonspecific for organic pesticides, that is, any organic compound that captures an electron gives a response. For example, Schafer et al. (1969) examined 500 grab samples of finished drinking water and raw water for ten chlorinated hydrocarbons between March 1964 and June 1967. These samples were extracted directly with hexane and were injected directly into a gas chromatograph equipped with an electron capture detector. Six stationary phases were employed. "Identification" was attempted by comparison of relative retention times to standards on the six columns.

Westlake (1971) made some interesting comments about use of the electron capture detector. "It is also the worst possible choice for identifying the detected compounds. Introduced by Lovelock and Lipsky in 1960, the use of this detector has revolutionized pesticide residue determinative procedures. This detector, however, has the unfortunate property of giving responses to a host of compounds other than pesticides and is, therefore, virtually useless for identification." Westlake continues: "Efforts toward using the electron capture gas chromatograph as a qualitative tool have centered first on the use of two or more different columns that hopefully will give different relative retention times for the various eluting peaks and, with the judicious use of standards some assurance of the identity of unknowns. This procedure can be time consuming and, while helpful, still may not yield anything positive."

Elgar (1971) shares the same concern as Westlake: "It is important to confirm the identity of pesticide residues convincingly. Some methods, such as TLC, paper chromatography, or p-values share the same physical property of partition in achieving separations of mixtures. They do not give independent evidence in achieving separations of mixtures. Similarly, GLC retention times for a compound on different stationary phases are often highly correlated. Thus the choice of confirmatory techniques should be carefully made."

Some attempt has been made in the United States (Cohen and Pinkerton, 1966; Stanley et al., 1971; Weekel et al., 1966) and in England (Wheatley and Harman, 1965; Abbott et al., 1965, 1966; Tarrant and Tatton, 1968) to determine the extent of atmospheric transport of organic pesticides. Dieldrin, 0.003 ppm, was found in dust fallout collected in Cincinnati, Ohio (Cohen and Pinkerton, 1966; Weekel et al., 1966). In an analysis of the atmosphere, samples were collected at nine locations in the United States (urban, rural, etc.). Dieldrin was found in only one location, Orlando, Florida, and in 50 of 99 samples. The maximum concentration was 29.7 ng/$\underline{M}^3$. Aldrin was found in only one sample in only one location - Iowa City (8.0 ng/$\underline{M}^3$). In both of these surveys no confirmation of the GLC peaks was attempted. Identification was made by a relative retention time on two or three GLC columns. The English analysts were more careful than the United States analysts. They examined rain water samples followed by gas-liquid chromatography and still called the results "apparent organochlorine insecticides" because not enough pesticide was available for infrared confirmation (Wheatley and Hardman, 1965).

A rather interesting study was reported by Frazier et al. (1970) in which "apparent organochlorine insecticides" were found in sealed 1910 soil samples. This study was, of course, an attempt to demonstrate the analytical uncertainties in soils. Thirty-four samples were analyzed for chlorinated hydrocarbons on three GLC columns commonly employed for this purpose. On one column, QF-1/OV-17, five of the 34 samples yielded aldrin in concentrations greater than 1 ppb, whereas one sample yielded dieldrin. On another column, QF-1/DC-200, aldrin was found in 20 of 34 samples with concentrations ranging from 3 to 294 ppb, whereas only one sample gave dieldrin. The significance of the latter column is that it is used in a rather routine fashion for aldrin and dieldrin in various environmental samples, especially soils.

There is one additional note concerning analytical difficulties with organic contaminants in environmental samples namely, the interference of PCBs. Reynolds (1971) has documented this problem reasonably well. Under GLC conditions commonly employed for the detection of chlorinated hydrocarbons, it was shown that Aroclor 1254 with its 14 major peaks gave relative retention times the same as aldrin and dieldrin. A relatively nonpolar stationary phase was employed: 4% SE-30/6% QF-1 on 60/80 mesh chromosorb W. Data were presented from the analysis of bird tissue, brain and liver tissue, and eggs to demonstrate the interference of PCB with DDD and DDT. It may well be that some of the early reports of the ubiquitous distribution of DDT in animal tissue may have been PCBs. The latter is, of course, a serious environmental problem.

B. Recovery Difficulties

One of the major problems with identification or organic contaminants in environmental samples lies with the initial recovery step. Many techniques and innovations have been attempted. The recovery step is extremely important since it provides separation and concentration (in most cases) from the water phase. On the other hand, many recovery steps are inefficient and destroy or modify the original molecule.

Gunther and Blinn (1955) concluded: "Environmental samples are weathered (i.e., transformations due to physical, chemical and metabolic forces) whereupon the molecule may be modified or may occur in aggregates. Consequently, the organic contaminant may not be in the same form as the fortified sample. Recovery efficiencies from environmental samples will approach laboratory fortification if the organic is dissolved completely, not absorbed or particulate matter, and not dissociated, and if the natural water characteristics of pH, temperature, and ionic strength can be duplicated."

IV. TRANSPORT MECHANISMS FOR DISTRIBUTION THROUGHOUT THE ENVIRONMENT

It may be stated rather succinctly that organic compounds do, indeed, contaminate man's environment. Furthermore there is some evidence that there is widespread distribution. What, then, are the forces affecting the occurrence, persistence, distribution, and transport in natural aquatic environments? These factors are brought together under one term that is labeled "fate". We will examine this question largely from the organic pesticide literature. These compounds represent the major classes of organics, namely, the pesticide may be ionic, basic, acidic, and nonionic or neutral. Also, considerably more information is known about pesticides in natural systems than any other type of organic contaminant. Pesticides may undergo surface reactions, may hydrolyze, or may be photochemically or chemically degraded.

A. Pesticide Transport to Aquatic Environments

An organic pesticide applied into an agricultural situation can be transported from a watershed to a receiving body of water by several mechanisms: (a) from the atmosphere originating from application-associated losses, volatilization, and wind erosion, (b) from ground water, and (c) from runoff by movement in solution as the soil-pesticide complex (Pionke and Chesters, 1973). The latter mechanism is examined in some detail below.

1. Losses to the Atmosphere. Inadvertent losses from application are greatly variable and depend upon climatic conditions and methods of application and formulation. Pesticides entering into local drainage waters from aerial sprays may be associated with "carriers" and emulsifiers that can temporarily attain abnormally high concentrations relative to their actual solubilities. This has been observed with DDT (Cole et al., 1970; Grzenda et al., 1964).

Volatilization of soil-applied pesticides is possible and may be appreciable if the compound is not incorporated into the soil. This has been documented in detail by Spencer within this treatise (Part 1, Chapter 5).

It has been suggested that most airborne pesticides are associated with dust particles (Cohen and Pinkerton, 1966; Weebel et al., 1966). Whether the volatilized pesticide is adsorbed by airborne dust or is from dislodgement of the soil particle-pesticide complex due to wind erosion is not certain. In some geographical areas, wind erosion provides an important aerial source of pesticides that are redeposited into water bodies (Cohen and Pinkerton, 1966). Pesticide contents in the air provide the potential amounts that could be returned to a watershed and associated water systems from rainfall (see foregoing discussion).

Pionke and Chesters (1973) provide an excellent analysis of the question of aerial transport of pesticides. "Although concentrations are low-level, the return of these compounds in solution with the 'host' particulate matter to the earth's surface is continuous and contribution of pesticides through this pathway into the aquatic system is difficult to evaluate. Dissolved in rainfall, these compounds are probably well mixed throughout the aquatic system if they gain direct entry, or they probably react with soil colloids or suspended materials if deposited on the watershed. Desorption characteristics of pesticides adsorbed to airborne particulate control solution concentration from this source, particularly if the pesticide is strongly adsorbed. Furthermore, the eventual distribution of these colloidal particles in the aquatic system affects the distribution of the pesticide." In this treatise, Junge has proposed global atmospheric residence times for chlorinated hydrocarbon pesticides based upon present knowledge.

2. Losses to Ground Water. The potential leaching of pesticides from soils to ground waters depends primarily on the adsorbtivity of the compound on soil colloids. The general nonleachability of most chlorinated hydrocarbons has been well documented (Carter and Stringer, 1970; Terriere et al., 1966;

Voerman and Besemer, 1970) for many soil types including sands and aquifer sands and at various concentrations that exceeded, on occasion, 1000 μg/g (Carter and Stringer, 1970). Cationic pesticides are adsorbed very strongly by soil. For example, the dipyridinium herbicides (diquat and paraquat) are highly resistant to leaching (Dixon et al., 1970; Weed and Weber, 1969). The s-triazines, thiocarbamates, and urea derivatives are somewhat mobile (Harris, 1969), but high contents of soil organic matter will reduce this movement (Pionke and Chesters, 1973). Hague and Freed (1974) have provided some very valuable information on the environmental behavior of organic pesticides in soil by developing vaporization and leaching indexes in soil (Table 3).

3. Pesticide Losses in Surface Runoff. Runoff (defined as the combination of runoff waters with suspended soil particles dislodged by erosion) from agricultural lands has been implicated as one of the primary mechanisms of introducing chlorinated hydrocarbons and organophosphates into aquatic environments Lauer et al., 1966). In one investigation, diazinon and DDT concentrations, in runoff generated by irrigation, increased 10^3 to 10^5 times the antecedent concentrations (<0.16 mg/liter) 24 hr after aerial application of 2 kg/ha DDT and 1 kg/ha diazinon (Hindin et al., 1966). The maximum contents of DDT and diazinon in the runoff water that remained in the aqueous portion after three hours of settling were 506 and 975 mg/liter, respectively. A number of studies on small agricultural watersheds under reasonably controlled conditions has been made and the results were detailed by Browman and Chesters (Part 1, Chapter 4, within this treatise). These data suggest that losses of most chlorinated hydrocarbons relative to the amount applied are "low" even though the insectides were applied to the soil surface. In one case, 30 times more dieldrin was transported by suspended sediment than by runoff water (Caro and Taylor, 1971). Generally, it appears that runoff losses of most surface applied pesticides are greater in the period immediately following application than at some later time (Caro and Taylor, 1971).

B. Surface Reactions

Adsorption and retention of organic pesticides on various environmental surfaces may represent the major natural "force" affecting their fate in aquatic systems. The pesticidal molecule encounters surfaces of: soil particles, clay minerals, carbon, oxides or iron, aluminum, abiotic organic matter, and such biotic organic matter as tree leaves and aquatic vegetation. Pionke and Chesters (1973) indicate the "the relationship between sediment

TABLE 3. Vaporization and Leaching Indexes of Organic Pesticides in Soil.[a]

Compound	Vaporization[c] index (from soil)	Leaching[d] Index
Herbicides		
Alachlor	3.0	1.0-2.0
Propanil	2.0	1.0-2.0
Trifluralin	2.0	1.0-2.0
Dalapon-Na	1.0	4.0
MCPA (acid)	1.0	2.0
2,4-D (acid)	1.0	2.0
2,4,5-T (acid)	1.0	2.0
Insecticides		
Carbaryl	3.0-4.0	2.0
Malathion	2.0	2.0-3.0
Naled	4.0	3.0
Dimethoate	2.0	2.0-3.0
Fenthion	2.0	2.0
Diazinon	3.0	2.0
Ethion	1.0.2.0	1.0-2.0
Oxydemeton-methyl	3.0	3.0-4.0
Azinophosmethyl	-	1.0-2.0
Phosphamidon	2.0-3.0	3.0-4.0
Mevinphos	3.0-4.0	3.0-4.0
Methyl parathion	4.0	2.0
Parathion	3.0	2.0
DDT	1.0	1.0
BHC	3.0	1.0
Chlordane	2.0	1.0
Heptachlor	3.0	1.0
Toxaphene	4.0	1.0
Aldrin	1.0	1.0
Dieldrin	1.0	1.0
Endrin	1.0	1.0
Fungicides		
Captan	2.0	1.0
Benomyl	3.0	2.0-3.0
Zineb	1.0	2.0
Maneb	1.0	2.0
Mancozeb	1.0	1.0

[a]From Hague and Freed (1974). Reprinted with permission of Springer-Verlag Co., New York.
[b]Estimated from best available information for loam soil at 25°C under annual rainfall of 150 cm.
[c]A vaporization index number of 1 = vapor loss of less than 0.1 kg/ha/yr, 2 = from 0.2 to 3.0 kg/ha/yr or more, 3 = 3.5 to 6.5 kg/ha/yr or more, and 4 = 7 to 14 kg/ha/yr or more.
[d]A leaching index number indicates the approximate number of centimeters moved through the soil profile with an annual rainfall of 150 cm; thus, an index of 1 = <10 cm, 2 = <20 cm, 3 = >35 cm, and 4 = >50 cm.

and pesticides controls or influences the movement, distribution, toxicity, and persistence of many pesticidal compounds which might appear in lakes or streams. Often adsorption by soil will stabilize the pesticides against loss by volatilization, leaching, or microbiological degradation. Adsorption may result also in the formation of a pesticide-soil complex which may be removed intact to aquatic systems by erosive forces. Because the aquatic pollution hazard of an applied pesticide is only truly removed upon its degradation to a nontoxic product, perhaps the most important effect is that adsorption can alter the rate of biological and/or chemical degradation of pesticides."

1. Physicochemical Characters of Adsorbate and Adsorbent. Any surface reaction is dependent, of course, upon the chemical nature of the adsorbent and adsorbate. For example, ionic pesticides reacting with ionic surfaces are retained to a greater extent than nonionic pesticides reacting with nonionic surfaces. Adsorption and desorption of organics by and from various surfaces have been described by Browman and Chesters in Part 1, Chapter 4, after the work of Bailey and White (1970). For adsorbents, the major considerations are primarily the area and configuration and magnitude, distribution, and intensity of the electric field at the surface. For example, the clay mineral kaolinite has a low cation exchange capacity and a relatively low surface area which gives it a limited adsorption capacity. On the other hand, the 2:1 expandable clay minerals, montmorillonite, and vermiculite have high cation exchange capacities and surface areas. Their adsorption capacities are very much greater than kaolinite. Porosity is another important property of the adsorbent. Carbon provides an excellent example. Very little of the adsorption on granular carbon occurs on the external surface with most occurring in the internal pores. Another factor lies in the exposed functional groups on the surface. Carbon presents a surface of various organic functional groups (-COOH, = O, -OH, etc.). The nature of these groups determines the mechanism by which adsorption occurs: (a) physical (due to van der Waal's forces), (b) hydrogen bonding, (c) coordination complexes, or (d) chemical bonds. All of these factors combine to determine the extent of adsorption and the adsorptive capacity.

2. Extent of Adsorption. One of the major questions concerning the transport and distribution of organic pesticides in aquatic environments is their adsorption onto soil particles, soil colloids, and suspended sediments. For example, dieldrin, a chlorinated hydrocarbon, is relatively insoluble in water, 0.1 to 0.25 mg/liter. This molecule is considered to be nonpolar, has no functional groups, and is a neutral or nonionic

TABLE 4. Adsorption of Dieldrin on Soil and Clay Minerals at 30 to 32°C.[a]

Soil type	Total Clay(%)	Organic Matter(%)	Cation Exchange (meg/100 g)	Dieldrin Adsorbed X_m values[b] (μg/g)
1	18.7	4.2	20.8	70.0
1a	17.9	4.3	21.1	116.0
1b	19.0	4.2	19.7	67.0
2	24.6	3.9	31.2	61.0
2a	24.5	4.0	32.4	75.0
2b	26.6	3.8	28.9	48.0
3	21.9	9.1	22.6	340.0
Peat	-	40.0	78.8	714.0
Montmorillonite	53.4	-	117.0	32.0
Halloy.	49.2	-	10.6	164.0
Atta.	-	-	28.0	58.5
Bento.	-	-	-	18.0
Illite	40.4	-	25.0	71.0

[a]From Eye, (1968). Reprinted with permission of the Water Pollution Control Federation.

[b]Langmuir's "monolayer" coverage.

molecule. That dieldrin is adsorbed to various extents on soil types may be seen in Table 4 (Eye, 1968). Organic matter content, usually expressed as percent, appears to be the major factor controlling the extent of adsorption. As the percentage of organic matter is increased, the Langmuirian monolayer coverage value, X_m, becomes greater. Cation exchange capacity (CEC) has little or no correlation with adsorption, especially among the clay minerals. Also, the clay minerals, in general, adsorbed less dieldrin than the soil types 1, 2, and 3.

Some indication of the extent of pesticide transport by suspended sediment may be obtained from the adsorption data in Table 4. For example, the Colorado River, Grand Canyon, Arizona, on October 3, 1948, carried 156,000 ton/day of suspended sediment (this was chosen at random and is a "high" number) (Anon., 1953). Assuming the datum of 75 μg dieldrin/g of suspended sediment to be an average value, 11.7 tons of this chlorinated hydrocarbon would be the daily rate of transport under these conditions. This calculation assumes, of course, that no desorption occurs and that all of the suspended sediment contains dieldrin.

Cationic pesticides are adsorbed to a greater extent than neutral molecules. Weber (1972) provided some excellent data from the adsorption of eight cationic pesticides (pyridinium or quarternary ammonium compounds). Their adsorption isotherms are seen in Figure 1 (Weber, 1972). Several generalizations may be made. For example, divalent cations (e.g., diquat) were adsorbed to a greater extent than the monovalent species (e.g., Et. Pry. Br.) on montmorillonite and larger cations were adsorbed more than smaller ones (top isotherm). In the case of cationic pesticides, adsorption usually occurs to the CEC of the adsorbent.

Type of surface is also a variable affecting the transport and distribution of organic pesticides in aquatic environments. Table 5 lists the Langmuir monolayer coverage values for several compounds on various surfaces. The extent of adsorption may be compared through the X_m values. Carbon provides a surface that is covered to a greater extent than the clay mineral surfaces.

3. Extent of Desorption. If suspended sediments are involved in the transport of organic compounds through natural aquatic environments, then the extent and rate of desorption are extremely important factors affecting their contents in the water phase. Once again, we look at the pesticide literature. Desorption of some pesticides from soils apparently occurs at a much slower rate than adsorption as illustrated by simazine

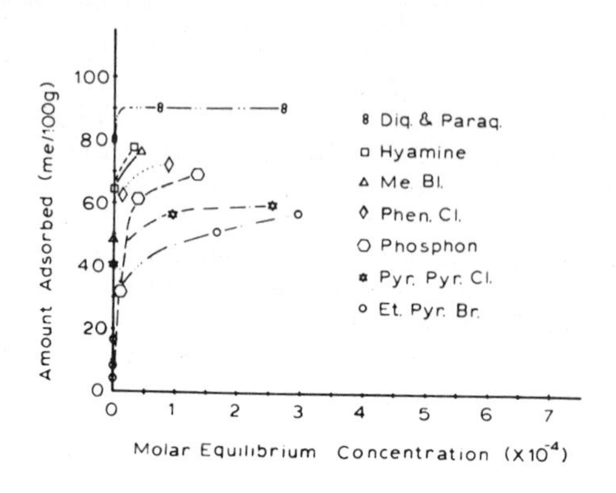

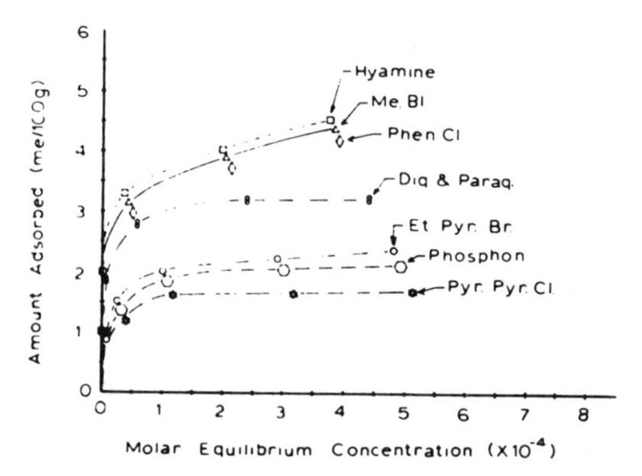

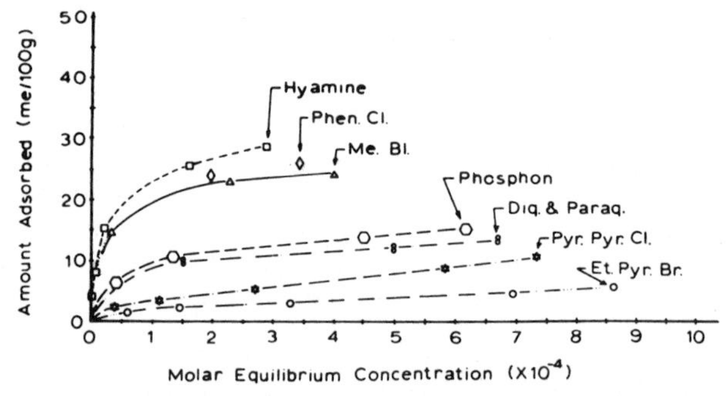

FIGURE 1. Isotherms for adsorption of several cationic pesticides on Na-montmorillonite (top), Na-kaolinite (center), soil organic matter (bottom). After Weber (1972). Reprinted with permission from the Adv. in Chem., Series No. 111. Copyright by the American Chemical Society, "Fate of Organic Pesticides in the Aquatic Environment."

TABLE 5. Langmuir X_m Values - "Monolayer" Coverage for Organic Pesticides on Various Surfaces.

Compound	Type	Surface	X_m(mg/g)
2,4-D	Anionic	Granular carbon	387.0
2,4,5-T	Anionic	Granular carbon	448.0
Silvex	Anionic	Granular carbon	464.0
Parathion	Neutral	Granular carbon	530.0
Phenol	Neutral	Granular carbon	103.0
1-Chloro-4-Nitrobenzene	Neutral	Granular carbon	400.0
Dieldrin	Neutral	Montmorillonite	0.032
Dieldrin	Neutral	Illite	0.071
Dieldrin	Neutral	Peat	0.714
Diquat	Cationic	Montmorillonite	1.80

(Williams, 1968); atrazine, linuron, monuron, and CIPC (Hance, 1967); and aldrin (Yaron et al., 1967). Several of these compounds that were almost completely adsorbed in 7 hr were not completely desorbed after 72 hr (Hance, 1967). Irreversible adsorption has been demonstrated for DDT (Grenzi and Beard, 1970), lindane (Adams and Li, 1971), aldrin (Yaron et al., 1967), and monuron (Hance, 1967) in soils, 2,4-D and amiben in muck (Harris and Warren, 1964), heptachlor and dieldrin on clay minerals (Huang and Kiao, 1970), and toxaphene on lake sediments (Veith and Lee, 1971). Thus there is some evidence that organic pesticides remain adsorbed upon surfaces they contact. Desorption is apparently limited in scope. For example, simazine was recovered from a soil to the extent of 26% after a "first" desorption period of 16 hr and 13% after a "second" desorption period of 16 hr (Williams, 1968).

C. Reactions with Natural Surfaces

"Natural" surfaces also play an important role in the distribution of organic compounds, in general, and organic pesticides, specifically, in aquatic environments. The distribution of toxaphene in several portions of a natural lake environment is seen in Table 6 (Kallman et al., 1962). In this case history study, toxaphene appeared to concentrate rather rapidly on the

TABLE 6. Distribution of 0.05 mg/liter Toxaphene in Clayton Lake, New Mexico[a]

Sample	Toxaphene (mg/liter)	Time (days)	Remarks
Water	0.010	4.25	Windward
Water	0.028	4.25	Lee
Water	0.001	269	
Trout[b]	5.4	2.0	Netted
Trout	4.2	2.54	Dead
Bullheads[c]	4.2	3.0	Netted
Bullheads	15.2	5.0	Severely affected
Trout	2.5	262	Alive
Aquatic plants[d]	2.3	2 hr	Lee
Aquatic plants	1.8	2 hr	Windward
Aquatic plants	4.1	1.0	Lee
Aquatic plants	0.4	1.0	Windward
Aquatic plants	18.3	9.0	Lee
Aquatic plants	11.0	9.0	Windward
Bottom sediments	0.04	0.25	Windward
Bottom sediments	0.13	9.0	Lee
Bottom sediments	Trace	9.0	Windward

[a]After Kallman et al. (1962). Reprinted with permission of the American Fisheries Society.

[b]Added after toxaphene treatment.

[c]Added before toxaphene treatment.

[d]ppm (dry weight) in aquatic plants and sediments.

waxy surface of the aquatic plants. Initial dosage to the lake was 0.05 mg/liter. It should be noted also that considerable quantities of toxaphene were adsorbed by the fish.

There is some evidence that naturally-occurring organic matter, that is, humic acid, may be a factor in the transport of pesticides in soil-water systems. Wershaw et al. (1969) indicated that the solubility of DDT in a 0.5% sodium humate solution was increased at least 20 times. DDT solubility in water is about 4 x 10^{-5} g/liter. Sodium humate is considered to be a colloid in micelle form. Data were presented also to show that 2,4,5-T is adsorbed strongly by humic acid (Wershaw et al., 1969). Similar observations were made by Poirrier et al. (1972) with (C^{14})DDT and naturally colored surface waters from Louisiana. Their data suggested a concentration factor of 15,800 from a 0.168-ppb solubility of DDT in water (by weight). Khan (1974) found that a fulvic acid-montmorillonite adsorbed about 6.5 5.2 μmol of 2,4-D/g of complex at 5 and 25°C, respectively. The implication of these three studies in the transport of pesticides from soils into fresh surfaces into estuaries and oceans is quite clear.

V. NONBIOLOGICAL DEGRADATIONS AND TRANSFORMATIONS

It may be stated categorically that all organic contaminants are thermodynamically unstable in natural aquatic environments. These so-called recalcitrant compounds have their carbon atoms in a reduced valence state. Consequently, there is an inherent tendency for the carbon to be oxidized to higher valence states in an aerobic state. If conditions are favorable CO_2(g) is the eventual product.

Many terms are employed to describe the instability of organic contaminants in nonbiolgocial systems: oxidation, transformation, degradation, and so on. Definition and interpretation of these terms are, in most cases, arbitrary for most investigators and authors. In this text, oxidation and/or degradation refers to a change in the valence state of the organic C. Usually this change occurs whereby the carbon goes to a higher valence state that requires, of course, an electron acceptor. Transformation refers to any alteration in the configuration of an organic molecule. Examples of transformations would be: the hydrolysis of an organic phosphorus, carbamates, and ureas compounds; replacement of the =S by =0 in organic phosphorus compounds; modification of the trichloroethane group in DDT; replacement of the Cl atom by OH in atrazine, and so on. These two definitions are somewhat at variance with Kearney and Kaufman (1969), who prefer the term "degradation" "to cover all transformations of organic herbicides without particularly trying to ascribe these to enzyme or particulate systems."

A. Chemical Systems

1. Thermodynamic Stability. A natural water in equilibrium with the atmosphere should be saturated with dissolved oxygen (neglecting for the moment aerobic biological transformations of organic matter). In this situation, this water has a well-defined redox potential of approximately +800 MV (P_{O_2} = 0.21 atm, pH = 7.0, 25°C). A simple equilibrium calculation shows that, at this E_h value, all organic carbon should be present as C(+IV) or as CO_2, HCO_3^-, or CO_3^{2-}, S and N should occur in the form of SO_4^{2-} and NO_3^-, respectively. That organic compounds are unstable in aquatic environments may be shown from a simple thermodynamic model:

$$
\begin{array}{lcl}
CH_2O + H_2O & = & CO_{2(g)} + 4H^+ + 4e \\
O_{2(g)} + 4H^+ + 4e & = & 2H_2O \\
\hline
CH_2O + O_{2(g)} & = & CO_{2(g)} + H_2O
\end{array}
$$

The free energy change of this reaction is -117.99 kcal/mol which gives a E_h^o value of 1.28 V. Under the European sign convention, a negative ΔG^o reaction value and a positive E_h^o value denotes that the left to right reaction is feasible. A similar model may be calculated for the organic herbicide, 2,4-D, in an oxygenated environment:

$$2C_8H_6O_3Cl_2 + 15O_{2(g)} = 16OCO_{2(g)} + 4H^+ + 4Cl^- + 4H_2O.$$

The free energy change for this reaction (25°C) was calculated to be -1600.78 kcal/mol, which indicated that it is feasible for the carbon in 2,4-D to be oxidized to CO_2(g) with O_2(g) as an electron acceptor. This model does not, however, indicate the reaction kinetics or if the reaction will even occur. Only laboratory experimentation will answer these two points. More and detailed information on the thermodynamic stability of organic pesticides in aquatic systems is given by Gomaa and Faust (1971a).

2. Artificial Systems. There are numerous electron acceptors available for oxidation of organic compounds, in general, and for organic pesticides, specifically, in artificially created environments. However, where the natural environment is concerned, the oxidative reactions are extremely slow and are incomplete in the sense that toxic reduction products may be formed. It is extremely desirable to effect "complete" degradation to CO_2, H_2O, and so on.

Historically, $KMnO_4$, ClO_2, Cl_2, and O_3 have been employed for the oxidation of organic compounds at water treatment plants. Consequently, these oxidants have been investigated for their capacity to degrade organic pesticides (Gomaa and Faust, 1971a; Cohen et al., 1960, 1961; Aly and Faust, 1965; Robeck et al., 1965). Several basic concepts have evolved from these five studies for the oxidation of organic compounds.

a. Stoichiometry. Oxidative processes at water treatment plants, for example, are often ineffective because inadequate stoichiometries, that is, chemical dosages, are employed. Gomaa and Faust (1971a) proposed the exact stoichiometric relationship for several organic pesticides and inorganic oxidants. For example, parathion and $KMnO_4$ gave:

$$C_{10}H_{14}O_5NSP + 20MnO_4^- + 14H^+ = 20MnO_2 + 10CO_2 + SO_4^{2-} + NO_3^- + PO_4^{3-} + 14H_2O$$

$$3C_{10}H_{14}O_5NSP + 50MnO_4^- = 50MnO_2 + 15C_2O_4^{2-} + 3NO_3^- + 3SO_4^{2-} + 3PO_4^{3-} + 20H_2O + 2\bar{O}H$$

These two reaction models suggest rather complex molar ratios of oxidant to compound. The first reaction represents acidic conditions in which the $KMnO_4$/parathion molar ratio is 20:1. In the second reaction, the molar ratio under alkaline conditions is 16.67:1. In order to propose these two reactions and others (Gomaa and Faust, 1971), several assumptions were made for the reaction products:

1. MnO_4^- goes to $MnO_{2(s)}$ and not to Mn^{2+} (Stewart, 1965).
2. Organic C goes to CO_2 in acid conditions (C + IV).
3. Organic C goes to $C_2O_4^{2-}$ in alkaline conditions (C + III).
4. Nitro group goes to NO_3^-, P group goes to PO_4^{3-}, and S group goes to SO_4^{2-}.

There is some credibility to the proposed reactions since the moles of $KMnO_4$ consumed per mole of compound was in good agreement with the calculated values (Gomaa and Faust, 1971a).

B. Reaction kinetics. Well-designed kinetic experiments yield data from chemical oxidation systems that will:

1. Confirm therodynamic reactions models with respect to feasibility and stoichiometry,
2. Order of the disappearance reaction, and
3. Rate and reaction times required for process design.

Determination of the order of the pesticidal oxidation reactions may be accomplished by fitting the data into several kinetic equations. These reactions are, of course, extremely complex, and an exact order is not expected over the entire course of events. Intermediate oxidation products will compete with the parent molecule for the oxidant. Nonetheless, Gomaa and Faust (1971b) found that an integrated form of a second-order expression provided reasonably constant rate constants for the $KMnO_4$ oxidation of two dipyridylium quarternary herbicides - diquat and paraquat. This expression was:

$$\left(\frac{2.303}{aC^o_{ox} - bC^o_{herb}}\right)\left(\log \frac{C^o_{herb}\, C_{ox}}{C^o_{ox}\, C_{herb}}\right) = K_{OB}t$$

where t is reaction time in minutes, C^o_{herb} is the initial herbicide molar concentration, C_{herb} is the herbicide concentration at time t, C^o_{ox} is the initial oxidant concentration, C_{ox} is the oxidant concentration at time t, a is the number of herbicide moles given in the model reaction, and b is the number of oxidant moles given in the model reaction. The second-order rate constants were confirmed by effecting the reactions at three separate orders of magnitude in initial concentrations of the reactants (Gomaa and Faust, 1971b). The stoichiometric proportions were the same in each system, however.

Kinetically, the oxidative reactions were observed to be "slow". Table 7 shows the percentage of "complete" oxidation within a specific reaction time. For example, at a pH value of 7.4, only 31% of the initial concentration of parathion was oxidized within a reaction time of 44 hr. Note the very "slow" reaction of 2,4-D toward $KMnO_4$.

Not all chemical oxidations of organic pesticides are slow, however. In the decomposition of diquat and paraquat by ClO_2 at pH values of 8.14, 9.04, and 10.15, the rates were extremely rapid and could not be measured with experimental techniques. These reactions were complete in less than 1 min (Gomaa and Faust, 1971b).

TABLE 7. "Slow" Oxidation Reactions of Pesticides and $KMnO_4$.[a]

Parathion[b]			Paraoxon[b]		
pH Value	Time (hr)	Oxidation (%)	pH Value	Time (hr)	Oxidation (%)
3.1	44	83	3.1	92	48
5.0	44	53	5.0	92	25
7.4	44	31	7.4	116	14.5
9.0	6	91	9.0	3	95
Diquat[c]			Paraquat		
5.1	20	21	5.1	24	26
9.1	2	88	9.1	6	98
2,4-D[d]					
3.1	24	0			
7.4	24	0			
10.1	24	0			

[a]Temp. = 20°C, I = 0.02 M.
[b]$[KMnO_4]_o$ = 8 x 10^{-4} M, $[\text{Parathion}]_o$ = 3.95 x 10^{-5} M, $[\text{Paraoxon}]_o$ = 4.81 x 10^{-5} M.
[c]$[KMnO_4]_o$ = 10^{-3} M, $[\text{Diquat}]_o$ = 4.16 x 10^{-5} M, $[\text{Paraquat}]_o$ = 3.65 x 10^{-5} M.
[d]$[KMnO_4]_o$ = 158.0 mg/liter, [2,4-D] = 25.0 mg/liter.

In the design of water treatment processes, sufficient time should be afforded so that the oxidative reaction goes to completion. In this case, completion means the conversion of the organic C to CO_2(g). The reaction or contact times required for 50, 90, and 99% disappearance of an initial concentration of diquat were calculated for a $KMnO_4$ system. A pseudo-first-order rate constant was generated by multiplication of the second-order rate constant by the initial oxidant concentration. An assumption was made whereby this initial oxidant concentration is in excess of the diquat concentration, and little oxidant was consumed in the course of the reaction. These contact times are given in Gomaa and Faust (1971a).

c. Effect of $[H_3^+O]$. Apparently, one of the reaction variables affecting the kinetics of chemical oxidation of pesticides is the hydronium ion activity. There is sufficient evidence in the previous studies to demonstrate this point. For one reason or another, these oxidative reactions are faster when alkaline pH values prevail. For example, the rate of reaction of diquat-$KMnO_4$ at pH 9.13 is about 77 times faster than at pH 5.12, whereas the paraquat-$KMnO_4$ reaction is about 62 times faster at pH 9.13 than at 5.12 (Gomaa and Faust, 1971b). Similar observations were made with ClO_2 and Cl_2 as the oxidant (Gomaa and Faust, 1971b). In some systems, the effect of increasing the $[H_3^+O]$ leads to no reaction between the oxidant and pesticide. For example, ClO_2 had no effect on diquat and paraquat at pH values of 5.06, 6.17, and 7.12 (20°C) (Gomaa and Faust, 1971b). It should be obvious that a pH adjustment into the alkaline range is required to effect the fastest reaction and the least contact time.

d. Effect of temperature. That the rate constants of these chemical oxidative reactions vary greatly with temperature has been shown (Gomaa and Faust, 1971b). It is the general observation that reaction rates increase at higher temperatures.

e. Formation of intermediate products. Ideally, any chemical oxidation of an organic contaminant should carry the degradation of carbon to $CO_2(g)$. In the course of the reaction, intermediate products may be formed that, if not completely oxidized, may be more toxic or physiologically more harmful than the parent compound. This is especially important whenever an oxidative process is employed at a potable water treatment plant. It is extremely undesirable to have an incomplete oxidation and to have organic toxicants in the potable water.

Throughout the course of a reaction of $KMnO_4$ with parathion and paraoxon (Gomaa and Faust, 1971a), it was observed that deviation occurred from the second-order kinetic equation. This suggested the formation of intermediate organic products whose rates of oxidation were, in turn, slower than the parent molecule. Consequently, the $KMnO_4$-parathion, -paraoxon, -p-nitrophenol, and -2,4 dinitrophenol were studied and were reported in great detail elsewhere (Gomaa and Faust, 1971a, 1972). Figure 2 shows the proposed pathway in the $KMnO_4$ oxidation of parathion (Gomaa and Faust, 1971a). Under acidic and neutral conditions in the $KMnO_4$-parathion system, paraoxon was detected. These observations led, of course, to kinetic studies of the $KMnO_4$ oxidation of paraoxon and p-nitrophenol. Under acidic and neutral pH values, the paraoxon system did not yield any oxidation products that would have been detected by the analytical techniques

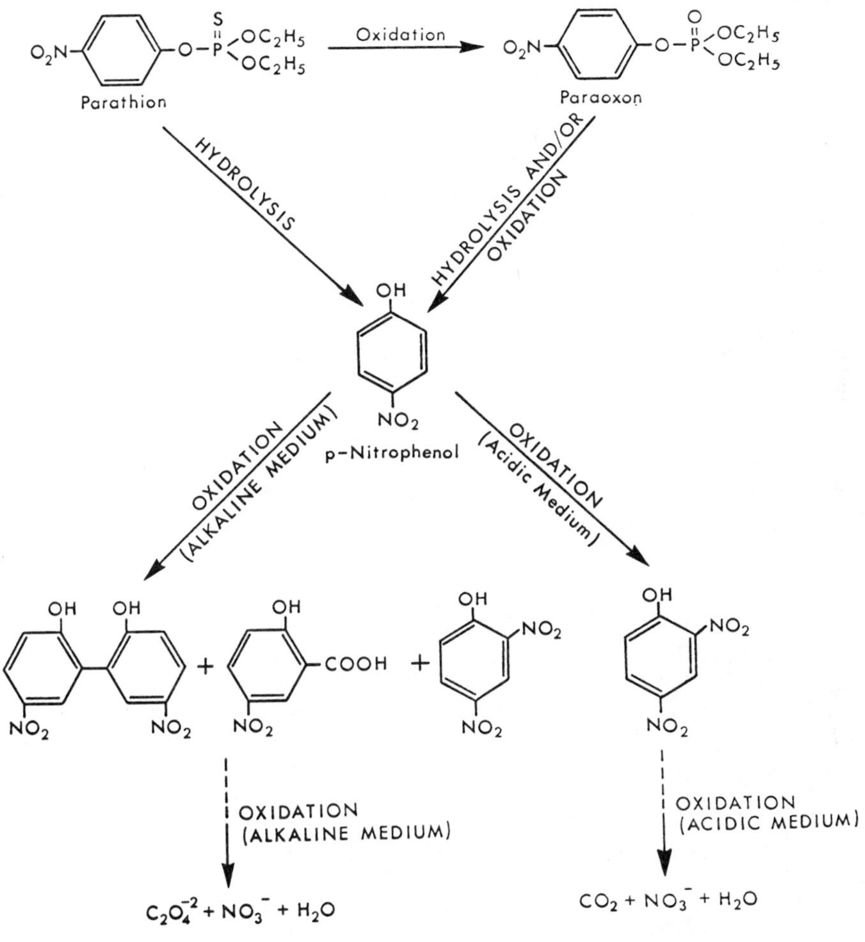

FIGURE 2. Proposed pathways for the $KMnO_4$ oxidation of parathion. Reprinted from Gomaa and Faust (1971a), p. 367, by courtesy of Marcel Dekker, Inc.

of GLC, TLC, and ultraviolet scan. Under alkaline conditions, however, p-nitrophenol was found. In the p-nitrophenol-$KMnO_4$ system, 2-4-dinitrophenol appeared as an intermediate product within the pH-value range of 3.1 through 9.0. Also two unknown spots appeared on the TLC plate at a pH value of 9.0. These unknowns were identified subsequently by infrared and mass spectroscopy as 2-hydroxy-5-nitrobenzoic acid and 2,2' dihydroxy-5,5' dinitrodiphenyl.

B. Photochemical Degradation

Among the environmental factors that influence the persistence of organic compounds in aquatic environments would be the decomposition of these compounds under the influence of solar radiation. Many organic pesticides undergo drastic changes upon exposure to ultraviolet light and artificial or natural sunlight (Mitchell, 1961; Weldon and Timmons, 1961; Jordan et al., 1964; Crosby et al., 1965; Crosby and Tutass, 1966; Eberle and Gunther, 1965; Abel-Wahah and Casida, 1967). Sunlight does not necessarily cause the degradation of pesticides into their constituent parts or into simpler compounds. In many cases, sunlight transforms pesticides into materials of similar or even greater structural complexity. The toxicology and persistence of these materials are, for the most part, unknown (Rosen, 1971). In spite of the difficulties involved in extending laboratory observations to natural environments, much useful information can be obtained from studying the photochemistry of pesticides.

Dieldrin was photoisomerized in sunlight and under laboratory conditions (Robinson et al., 1966; Parson and Moore, 1966) to photodieldrin. Similarly, aldrin was converted to photoaldrin in sunlight (Rosen and Sutherland, 1967). The reaction mixtures obtained from a 1-month exposure of an aldrin film contained 2.6% unaltered aldrin and 9.6% photoaldrin. The remainder consisted of 4.1% dieldrin, 24.1% photodieldrin, and 59.7% of the major product, an unidentified material with an average molecular weight of 482. The major product of aldrin photolysis was found subsequently to be approximately five times more toxic to flies than DDT. Yet, this material would not pass through gas chromatographic columns normally used. If this substance is an actual environmental product, it may go undetected.

Bell (1956) found 2,4-dichlorophenol when a distilled water solution of 2,4-D was exposed to ultraviolet light. On the other hand, no phenol was produced when a buffered solution at pH 7.0 was irradiated. It was proposed that ultraviolet decomposition of 2,4-D resulted in the formation of phenolic compounds. Also, rupture of the aromatic ring may result in the production of aliphatic products.

Aly and Faust (1964) researched the effect of ultraviolet irradiation on 2,4-D compounds (the Na salt, the isopropyl and butyl esters) in the laboratory. Apparently, the initial pH value of the irradiated solution influences the degree of decomposition of each compound. Decomposition was relatively slow at acidic pH values and became faster as the pH value was increased. Irradiation resulted in cleavage of the 2,4-D compounds at the ethereal linkage. 2,4-Dichlorophenol and some unknown free acids were produced as evidenced by a decrease in the pH value of the irradiated solutions. Under acidic conditions (pH 4.0), 2,4-dichlorophenol accumulated in the systems due to slow decomposition. On the other hand, no phenol was detected in the irradiated 2,4-D solutions at pH 9.0 because it is decomposed as fast as it is produced at this pH value.

Crosby and Tutass (1966) identified the intermediate products that resulted from ultraviolet irradiation of 2,4-D. In addition to fission of the ethereal bond, the compound underwent stepwise substitution of hydroxyl for chlorine until 1,2,4-trihydroxybenzene was formed. The latter was rapidly air-oxidized to the major product, a polymer whose infrared spectrum was similar to a sample of humic acid prepared from 2-hydroxybenzoquinone.

The herbicide 3-(p-bromophenyl)-1-methoxy-1-methyl urea (metobromuron) was converted to a 20% yield of a phenolic compound (compound 1) after exposure of a 225-mg/liter aqueous solution to sunlight for 17 days (Rosen and Strusz, 1968). Also, oxidation to compounds 2 and 3 occurred:

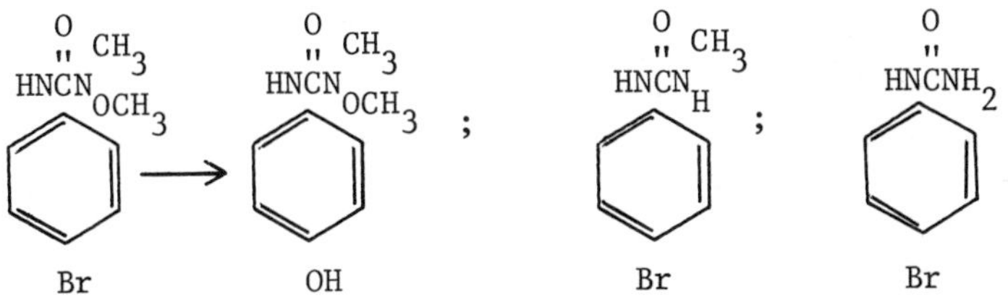

$HNC(=O)N(CH_3)OCH_3$ / Br → $HNC(=O)N(CH_3)OCH_3$ / OH ; $HNC(=O)N(CH_3)H$ / Br ; $HNC(=O)NH_2$ / Br

Metobromuron (1) (2) (3)

A material whose mass spectrum indicated that it was a condensation product of the parent compound (metobromuron) and product (1) was isolated also. The photolysis of linuron in aqueous solution proceeded in a manner similar to metobromuron as shown in the following reaction (Rosen and Strusz, 1968).

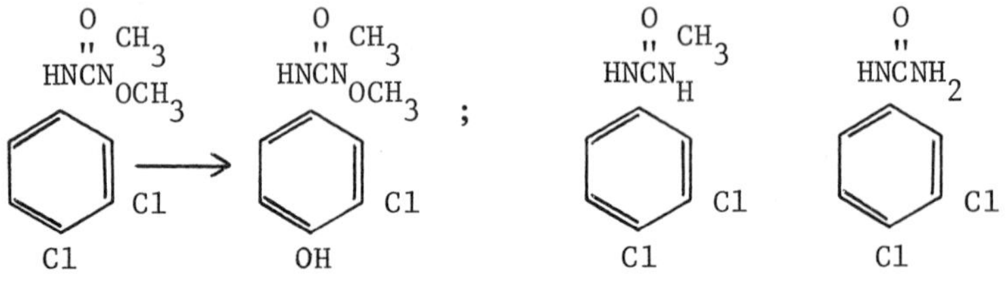

$$\text{Linuron (3-(3,4-dichlorophenyl)-1-methoxy-1-methylurea: } C_6H_3Cl_2\text{–NHC(O)N(CH}_3\text{)(OCH}_3\text{))} \longrightarrow (4)\ \text{HO, Cl–}C_6H_3\text{–NHC(O)N(CH}_3\text{)(OCH}_3\text{)}$$

$$(5)\ Cl_2C_6H_3\text{–NHC(O)N(CH}_3\text{)H} ; \quad (6)\ Cl_2C_6H_3\text{–NHC(O)NH}_2$$

Linuron (4) (5) (6)

The effect of ultraviolet irradiation upon the stability of the carbamate insecticides: 1-napthyl-*N*-methylcarbamate (sevin), *o*-isopropoxyphenyl-*N*-methylcarbamate (baygon), and 1-phenyl-3-methyl-5-pyrazolyl-dimethylcarbamate (pyrolan) in aqueous solutions was studied by Aly and El-Dib (1971): the pH value of the aqueous medium was an important factor in determining the rates of photolysis of sevin and baygon. The rates were slow at low pH values and tended to increase with an increase of pH value. However, the decomposition of pyrolan and dimetilan was not affected by the pH value of the irradiated medium. The primary effect of the ultraviolet light irradiation appeared to be the cleavage of the ester bond resulting in the production of the phenol or heterocyclic enol of the four carbamate esters. The effect of pH on the photolysis of sevin and baygon is analogous to that observed with 2,4-D (Aly and Faust, 1964) where the photodecomposition was much faster in alkaline than in neutral or acidic media. It was assumed that cleavage of the ester bond is not the only effect of the ultraviolet light but probably other modifications in the molecule tend to occur. Crosby et al. (1965) studied the photodecomposition of some carbamate esters. These investigators reported that the photodecomposition of sevin yielded, in addition to 1-naphthol, several cholinesterase inhibitory substances which indicate that these compounds retained the carbamate ester group intact. Also, irradiation resulted in changes at other positions in the sevin molecule.

It should be indicated that almost all studies of the photolyses of pesticides have been conducted by exposing pesticides to direct irradiation. Under actual environmental conditions, however, pesticides may be in intimate contact with materials that act as photosensitizers, which may absorb light energy and transfer this energy to the pesticide. This alters the products of direct irradiation or causing pesticides that do not absorb light to photolyze. For example, 3,4-dichloroaniline, a metabolite of linuron (Nashed and Ilnicki, 1967), diuron (Dalton et al., 1966), and propanil (Bartha and Pramer, 1967; Still and Kuzuiai, 1967), is farily stable to sunlight in aqueous solution. However, in the presence of riboflavin-5'-phosphate, a significant amount of photolysis occurs in a few hours. Several photoproducts of the sensitized reactions have been isolated, two of which have been identified as azo compounds. These materials are of interest in that they belong to a family of compounds that includes several members with carcinogenic properties (Weisburger and Weisburger, 1966).

Sunlight energy catalyzes the decomposition of organic compounds in natural waters. An excellent example is provided by Hedlund and Youngson (1972) who examined the sunlight-caused photodecomposition of picloram (4-amino-3,5,6-trichloropicolinic acid) under natural conditions. Table 8 gives a summary of the experimental variables and kinetics of the decomposition reaction. Pseudo-first-order kinetics was observed. The experimental conditions were:

1. Case 1: A hazy sunshine study conducted at Seal Beach, California. Thirty days of exposure, 26 of which were actual sunshine.

2. Case 2: Photodegradation was conducted in 3.65 m-deep containers at Walnut Creek, California. Solutions were circulated.

3. Case 3: Photodecomposition was conducted in distilled and canal water and in 2.54 cm-deep trays at Walnut Creek, California.

4. Case 4: Initial picloram concentrations were varied. Biological and radiochemical assays indicated that the photolysis products were not phytotoxic or in nondetectable concentrations.

5. Case 5: An effect of depth study. Columns were exposed 4 hr/day when the sun was most directly overhead.

It appears that picloram does disappear from water via photodecomposition under natural conditions and in a relatively short period of time.

TABLE 8. Summary of Experimental Variables and Kinetics - Photodecomposition of Picloram[a]

Case	Initial Concentration[b] ($M x 10^6$)	Solution depth	Time of year	Other variables	Rate constant	Half life
1	37.3	8 cm	March	Actual days	0.0738/day	9.4 days
				Estimated sunshine	0.0855/day	8.1 days
2	4.14	3.65 m	September-October	-	0.0168/day	41.3 days
3	20.7	2.54 cm	August	Distilled water	0.306/day	2.3 days
				Canal water	0.280/day	2.5 days
4[c]	414.1	4.6 cm	July	Bioassy	0.116/day	6.0 days
				$BuOH/NH_3$ Chromatography	0.0956/day	7.3 days
				$C_6H_6/C_2H_5COOH/H_2O$ Chromatography	0.0915/day	7.6 days
5[d]	0.409	0.292 m	August-September	-	$8.83 x 10^{-3}$/hr	78.5 hr
	0.401	1.82 m		-	$4.49 x 10^{-3}$/hr	154 hr
	0.421	3.65 m		-	$3.01 x 10^{-3}$/hr	230 hr

[a]After Hedlund and Youngson (1972). Reprinted with permission of the American Chemical Society.

[b]For conversion to other units, 4.14×10^{-6} M picloram is about equal to 1 ppm or 2.72 lb/acre-foot of water.

[c]Calculations based on construction method.

[d]Time values based on noncontinuous exposure, 4 hr/day at midday.

Photodecomposition products may be extremely signficant environmental problems. It appears that ultraviolet irradiation of organic pesticides is "incomplete". That is, the parent molecule is not degraded completely to $CO_2(g)$. Intermediate or even final decomposition products appear in many systems that, in turn, may have toxic properties or may impart some adverse organoleptic quality to the water and so on. For example, Crosby and Tutass (1966) examined the photodecomposition of 2,4-D (XII in Figure 3) and found several decomposition products. As seen in Figure 3, the major reaction is cleavage of the ether bond to

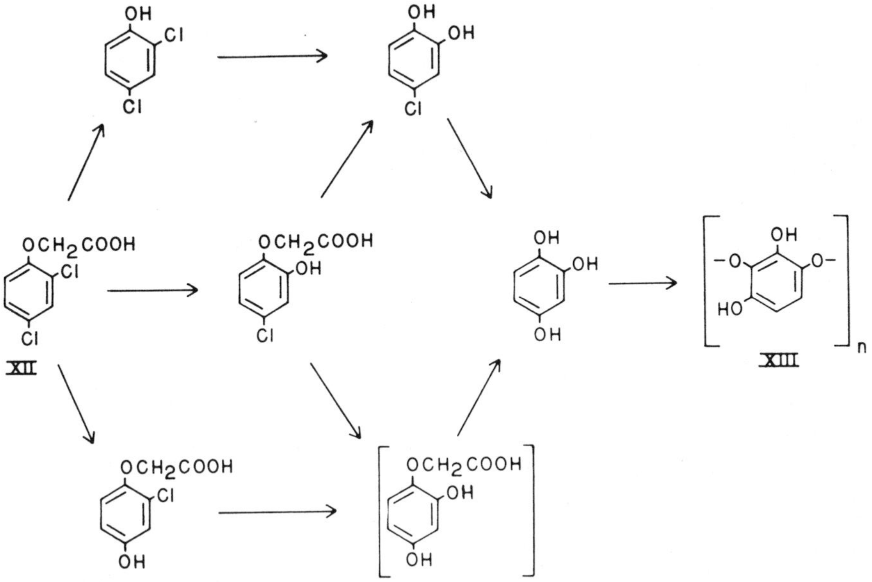

FIGURE 3. Photolysis of 2,4-dichlorophenoxy acetic acid. After Crosby and Tutass (1966). Reprinted with permission from the J. Agr. Food Chem. Copyright by the American Chemical Society.

produce 2,4-dichlorophenol, which is dehalogenated to 4-chlorocatechol and then to 1,2,4-benzenetriol. In turn, this compound is oxidized rapidly in air to a mixture of polyquinoid humic acids (XIII in Figure 3) by a light independent process. Crosby and Tutass subsequently stated: "the effects observed from irradiation in sunlight were qualitatively very similar to results from laboratory experiments." This research is an excellent example of potentially hazardous products being produced from partial photodecomposition.

It seems reasonable, therefore, to suggest that photodecomposition may account for some loss of the pesticide residues in clear surface waters exposed to long periods of sunlight. However, photolysis may be a minor factor in the decomposition of the pesticide residues in highly turbid waters where the penetration of light will be greatly reduced.

C. Hydrolytic Transformations

Many organic compounds, in general, and organic pesticides, specifically, may undergo hydrolytic transformations in aquatic environments. Many of these compounds have been synthesized as an organic ester of some sort. This is especially true for the pesticides that are phosphates, carbamates, ureas, and phenoxyacetates. For example, the organophosphorus pesticide is a tertiary phosphate or thiophosphate ester:

```
           S(O)
R - O       ||
     \
      P - O - R'
     /
R - O      (S)
```

R is usually a methyl or an ethyl group and R' is an organic moiety. Hydrolytic transformation of these compounds occur from rupture of either the P-O(S) bond (alkaline conditions) or the (S)O-R' bond (acidic conditions). Very seldom is the hydrolysis carried to the point where the R-O bond is ruptured (tertiary hydrolysis). Once again the nature of the R' group becomes important because the hydrolytic product may be environmentally hazardous.

The kinetics of hydrolysis was determined for parathion and paraoxon (Gomaa and Faust, 1972) and for diazinon and diazoxon (Gomaa et al., 1969). These two studies departed from previous studies wherein the hydrolytic reactions were effected in purely aqueous systems and at temperatures simulating natural water conditions. Also, the progress of the reactions was directly determined for disappearance of the parent molecule as well as appearance of the products. First-order kinetic behavior was observed for the hydrolytic transformation of these four organophorus compounds. Parathion and paraoxon are relatively stable in acidic and neutral conditions. The half-life values observed for parathion and paraoxon (174 to 108 days, 197 to 144 days) at pH values 3.1 to 7.4 are sufficient to permit environmental damage in a natural water body. On the other hand, diazoxon is relatively short-lived, especially under high acidic and alkaline conditions.

Temperature is an environmental factor that affects the rates at which hydrolysis reactions occur. The energy of activation values were calculated from the temperature effect. Those reactions that yield the higher Ea values will exhibit the greatest effect of temperature upon their rate of hydrolysis. For example, parathion has an activation energy of 16.4 kcal/mol at a pH value of 3.1, whereas diazinon has a value of 13.1 parathion shows the greater difference in the $t_{1/2}$ values at the two temperatures (Gomaa and Faust, 1972; Burnham et al., 1972). Two rather important environmental implications evolve from these temperature studies: (a) those compounds with the higher Ea values are more persistent at the lower temperatures of 0 to 20°C, and (b) hydrolysis is generally slower under acidic conditions.

Esters of carbamic acid (the carbamate pesticides) also hydrolyze, but mostly under alkaline conditions:

$$R - O - \overset{O}{\overset{\|}{C}} - \overset{R'}{\overset{|}{N}} - R'' + H_2O \xrightarrow{(OH^-)} ROH + R' - \overset{H}{N} - R'' + CO_2(g)$$

where the products are a hydroxy compound (phenol), an amine, and CO_2(g). Aly and El-Dib (1971) reported the hydrolytic stability of four carbamates: sevin, baygon, pyrolan, and dimetilan over the pH value range of 2 to 10. First-order kinetics of hydrolysis was observed from which the half life values were calculated (Aly and El-Dib, 1971). Pyrolan and dimetilan did not hydrolyze within this pH range and at 20°C. Baygon resisted hydrolysis at pH values 3 through 7, but did decay under alkaline conditions. Sevin was the least stable of the four carbamates with hydrolysis occurring at pH values 7.0 and above with the rate increasing as the $[OH^-]$ was increased.

In an attempt to simulate natural conditions, Bailey et al. (1970) studied the hydrolysis of the propylene glycol butyl ether ester of 2-(2,4,5)TP (silvex) in three pond waters. There was an apparent rapid decay of this ester as seen by the $t_{1/2}$ values of 5, 7, and 8 hr, respectively.

Hydrolysis is, indeed, an important variable affecting the fate of the appropriate organic compound in aquatic environments. The rate at which hydrolysis occurs is, of course, unique to the individual compound and is dependent, also, upon such environmental factors as $[H_3O^+]$ and temperature (Faust and Gomaa, 1972). Hydrolytic stability should be viewed with concern about the length of the time period required for complete hydrolysis and about the products that may be more toxic than the parent molecule.

VI. POSTSCRIPT - CHLORINATED HYDROCARBONS

After initiation of preparation of this manuscript, newspaper headlines and television programs featured the "discovery" of chlorinated hydrocarbons in drinking water (Anon., 1974). It is charged that halogenated hydrocarbons, especially trichloromethane (chloroform), are formed in the chlorination process for disinfection. Furthermore, it is suggested that these organic compounds may be physiologically harmful to humans because chloroform, for example, may be carcinogenic to laboratory animals (mice). Let us consider the evidence to date.

Rook (1974) of the Rotterdam Waterworks has observed the formation of several haloforms in natural surface waters during chlorination. Gas-liquid chromatographic and mass spectrometric techniques were employed to identify $CHCl_3$, CCl_4, $CHCl_2Br$, $CHClBr_2$, and $CHBr_3$ and traces of CH_2Cl_2, CH_2Br_2, and $C_2H_2Cl_2$. The observation of these compounds led to two perplexing questions: (a) What is the precursor? and (b) What is the source of the bromine? For the latter question, no satisfactory explanation was given for a source of bromine. Two possibilities exist: (a) contamination of the chlorine gas with elemental bromine and (b) formation of Br_2 from naturally-occurring Br^- in the surface water during chlorination. No evidence was offered, however, to support either source.

Rook's search for the precursor to the haloforms was more fruitful. The possibility for acetone (0.04 mg/liter) to serve in the reaction was confirmed by experiments in which $CHCl_3$, CH_2Cl_2Br, $CHClBr_2$, and $CHBr_2$ were found after 2-hr contact with 0.02 mg/liter Cl_2 and 1.4 mg/liter NaBr. Attempts to find acetone in the natural river waters at concentrations high enough (2 to 9 mg/liter) to yield significant quantities of the haloforms were unsuccessful. Instead, Rook proposed and presented compelling evidence that the polyhydroxybenzene building blocks of natural color molecules are responsible for the haloform reaction. These four compounds were formed from the chlorination of an aqueous infusion of peat in the presence of Br^-: $CHCl_3$, $CHCl_2Br$, $CHClBr_2$, and $CHBr_3$. In addition, haloforms were formed from the chlorination of several polyhydric phenols: pyrogallol; phloroglucinol; the *o*-, *m*, and *p*-dihydroxybenzenes and from two natural polyphenols, hesperidine and phlorizine.

Bellar et al. (1974) reported the occurrence of organohalides in raw surface waters, drinking water, and waste water. Chloroform (94.0 μg/liter), bromodichloromethane (20.8 μg/liter), dibromochloromethane (2.0 μg/liter), and ethyl alcohol were recovered and identified in finished drinking water by gas-liquid chromatographic-mass spectrometric techniques. Chloroform was found in the raw water source at a rather constant value (with time), 0.9 ± 0.2 μg/liter. This implies, of course, that chloroform is formed in the water treatment plant from the chlorination process as seen in Table 9. Several grab samples of waste water (mixture of industrial and domestic) collected from a local treatment plant contained these compounds: methylene chloride; chloroform; 1,1,1-trichloroethane; 1,1,2-trichloroethylene; 1,1,2,2-tetrachloroethylene; 1,1,2,2-tetrachloroethylene; and di- and trichlorobenzenes. Of course, some of these compounds could have industrial origins rather than being formed in the water. Again bromo-compounds were found for which no source was reported.

TABLE 9. Trihalogenated-Methane Content of Water from Water Treatment Plant[a]

Sample source	Sample point	Free Cl (ppm)	Chloroform	Bromodichloromethane	Dibromochloromethane
Raw river water	1	0.0	0.9	b	b
River water treated with chlorine and alum-chlorine contact time - 80 min	2	6	22.1	6.3	0.7
3-day-old settled water	3	2	60.8	18.0	1.1
Water flowing from settled area to filters[e]	4	2.2	127	21.9	2.4
Filter effluent	5	d	83.9	18.0	1.7
Finished water	6	1.75	94.0	20.8	2.0

[a]After Bellar (1974). Reprinted from Journal American Water Works Association, Volume 66, by permission of the Association. Copyrighted 1974 by the American Water Works Association, Inc., 6666 West Quincy Avenue, Denver, Colorado 80235.

[b]None detected. If present, the concentration is <0.1 μg/liter.

[c]Carbon slurry added at this point

[d]Unknown.

Similar results and compounds were reported by Dowty et al. (1975) for the drinking water of New Orleans, Louisiana. In addition, five halogenated hydrocarbons were isolated from blood plasma pooled from eight subjects. These compounds were found: 1-chloropropene, chloroform, carbon tetrachloride (plasma also), dichloroethane, trichloroethylene, dichloropropane, dichloropropene, bromodichloromethane, tetrachloroethylene (plasma also), and dibromochloromethane. In addition, three isomers of dichlorobenzene were found in the blood plasma. Speculation was offered that drinking water was the source of the halogenated hydrocarbons for the blook plasma.

In this treatise Kopfler et al. (1977) (Section V, Part 2) report 187 organic compounds were found in United States drinking water.

VII. ENVIRONMENTAL SIGNIFICANCE OF ORGANIC CONTAMINANTS

The scientific community must provide society with objective information about the potential hazards of organic contaminants in aquatic environments. There was a considerable public controversy as this chapter was being prepared about the detection of halogenated hydrocarbons in public drinking waters, especially in New Orleans, Louisiana. The reports (Rook, 1974; Bellar, 1974; Dowty, et al., 1975; Kleopfer and Fairless, 1972) are examples of excellent analytical detection. What is, however, the physiological threat of these organics to man? Is there sufficient evidence, to date, to indicate the water treatment process of chlorination as the cause of these compounds? Should the water utilities rush to alternative means of disinfection? These and many other questions must be asked and answered before environmental panic reigns. In discussion of these points, it is assumed that man is the target organism.

The effects of organic compounds on the water quality of man is reasonably well documented. Namely, the aesthetic effects of clarity, color, taste, and odor are well known. On the other hand, is there a physiological threat to man by individual or combinations of organic compounds? The obvious concern and threat is the one of cancer in man. Let us examine some of the evidence.

Whenever the cancer-in-man charge is made, the data should be reviewed with several points in mind. Carcinogenic evidence usually comes from reasonably well-controlled laboratory animal studies. First, can the information be extrapolated from laboratory animals to man? Toxicologists disagree on this point. Some say yes, whereas others say no. Second, a large proportion of the studies to date have been conducted with the CF-1 inbred strain of mouse. This strain has a high incidence of naturally-occurring, benign, hepatic nodules and to a lesser extent, hepatocellular carcinoma (a malignant growth). It is extremely difficult to separate statistically the cause and effect relationship against a high background of tumors in the control group. Third, there is the question of benign tumors becoming malignant. Many compounds produce a high incidence of these tumors in test animals. Again, the toxicologists disagree upon the transformation of benign into malignant tumors. Fourth, there is the question of semantics. Hepatonia is a term commonly employed to describe carcinogenic effects. It is a noncommittal term that is used for benign or malignant tumors. It is frequently, however, interpreted by the lay public to describe malignant situations. Let us examine two specific situations namely, aldrin and dieldrin, and CCE and CAE. There is evidence that dieldrin enhanced the incidence of naturally-occurring hepatic nodules in the CF-1 strain of mice (Walker et al., 1971) and significantly increased the incidence of histologically benign liver tumors (Davis and Fitzhugh, 1962). No carcinogenic action has been demonstrated for dieldrin in rats, dogs, or primates.

The largest single study of the effects of aldrin and dieldrin in man, frequently cited, was by Jager (1970). This was based on the occupational exposure of more than 800 workers at an insecticide plant over a period of about 15 yr. From these workers, a group of 233 was selected for long-term exposure studies and consisted of personnel who had been exposed at least 4 yr and up to 13.25 yr not only to aldrin and dieldrin, but to endrin and telodrin as well. The average blood level of dieldrin in this group was 0.035 μg/ml. This was equivalent to an average oral intake of 407 μg/man/day and represents over 50 times the daily intake of the general population of the United States. Medical examinations of this group of workers revealed no adverse effects from this type of exposure.

The Water Quality Criteria - 1972 (Anon., 1972a), makes the recommendation: "Because large values of CCE and CAE are aesthetically undesirable and represent unacceptable levels of unidentified organic compounds that may have adverse physiological effects, and because the defined treatment process has little or no effect on the removal of these organics, it is recommended that organics-carbon adsorbable as measured by the Low-Flow Sampler not exceed 0.3 mg/l CCE and 1.5 mg/l CAE in public water supply sources." There is conflicting evidence on the "adverse physiological effects." As mentioned previously, Dunham et al. (1967) concluded that CCE and CAE pollutants were not responsible for inducement of tumors in mice during an experimental period of 78 weeks. On the other hand, Hueper and

Payne (1963) stated "chloroform eluates, as well as alcohol eluates of such adsorbates on activated carbon prepared from raw and finished water, when subcutaneously injected or cutaneously applied, were probably the cause of leukemic reactions in some mice." In more recent studies, Tardiff and Deinzer (1973) reported the LD_{50} acute toxicity values for three CCE and one CAE toward male Carworth CF1 mice. The LD_{50} values for the three CCE were 32, 35, and 89 mg/kg and for the CAE, 84 mg/kg. These data are difficult to interpret, since no control data were given for comparison. In fact, the authors suggest, "The interpretation of the toxicity of mixtures of organic compounds as in CCE and CAE must be viewed with caution, particularly if conclusions are to be drawn about the toxicity of the source water. The process of eluting compounds from activated carbon is very harsh and undoubtedly creates many alterations in the structural configuration of the molecules. Such changes will affect the toxicity of the entire composition, but in a direction that is presently unpredictable."

Thus, we have an incomplete, conflicting, and controversial problem of potentially hazardous organic compounds in aquatic environments. These conflicts must be resolved. There is sufficient evidence and concern to justify a concerted research effort. Answers must be obtained so that man's aquatic environment is, indeed, potable.

ACKNOWLEDGMENTS

Paper of the Journal Series, New Jersey Agricultural Experiment Station, Department of Environmental Sciences, Cook College, Rutgers, The State University, New Brunswick, New Jersey 08903.

REFERENCES

1. Abbott, D. C., et al. 1965. Nature 208 (5017), 1317.

2. Abbott, D. C., et al. 1966. Nature 211 (5046), 259.

3. Abdel-Wahab, A. M. and Casida, J. E. 1967. J. Agr. Food Chem. 15, 479.

4. Adams, R. S., Jr. and Li, P. 1971. Soil Sci. Soc. Amer. Proc. 35, 78.

5. Ahling, B. and Jensen, S. 1970. Anal. Chem. 42 (13), 1483.

6. Aly, O. M. and El-Dib, M. A. 1971. In Organic Compounds in Aquatic Environments (S. D. Faust and J. V. Hunter, Eds.), New York, Chapter 20.

7. Aly, O. M. and Faust, S. D. 1964. J. Agr. Food Chem. 12, 541.

8. Aly, O. M. and Faust, S. D. 1965. J. Amer. Water Works Assoc. 57, 221.

9. Andelman, J. B. and Snodgrass, J. E. 1974. Crit. Rev. Environ. Cont. 4 (1), 69.

10. Anon. 1970. European Standards for Drinking Water, 2nd Ed., World Health Organization, Geneva, Switzerland.

11. Anon. 1953. Geological Survey Water-Supply Paper 1163, U.S. Dept. Interior, Washington, D.C.

12. Anon. 1972a. Water Quality Criteria-1972, Committee on Water Quality Criteria, Environmental Studies Board, Nat. Acad. Sci., Washington, D.C.

13. Anon. 1972b. Control of Oil and Other Hazardous Materials, Training Manual, U.S. Environmental Protection Agency, Office of Water Programs, Washington, D.C., September.

14. Anon. 1974. Willing Water (AWWA) 18 (12), 4.

15. Bailey, G. W. and White, J. L. 1970. Residue Rev. 32, 29.

16. Bailey, G. W., et al. 1970. Weed Sci. 18, 413.

17. Bartha, R. and Pramer, D. 1967. Science 156, 1617.

18. Bell, G. R. 1956. Bot. Gaz. 118, 133.

19. Bellar, T. A. and Lichtenberg, J. J. 1974. J. Amer. Water Works Assoc. 66, 703.

20. Breidenbach, A. W., et al. 1967. Pub. Health Rept. 82 (2), 139.

21. Burnham, A. K., et al. 1972. Anal. Chem. 44, 139.

22. Burnham, A. K., et al. 1973. J. Amer. Water Works Assoc. 65, 722.

23. Caro, J. H. and Taylor, A. W. 1971. J. Agr. Food Chem. 19, 379.

24. Carter, F. L. and Stringer, C. A. 1970. Bull. Environ. Contam. Toxicol. 5, 422.

25. Caruso, S. C., et al. 1966. Air Water Pollut. 10, 41.

26. Christman, R. F. and Ghassemi, M. 1966. J. Amer. Water Works Assoc. 58 (6), 723.

27. Cohen, J. M. and Pinkerton, C. 1966. In Organic Pesticides in the Environment (A. A. Rosen and H. F. Kraybill, Symposium Chairmen), Adv. Chem. Ser. 60, Amer. Chem. Soc., Washington, D.C., Chapter 13.

28. Cohen, J. M., et al. 1960. J. Amer. Water Works Assoc. 52, 1551.

29. Cohen, J. M., et al. 1961. J. Amer. Water Works Assoc., 53, 49.

30. Cole, H., et al. 1970. Bull. Environ. Contam. Toxicol. 2, 127.

31. Crosby, D. C. and Tutass, H. O. 1966. J. Agr. Food Chem. 14, 596.

32. Crosby, D. C., et al. 1965. J. Agr. Food Chem. 13, 204.

33. Cueto, C. and Biros, F. J. 1967. Toxicol. Pharmacol. 10, 261.

34. Dalton, R. L., et al. 1966. Weeds 14, 31.

35. Davis, K. J. and Fitzhugh, O. G. 1962. Toxicol. Appl. Pharmacol. 4, 187.

36. Dixon, J. B., et al. 1970. Soil Sci. Soc. Amer. Proc. 34, 805.

37. Dorn, H. F. and Cutler, S. J. 1959. Morbidity from Cancer in the United States, PHS Monogr. No. 56, Govt. Printing Office, Washington, D.C.

38. Dowty, B., et al. 1975 Science 187, 75.

39. Duke, T. W., et al. 1970. Bull. Environ. Contamin. Toxicol. 5 (2), 171.

40. Dunham, L. J., et al. 1967. Amer. J. Pub. Health 57 (12),

41. Eberle, D. O. and Gunther, F. A. 1965. J. Assoc. Off. Agr. Chem. 48, 927.

42. Elgar, K. E. 1971. In Pesticides Identification of the Residue Level, (F. J. Biros, Symposium Chairman), Adv. Chem. Ser. 104, Amer. Chem. Soc., Washington, D.C., Chapter 10.

43. Eye, J. D. 1968. J. Water Pollut. Control Fed. 40, (8), R316.

44. Faust, S. D. and Gomaa, H. M. 1972. Environ. Letts. 3 (3), 171.

45. Feltz, H. R., et al. 1971. Pestic. Monit. J. 5 (1), 54.

46. Frazier, B. E., et al. 1970. Pestic. Monit. J. 4, 67.

47. Gomaa, H. M. and Faust, S. D. 1971a. In Organic Compounds in Aquatic Environments, (S. D. Faust and J. V. Hunter, Eds.), New York. Chapter 15.

48. Gomaa, H. M. and Faust, S. D. 1971b. J. Agr. Food Chem. 19, 302.

49. Gomaa, H. M. and Faust, S. D. 1972. In Fate of Organic Pesticides in Aquatic Environments, (S. D. Faust, Symposium Chairman), Adv. Chem. Ser. 111, Amer. Chem. Soc., Washington D.C., Chapter 10.

50. Gomaa, H. M. et al. 1969. Residue Rev. 29, 171.

51. Green, R. S. and Love, S. K. 1967. Pestic. Monit. J. 1 (1), 13.

52. Grzenda, A. R., et al. 1964. J. Econ. Entomol. 57, 615.

53. Guenzi, W. D. and Beard, W. E. 1970. Soil Sci. Soc. Amer. Proc. 34, 443.

54. Gunther, F. A. and Blinn, R. C. 1955. Analysis of Insecticides and Acaricides. Vol. 6, Wiley (Interscience), New York, p. 476.

55. Hague, R. and Freed, V. H. 1974. Residue Rev. 52, 89.

56. Hance, R. J. 1967. Weed Res. 7, 29.

57. Harris, C. I. 1969. J. Agr. Food Chem. 17, 80.

58. Harris, C. I. and Warren, G. F. 1964. Weeds 12, 120.

59. Hedlund, R. T. and Youngson, C. R. 1972. In Fate of Organic Pesticides in Aquatic Environments, (S. D. Faust, Symposium Chairman), Adv. Chem. Series 111, Amer. Chem. Soc., Washington, D.C., Chapter 8.

60. Heuper, W. C. and Payne, W. W. 1963. Amer. J. Clin. Pathol. 39, 475.

61. Hindin, E., et al. 1966. In Organic Pesticides in the Environment. (A. A. Rosen and H. F. Kraybill, Symposium Chairmen), Adv. Chem. Ser. 60, Amer. Chem. Soc., Washington, D.C., Chapter 11.

62. Huang, J. C. and Kiao, C. S. 1970. J. Sanit. Eng. Div. (ASCE), Oct., 1057.

63. Jager, K. W. 1970. Aldrin, Dieldrin, Endrin, and Telodrin. An Epidemiological and Toxicological Study of Long-Term Occupational Exposure, Elsevier, Amsterdam.

64. Jordan, L. S., et al. 1964. Weeds 12, 1.

65. Kallman, B. J., et al. 1962. Trans. Amer. Fish Soc. 91, 14.

66. Kearney, P. C. and Kaufman, D. D. 1969. Degradation of Herbicides. Dekker, New York.

67. Khan, S. U. 1974. Environ. Sci. Tech. 8, 236.

68. Kleopfer, R. D. and Fairless, B. J. 1972. Environ. Sci. Tech. 6 (12), 1036.

69. Lauer, G. J., et al. 1966. Amer. Fish. Soc. Trans. 95, 310.

70. Lichtenberg, J. J., et al. 1970. Pestic. Monit. J. 4 (2), 71.

71. Lovelock, J. E. and Lipsky, S. R. 1960. J. Amer. Chem. Soc. 82, 431.

72. Middleton, F. M. and Lichtenberg, J. J. 1960. Indust. Eng. Chem. 52, 99A.

73. Mitchell, L. C. 1961. J. Assoc. Off. Agr. Chem. 44, 643.

74. Nashed, R. B. and Ilnicki, R. D. 1967. Proc. Northeast Weed Control Conf. 21, 564.

75. Parson, A. M. and Moore, D. J. 1966. J. Chem. Soc. 2026.

76. Pionke, H. B. and Chesters, G. 1973. J. Environ. Qual. 2 (1), 29.

77. Poirrier, M. A., et al. 1972. Environ. Sci. Technol. 6, 1033.

78. Reynolds, L. M. 1971. Residue Rev. 34, 27.

79. Robeck, G., et al. 1965. J. Amer. Water Works Assoc. 57, 181.

80. Robinson, J., et al. 1966. Bull. Environ. Contam. Toxicol. 1, 127.

81. Rook, J. J. 1974. Soc. Water Treat. Exam 23 (2), 234.

82. Rosen, A. A., et al. 1962. J. Water Pollut. Control Fed. 34, 7.

83. Rosen, A. A., et al. 1963. J. Water Pollut. Control Fed. 35, 777.

84. Rosen, A. A. 1969. Proc. 11th Sanit. Eng. Conf., Univ. Illinois, Urbana, p. 59.

85. Rosen, J. D. and Strusz, R. F. 1968. J. Agr. Food Chem. 16, 568.

86. Rosen, J. D. and Sutherland, D. J. 1967. Bull. Environ. Contam. Toxicol. 2, 1.

87. Schafer, M. L, et al. 1969. Environ. Sci. Technol. 3 (12), 1261.

88. Stanley, C. W., et al. 1971. Environ. Sci. Technol. 5 (5), 430.

89. Stewart, R. 1965. In Oxidation in Organic Chemistry. (K. Weberg, Ed.), Academic, New York, p. 36.

90. Still, C. C. and Kuzuian, O. 1967. Nature 216, 799.

91. Tarrant, K. R. and Tatton, J. O'G. 1968. Nature 219, 725.

92. Terriere, L. C., et al. 1966. In Organic Pesticides in the Environment. (A. A. Rosen and H. F. Kraybell, Symposium Chairmen). Adv. Chem. Ser. 60, Amer. Chem. Soc., Washington, D.C., Chapter 21.

93. Veith, G. D. and Lee, G. F. 1971. Environ. Sci. Technol. 5, 230.

94. Voerman, S. and Besemer, A. F. H. 1970. J. Agr. Food Chem. 18, 717.

95. Walker, A. I. T., et al. 1971. The Toxicology of Dieldrin: Long-Term Oral Toxicity Experiments in Mice, unpublished data.

96. Weaver, L., et al. 1965. Pub. Health Rept. 80, 481.

97. Weber, J. B. 1972. In Fate of Organic Pesticides in Aquatic Environments. (S. D. Faust, Symposium Chairman), Adv. Chem. Ser. 111, Amer. Chem. Soc., Washington, D.C., Chapter 4.

98. Weebel, S. R., et al. 1966. J. Amer. Water Works Assoc. 58, 1075.

99. Weed, S. B. and Weber, J. B. 1969. Soil Sci. 29, 379.

100. Weisburger, J. H. and Weisburger, E. K. 1966. Chem. Eng. News 124 (February 7.)

101. Weldon, L. W. and Timmons, F. L. 1961. Weeds 9, 111.

102. Wershaw, R. L., et al. 1961. Environ. Sci. Technol. 3, 271.

103. Westlake, W. E. 1967. In Pesticides Identification at the Residue Level. (F. J. Biros, Symposium Chairman). Adv. Chem. Ser. 104, Amer. Chem. Soc., Washington, D.C., Chapter 5.

104. Wheatley, G. A. and Hardman, J. A. 1965. Nature 207 (4996), 486.

105. Wheeler, W. B. and Frear, D. E. H. 1966. Residue Rev. 16, 86.

106. Williams, J. D. H., 1968. Weed Res. 8, 237.

107. Yaron, B., et al. 1967. J. Agr. Food Chem. 15, 671.

108. Zogorski, J. S. 1975. Ph.D. thesis, Rutgers University.

Interaction of Humic Substances with Herbicides in Soil and Aquatic Environments*

SHAHAMAT U. KHAN
Chemistry and Biology Research Institute
Canada Agriculture
Ottawa, Ontario, K1A 0C6, Canada

I. INTRODUCTION

Herbicides are being used in increased quantity for weed control in soils and waters, and new compounds are being added to the list every year. An understanding of their interactions with humic substances in soil and aquatic environments is required for effective use. Humic substances are widely distributed in nature, occurring in soils (Schnitzer and Khan, 1972), in lakes (Ishiwatari, 1969), in rivers (Lamar, 1968), and in the sea (Rashid and King, 1969). They are remarkable materials in that the small amounts present in soils and waters may influence the behavior of herbicides significantly. The importance and role of humic substances has been indicated in studies conducted over the past decade for a wide variety of herbicides. Numerous examples where binding, persistence, chemical- and biodegradation, leachability, and translocation have been shown to bear a direct relationship to humic substances content can be found in recent review articles by Hayes (1970), Burns and Hayes (1974), Stevenson (1972a,b), and Khan 1972, 1974a). We have directed our efforts for the last few years in the study of the interaction of herbicides with humic substances in order to understand this problem. It is the purpose of this chapter to review and discuss some of the work we have done in this field.

*Contribution No. 855.

II. HUMIC SUBSTANCES

Of fundamental importance to any such study is a basic understanding of humic substances. Organic matter in soils, sediments, and waters contains a wide variety of organic compounds, which may be classified conveniently into two main groups, nonhumic substances and humic substances. Nonhumic substances include those with still recognizable physical and chemical characteristics such as carbohydrates, proteins, peptides, amino acids, fats, waxes, alkanes, and low-molecular-weight organic acids. Most of these substances have a short survival rate in soil and aquatic environments as they can be relatively easily attacked by microorganisms. The major portion of the organic matter in most soils and waters, however, consists of humic substances. These are amorphous, dark colored, hydrophilic, acidic, partly aromatic, chemically complex organic substances that range in molecular weights from a few hundred to several thousand.

Based on their solubility in alkali and acid, humic substances are partitioned into three main fractions: (a) humic acid (HA), which is soluble in dilute alkali, but is precipitated by acidification of the alkaline extract; (b) fulvic acid (FA), which is that humic fraction that remains in solution when the alkaline extract is acidified, that is, it is soluble in both dilute alkali and dilute acid, and (c) humin, which is that humic fraction that cannot be extracted from the soil or sediment by dilute base and acid. The principal humic fractions are HA and FA, which have been extensively investigated during the past decade in several laboratories. From the analytical data published in the literature (Schnitzer and Khan, 1972), it becomes apparent that structurally, HA and FA fractions are similar, but differ in molecular weight, ultimate analysis, and functional group content, with FA having a lower molecular weight but higher content of oxygen-containing functional groups per unit weight. Important characteristics exhibited by the humic fractions are: resistance to microbial degradation (Kononova, 1966); the ability to form water-soluble and water-insoluble complexes with metal ions and hydrous oxide (Schnitzer and Skinner, 1965); and the ability to interact with clay minerals (Schnitzer and Kodama, 1966), and hydrophobic organic compounds such as alkanes, fatty acids, and dialkyl phthalates (Ogner and Schnitzer, 1970a,b; Schnitzer and Ogner, 1970; Khan and Schnitzer, 1971, 1972a), and pesticides (Hayes, 1970; Burns et al., 1974; Stevenson 1972a,b; Khan 1972a,b, 1974a). Table 1 shows some analytical characteristics of much researched HA and FA. The HA originated from the Ah horizon of a black chernozemic soil of western Canada and

FA extracted from Bh horizon of a Podzol from eastern Canada. Elementary and functional group analyses of HA differ from that for FA in the following respects: (a) HA contains more C, H, N, and S but less O than does FA; (b) the total acidity and COOH-content of FA are approximately twice as great as those of HA; (c) the ratio of COOH to phenolic OH groups is about 3 for FA but only approximately 2 for HA; and (d) E_4/E_6 ratios and ESR data also indicate differences between HA and FA.

Spectroscopic methods have been used for qualitative and quantitative investigations on humic substances. These methods have a number of attractive features: (a) they are nondestructive; (b) only small sample weights are needed; (c) they are experimentally simple and do not require special manipulative skills; and (d) they often provide valuable information on molecular structure and on chemical interaction.

Humic substances, like many relatively high-molecular-weight materials, yield generally uncharacteristic spectra in the visible region (400 to 800 nm). Absorption spectra of neutral, alkaline, and acidic aqueous solutions of HA and FA are featureless, showing no maxima or minima; the optical density decreases as the wavelength increases. Ultraviolet spectra (200 to 400 nm) of most humic substances are also featureless, with the optical density decreasing as the wavelength increases.

Infrared spectroscopy (4000 to 650 cm^{-1}) has been found useful in humic research. However, the assignment of absorption bands to certain groupings with the aid of correlation charts is still fraught with considerable uncertainty. It is, therefore, always advisable to corroborate spectral data with information obtained by other methods. Infrared spectra of the HA and FA are shown in Figure 1. The absorption bands are broad because of extensive overlapping of individual absorption. The main absorption bands are listed in Table 2. The most striking difference between the two spectra in Figure 1 lies in the intensities of the bands in the 2900 to 2800 cm^{-1} region and in the 1725 cm^{-1} band. The HA contains more aliphatic C-H groups than does the FA. The 1725 cm^{-1} band is very strong in the case of FA, but only a shoulder for HA, and so substantiate the chemical data in Table 1, which show that the FA contains considerably more COOH groups than does the HA. In general, infrared spectra of humic substances of diverse origins are very similar, which may indicate the presence of essentially similar chemical structures, differing mainly in the contents of functional groups.

HA and FA can be degraded by oxidation under alkaline as well as under acidic conditions and also by base hydrolysis (saponification) into aliphatic, phenolic and benzenecarboxylic

TABLE 1. Analytical Characteristics of HA and FA (Schnitzer and Khan, 1972)[a]

Characteristics	HA	FA
Elementary composition (%, on dry ash-free basis)		
C	56.4	50.9
H	5.5	3.3
N	4.1	0.7
S	1.1	0.3
O	32.9	44.8
Oxygen-containing functional groups (meq/g, on dry ash-free basis)		
Total acidity	6.6	12.4
Carboxyl	4.5	9.1
Total hydroxyl	4.9	6.9
Phenolic hydroxyl	2.1	3.3
Alcoholic hydroxyl	2.8	3.6
Total carbonyl	4.4	3.1
Quinone	2.5	0.6
Ketonic carbonyl	1.9	2.5
Methoxyl	0.3	0.1
E_4/E_6 ratio[b]	4.3	7.1
Free radicals (spins/g x 10^{-18})	0.8	0.2
Line width (G)	3.5	5.0
g - value	2.0029	2.0031

[a]Reprinted from "Humic Substances in the Environment", Marcel Dekker, Inc., New York, 1972, by courtesy of Marcel Dekker, Inc.

[b]Ratio of optical densities of 465 and 665 nm.

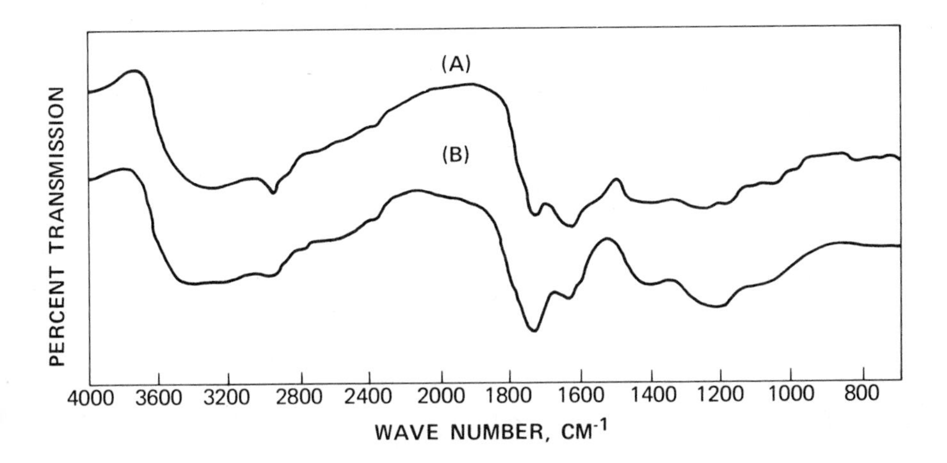

FIGURE 1. Infrared spectra of (A) HA and (B) FA. (After Schnitzer and Khan, 1972. Reprinted from "Humic Substances in the Environment", Marcel Dekker, Inc., New York, 1972, by courtesy of Marcel Dekker, Inc.)

TABLE 2. Main Infrared Absorption Bands of Humic Substances (Schnitzer and Khan, 1972)[a]

Frequency (cm^{-1})	Assignment
3400	Hydrogen-bonded OH
2900	Aliphatic C-H stretch
1725	C = O of COOH
	C = O stretch of ketonic caronyl
1630	Aromatic C = C (?), Hydrogen-bonded C = O of carbonyl (?)
	Double bond conjugated with carbonyl, COO^-
1450	Aliphatic C-H
1400	COO^-
	Aliphatic C-H
1200	C-O stretch
	OH deformation of COOH
1050	Si-O of silicate impurity

[a]Reprinted from "Humic Substances in the Environment", Marcel Dekker, Inc., New York, 1972, by courtesy of Marcel Dekker, Inc.

components. Types and yields of products resulting from the oxidation of HA and FA with different reagents are compared in Tables 3 and 4. The data show that CuO-NaOH and CuO-NaOH + $KMnO_4$ oxidations of HA and FA produced greater amounts of aliphatic and phenolic compounds than did any of the other reagents that were employed, and which also included hydrolysis with 2N NaOH. Peracetic acid oxidation of HA at 40 and 80°C (Table 3) and $KMnO_4$ oxidation of FA (Table 4) were the most effective procedures for the formation of benzenecarboxylix acids. A combination of CuO-NaOH + $KMnO_4$ was the most efficient oxidative degradation proceudre for obtaining the highest yields of products per g of HA and FA.

The major problem confronting humic materials specialists at this time is to determine how the major components combine or fit together to form the type of structural arrangements that we encounter in soils and waters. It has become more apparent that humic substances are not simple molecules but rather associations of molecules of biological, polyphenolic, lignin, or condensed lignin origins. These molecules are somehow held together and organized into polymeric structures of considerable stabilities by as yet unknown mechanisms.

For a more complete account of the chemistry and reactions of humic substances the reader is referred to a recently published book entitled "Humic Substances in the Environment" (Schnitzer and Khan, 1972).

III. HERBICIDES

The variety of herbicides used presently represents different classes of organic compounds. The types of interactions of these compounds with humic substances in soil and aquatic environments are enormous. This paper deals only with a few selected herbicides used in our studies. The common and chemical names, structure and some properties of these herbicides are shown in Table 5.

A. Humic Substances - Herbicides Interactions

Several mechanisms or combination of mechanisms have been suggested for the adsorption of herbicides by humic substances. These include hydrogen bonding, ion exchange, protonation, Van der Waal's forces, charge transfer, ligand exchange, and coordination through an attached metal ion.

1. Paraquat and Diquat. Paraquat and diquat readily dissolve and dissociate in aqueous solutions to form divalent cations.

TABLE 3. Compounds (mg) Produced by the Oxidation of 1 g of HA with Various Reagents and by Hydrolysis with 2 N NaOH

Compounds	Oxidation with					Hydrolysis with[e] 2 N NaOH
	Peracetic acid		$KMnO_4$[c]	CuO-NaOH[d]	CuO-NaOH[d] $KMnO_4$	
	at 40°C[a]	at 80°C[b]				
Alphatic	34.0	13.4	15.8	109.0	120.0	113.6
Phenolic	84.2	43.4	96.5	94.9	101.3	72.1
Benzene-carboxylic	160.1	152.4	137.4	36.2	74.6	17.1
TOTAL	278.3	209.2	249.7	240.1	295.9	202.8

[a]Schnitzer and Skinner (1974a).

[b]Schnitzer and Skinner (1974b).

[c]Khan and Schnitzer (1972b).

[d]Neyroud and Schnitzer (1974a).

[e]Neyroud and Schnitzer (1975).

TABLE 4. Compounds (mg) Produced by the Oxidation of 1 g of FA with Various Reagents and by Hydrolysis with 2 *N* NaOH

Compounds	Oxidation with					Hydrolysis with[e] 2 *N* NaOH
	Peracetic Acid		$KMnO_4$[c]	CuO-NaOH[d]	CuO-NaOH[d] + $KMnO_4$	
	at 40°C[a]	at 80°C[b]				
Aliphatic	25.8	23.1	1.0	104.7	111.1	103.8
Phenolic	115.4	40.6	88.5	146.7	150.8	133.8
Benzenecar-boxylic	83.7	72.5	114.9	35.1	48.8	27.0
TOTAL	224.9	136.2	204.5	286.5	310.7	364.6

[a]Schnitzer and Skinner (1974a).
[b]Schnitzer and Skinner (1974b).
[c]Khan and Schnitzer (1972b).
[d]Neyroud and Schnitzer (1974a).
[e]Neyroud and Schnitzer (1975).

TABLE 5. Chemical Designation, Structure, and Some Properties of the Selected Herbicides.

Common name	Trade name	Chemical name	Structure	Molecular Weight	Water solubility (20-25°C, ppm)	LD_{50}[a] (mg/kg)
Paraquat	Gramoxone	1,1'-Dimethyl-4,4'-bipyridylium dichloride	$H_3C\text{-}N^+$ … $N^+\text{-}CH_3$ $2Cl^-$	257	Soluble	150
Diquat	Reglone	1,1'-Ethylene-2,2-bipyridylium dibromide	N^+ N^+ H_2C-CH_2 $2Br^-$	344	Soluble	400-440
2,4-D	Weedone 638	2,4-dichlorophenoxyacetic acid	Cl Cl $OCH_2 \cdot COOH$	221	650	300-1000
Picloram	Tordon	4-amino-3,5,6,-trichloropicolinic acid	H_2N Cl COOH N Cl Cl	241.5	430	8,200
Linuron	Lorox	3-(3,4-dichlorophenyl)-1-methoxyl-1-methylurea	Cl Cl $NH\overset{O}{C}N<^{CH_3}_{OCH_3}$	249.1	75	1,500

[a]Acute oral toxicity for rats.

They have high solubility in water and behave as strong electrolytes in solution, even under acid conditions. Therefore, hydrophobic-hydrophilic interactions, and their dissociation constants are of no direct relevance to their adsorption characteristics (Burns and Hayes, 1974). They can react with more than one negatively charged site on humic colloid. It has been demonstrated that adsorption of paraquat and diquat by humic substances is always accompanied by the release of significant concentration of hydrogen ions (Best et al., 1972; Burns, Hayes, and Stacey, 1973 a,b,c; Khan, 1974b). This suggests that an ion exchange process is involved. An ion exchange mechanism is of special significance because the herbicide adsorbed in this manner may become ineffective in controlling weeds.

Infrared spectroscopy has been utilized to demonstrate that ion exchange is the predominant mechanism for adsorption of diquat and paraquat by humic substances (Burns, Hayes, and Stacey, 1973c; Khan 1974b). This situation is illustrated in Figure 2. On addition of the herbicide, the main changes occurred in the 1500 to 1800 cm^{-1} region, so only this part of the spectrum is presented herein (Khan, 1974b). In the spectrum of HA (curve A), the 1720 cm^{-1} band (carbonyl of carboxylic acid) was more prominent than that at 1610 cm^{-1} (carboxylate). Upon addition of herbicides, the intensity of 1720 cm^{-1} band diminished while that at 1610 cm^{-1} increased. This indicated conversion of COOH to COO^- groups, which react with the cationic herbicide to form carboxylate bonds. It was not possible to deduce from the spectra whether OH groups participated in the reaction, as the intensity of OH absorption near 3400 cm^{-1} (not shown here) remained virtually unchanged. Notice that a considerable proportion of H^+ in COOH remained inaccessible to the larger herbicide cations. It has been observed that humic substances adsorb paraquat and diquat in amounts that are considerably less than their exchange capacity (Burns, Hayes, and Stacey, 1973b; Khan 1973 a,b,c, 1974b). The large size of the herbicide cations seems to result in steric hindrance so that they are not exchanged with ionizable H^+ as effectively as the smaller inorganic cations.

Further evidence for the ion exchange mechanism was procured by the potentiometric titrations of HA and herbicide-HA complexes (Figure 3). The decrease in consumption of alkali for herbicide-HA complexes titration (curves B, C, D versus A) suggests that ionization of acid functional groups was involved in paraquat and diquat interaction with HA (Khan, 1974b).

It has been shown that paraquat is adsorbed on the nonpolar polystyrene molecules, Amberlite XAD-2 (McCall et al., 1972; Burns, Hayes, and Stacey, 1973b) indicating nonexchange processes.

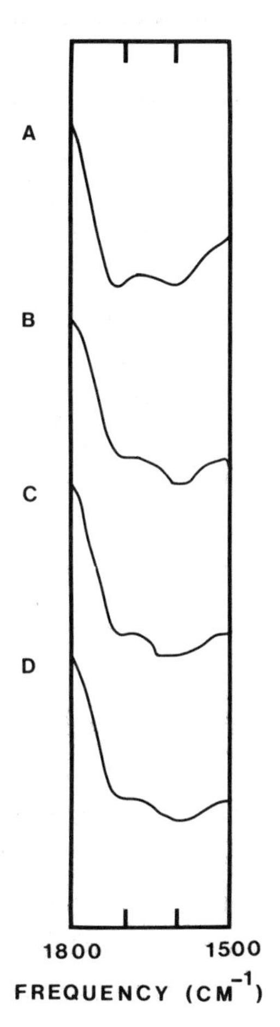

FIGURE 2. Infrared spectra in the region 1500 to 1800 cm^{-1} of (A) HA, (B), HA-diquat, (C) HA-paraquat, and (D) HA-diquat + paraquat. (After Khan, 1974b. Reproduced from Journal of Environmental Quality, Vol. 3, 1974, pages 202-206, by permission of the American Society of Agronomy.)

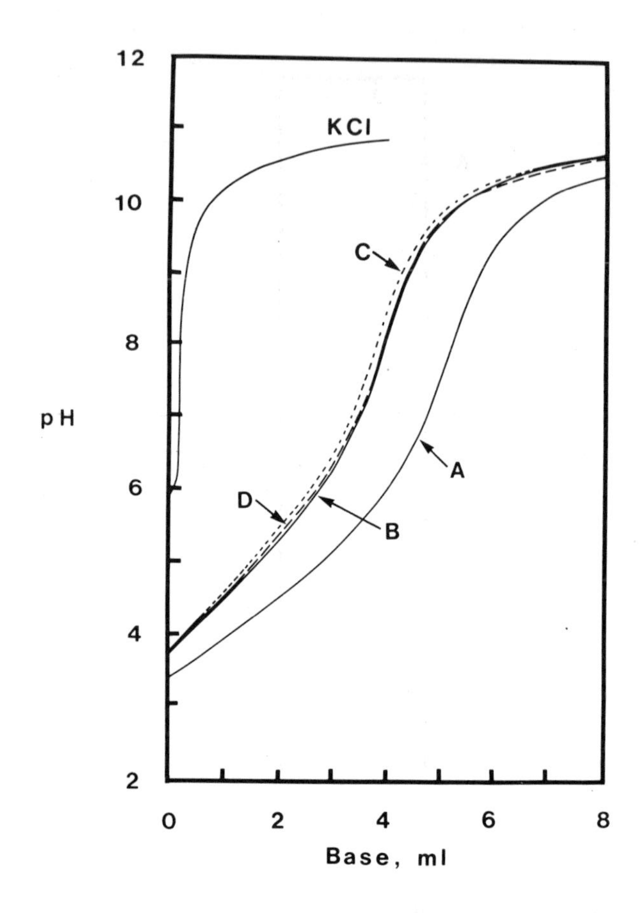

FIGURE 3. Potentiometric titration curves of (A) HA, (B) HA-diquat, (C) HA-paraquat, and (D) HA-diquat + paraquat. (After Khan, 1974b. Reproduced from Journal of Environmental Quality, Vol. 3, 1974, pages 202-206, by permission of the American Society of Agronomy.)

Thus although adsorption of paraquat and diquat by humic substances is mainly by ion exchange, it appears that other mechanisms may also be involved as well. The bipyridylium cations will readily accept and stabilize an additional electron (Homer and Tomlinson, 1959). Paraquat is known to form charge transfer complexes with various anionic and neutral donor species (Foster, 1969; Haque and Lilley, 1972). White (1969) provided evidence for the charge transfer processes in paraquat complexes with a wide variety of organic donor molecules including catechol, phloroglucinol, 3,4-dihydroxybenzoic acid, *p*-aminobenzoic acid, and hydroquinone, all of which may be present in humic substances. Humic substances possess measurable free-radical content (Schnitzer and Khan, 1972). According to Hamaker and Thompson (1972), it is likely that charge transfer mechanisms are involved in the binding of many chemicals to humic materials. Burns, Hayes, and Stacey (1973b) postulated the involvement of charge transfer mechanisms in paraquat adsorption of HA. However, in a later study these workers (Burns, Hayes, and Stacey, 1973c) failed to obtain evidence for such mechanisms in paraquat-HA complexes in aqueous system. They based their conclusions on the failure of paraquat adsorption maximum shifts to longer wavelengths in the paraquat-HA complexes as determined by ultraviolet spectroscopy. According to these workers, spectral shifts would be expected when charge transfer processes are involved in the adsorption (Burns, Hayes, and Stacey, 1973 a,b,c). While the ultraviolet spectroscopic method has been employed as a tool for identifying charge transfer interactions, the technique lacks sensitivity for determining the charge transfer bonding between bipyridylium compounds and humic substances as a result of light-scattering losses.

Infrared spectroscopic techniques have been employed in our laboratory to obtain evidence for the involvement of charge transfer mechanisms in the adsorption of paraquat and diquat by humic substances (Khan, 1973a, 1974b). The infrared spectra of diquat dibromide and paraquat dichloride show strong bands, at 792 and 815 cm^{-1}, respectively (Fig. 4). These bands are assigned to the out-of-plane vibration mode of C-H (Bellamy, 1955). It has been reported that the bipyridinium ions show characteristic changes in the out-of-plane C-H deformation modes on charge transfer complex formation with various anionic species and clay minerals (Haque, Lilley, and Coshow, 1970; Haque and Lilley, 1972). The interaction of diquat with humic substances resulted in a shift of the band to about 765 cm^{-1} whereas the corresponding band in paraquat showed a shift to about 825 cm^{-1} (Figure 5.) This indicated the formation of charge transfer complexes between humic materials and bipyridylium herbicides. The infrared spectra shown in Figure 5 were recorded employing five times expansion of both ordinate and abscissa scales. This is definite evidence for the shifts in these bands on charge transfer complex formation between bipyridylium herbicides and humic substances.

The picture that emerges from the foregoing discussion clearly indicates the involvement of ion exchange and charge transfer

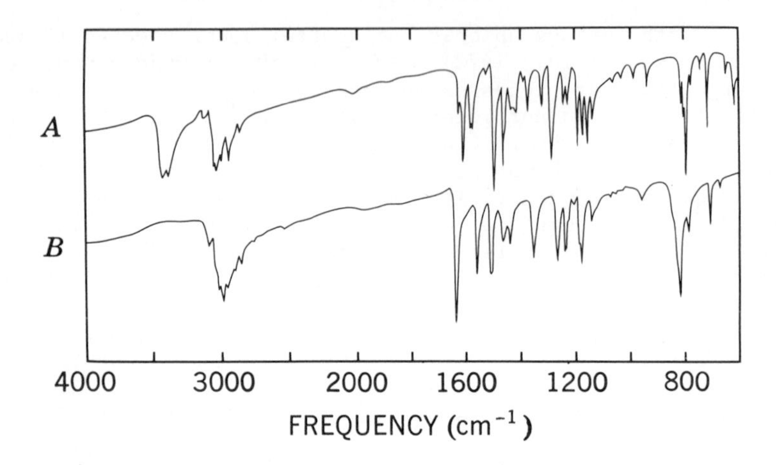

FIGURE 4. Infrared spectra of (A) diquat dibromide and (B) paraquat dichloride. (After Khan, 1974b. Reproduced from Journal of Environmental Quality, Vol. 3, 1974, pages 202-206, by permission of the American Society of Agronomy).

mechanisms in the adsorption of paraquat and diquat by humic substances. An estimate of the relative importance of charge transfer and ion exchange mechanisms in the adsorption of paraquat and diquat by HA and FA will remain a matter of conjecture until more information is available. However, judging from the available data it may be assumed that the ion exchange mechanism plays a dominant role in the adsorption processes. It is also possible that other processes such as hydrogen bonding and Van der Waal's forces also contribute to the adsorption mechanisms.

The adsorption of paraquat and diquat is influenced by the nature of the cation present initially on humic materials (Table 6). The cation order for increasing adsorption for the two herbicides was nearly the same and followed the sequence: $Al^{3+} < Fe^{3+} < Cu^{2+} < Ni^{2+} < Zn^{2+} < Co^{2+} < Mn^{2+} < H^{+} < Ca^{2+} < Mg^{2+}$ (Khan, 1974b). For the polyvalent cations, the stability of metal-HA complexes or the strength of cation binding as determined by the relative positions of the titration curves and the magnitude of the pH drop have been shown to follow the sequence: $Fe^{3+} > Al^{2+} > Zn^{2+} \geqslant Ni^{2+} > Co^{2+} > Mn^{2+} > Ca^{2+} \geqslant Mg^{2+}$ (Khan, 1969; Van Dijk, 1971). Thus it appears that the extent of paraquat and diquat adsorption on cation saturated HA's is inversely proportional to the relative strength of cations binding to the HA or the stability of the metal-HA complexes. In soils and waters the

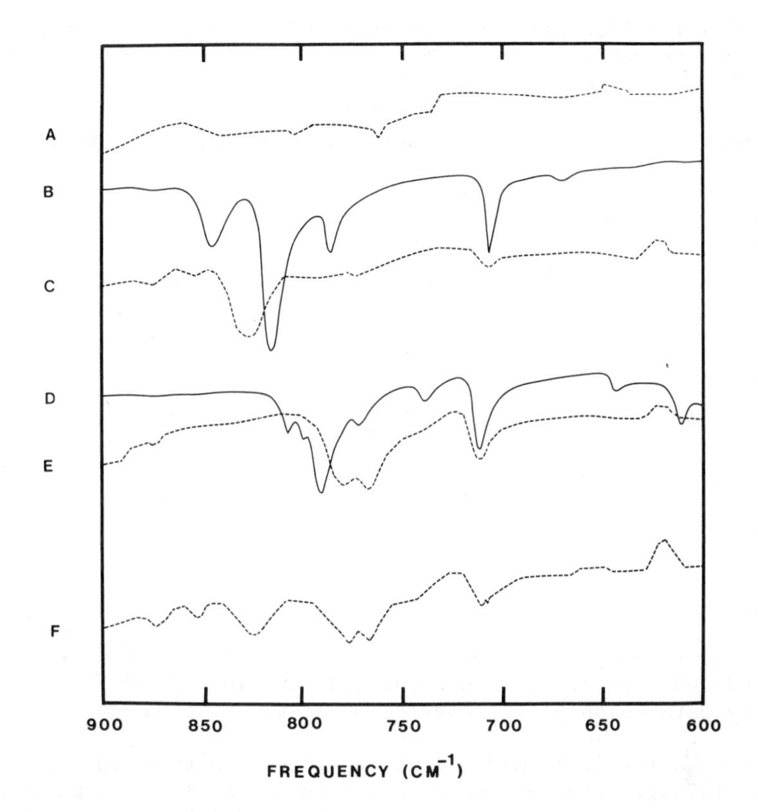

FIGURE 5. Infrared spectra in the region 600 to 900 cm^{-1} of (A) HA, (B) paraquat, (C) HA-paraquat, (D) diquat, (E) HA-diquat, and (F) HA-paraquat + diquat. (After Khan, 1974b. Reproduced from Journal of Environmental Quality, Vol. 3, 1974, pages 202-206 by permission of American Society of Agronomy).

occurrence of humic substances in the free state is unlikely as they are always associated with the inorganic cations present in the systems. It follows, therefore, that the adsorption of paraquat and diquat on humic substances in soil and aquatic environments may be roughly determined by the kind of cation present on the adsorbent.

The adsorption or complexing behavior of humic materials for bipyridylium herbicides is affected by the pH of the system. Studies carried out in our laboratory have shown that when paraquat and diquat were equilibrated in aqueous solution separately

TABLE 6. The Adsorption of Paraquat and Diquat from Aqueous Solutions on Various Cation-Saturated Humic Acids (Khan, 1974b).[a]

Adsorbent	Herbicide adsorbed % of initial amount added per 100 g of humic acid	
	Paraquat	Diquat
Al-HA	22.0	21.6
Fe-HA	26.2	25.8
Cu-HA	45.9	47.3
Ni-HA	67.9	74.9
Zn-HA	70.6	75.5
Co-HA	71.3	76.5
Mn-HA	72.7	78.0
H-HA	77.6	78.1
Ca-HA	78.8	80.4
Mg-HA	81.6	87.2

[a]Reproduced from J. Environmental Quality, Vol. 3, 1974, pp. 202-206 by permission of the American Society of Agronomy.

with the HA (pH 3.3), more diquat was adsorbed than paraquat (Khan, 1974b). However, HA in solution (pH 6.9) complexed paraquat in larger amounts than diquat (Khan, 1973a). It should be noted that HA in solution, such as at pH 6.9, will likely adopt a stretched configuration (Ong and Bisque, 1968). Upon addition of metal salts or acid, the cations attach themselves to the negatively charged functional groups, which result in a reduction of the intramolecular repulsion in the polymer chain and favor its coiling (Ong and Bisque, 1968). Thus the addition of excess salt or acid to humic polyelectrolyte would cause precipitation. In the solid state, HA is considered to have laminated, textured makeup of particles 130 Å in diameter (Orlov and Glebova, 1972). The bipyridylium cations differ in charge location or distribution on the structure, dimension, and flexibility of the molecule. Paraquat is a more flexible molecule than diquat, and more of the constituent atoms can approach an interacting material. Thus it appears that the stretched configuration of HA molecule presents a favorable correspondence between the distance of charge separation of paraquat and the distance between the charge sites on the HA molecule. Furthermore, Van der Waal's forces also probably will make a greater contribution to the overall adsorption of paraquat than is likely in the case of diquat. However, the precipitate or coiled geometry of humic colloid assumes a more compact nature and the distance between the charge sites may become shorter. Under these conditions diquat adsorption exceeds in amounts as compared to paraquat probably due to the smaller size of the cation that results in less steric hindrance, and a close match of the charge separation of diquat with the distance between charge site on HA surface.

In soils and sediments, the clay fraction is closely associated with organic matter and exist in the form of organo-clay complexes (Greenland, 1965). It is likely that the intimate association of clay and organic matter may cause a considerable change in the clay adsorptive capacity. Therefore, we investigated the adsorption of diquat and paraquat by an organo-clay complex (Khan, 1973b). The organo-clay complex was prepared by treating montmorillonite with FA as described by Kodama and Schnitzer (1969). It was observed that the adsorption of diquat and paraquat by the organo-clay complex followed nearly the same pattern as has been reported for the pure clay or organic matter. The two herbicides were completely adsorbed at the lower concentration but in the presence of an excess of the herbicide a level of maximum adsorption was observed. It was also observed that diquat and paraquat were adsorbed in considerably greater amounts by 1 g of montmorillonite when present in the form of organo-clay complex. Thus it appears that FA, which is the most prominent humic compound in soil solution, on interacting with clay minerals may facilitate the adsorption of herbicides on clays in soils and sediments.

2,4-D and Picloram. Adsorption studies with acidic herbicides and organic matter, humic substances, or charcoal showed that the compounds are readily adsorbed in moderate amounts. Adsorption is pH dependent, being greater under acid conditions where the herbicides are adsorbed in the molecular form (Weber, 1972). Stevenson (1972a) suggested that the adsorption mechanisms for these anionic herbicides in acidic medium involve hydrogen bonding, with multiple sites being available on both the herbicide and the organic matter surface. In neutral or alkaline soils, anionic herbicides can be bound through a salt linkage.

We attempted to obtain information on the equilibrium and kinetics of adsorption of 2,4-D and picloram on HA (Khan, 1973c). The two herbicides have been used extensively in Canada for the control of broad-leaved weeds in a variety of crops.

The empirically derived Freundlich equation, $X = KC^n$, was used to describe the adsorption of the herbicides by HA. X is the ratio of the herbicide to HA mass, C is the herbicide concentration in solution upon achieving equilibrium, and K and n are constants. The data obtained gave reasonably good straight lines by plotting log X against log C. A typical Freundlich plot for adsorption of 2,4-D on HA at 5°C is shown in Fig. 6. The values of n and log K (C = 1 ppm) were estimated from the Freundlich plots by using the method of least-square fit (Table 7). In comparing adsorptivity of the two herbicides by HA, the K value was used as an index for relating the degree of adsorption. The necessary conditions are that n values (slope) be approximately similar (Hance, 1965) and determination be made at the same C values. Examination of K values in Table 7 shows that the adsorptive capacity of HA for picloram was slightly greater than 2,4-D at both temperatures.

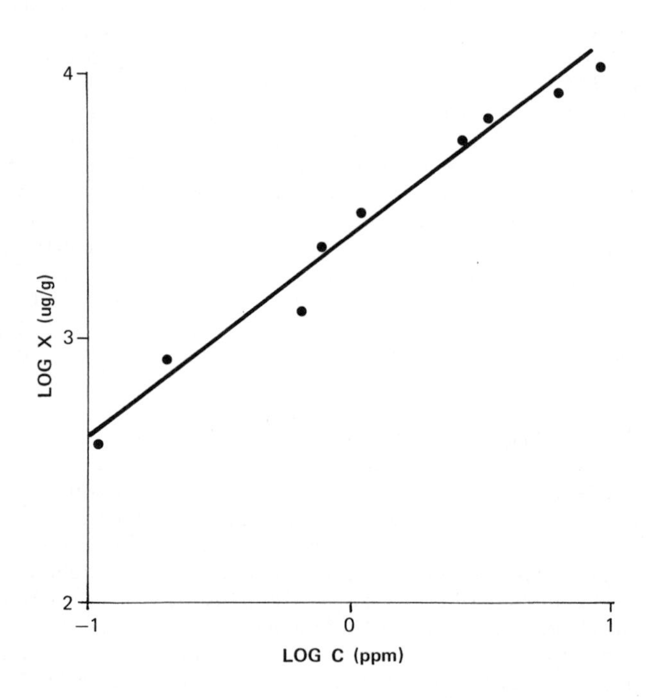

FIGURE 6. Freundlich plot for the adsorption of 2,4-D on HA at 5°C. (After Khan, 1973c. Reproduced from the Canadian Journal of Soil Science, 53, 429, 1973, by permission of the Agricultural Institute of Canada).

TABLE 7. Freundlich Isotherm Constants for the Adsorption of 2,4-D and Picloram on HA[a]

Herbicide	Intercept log K		Slope N	
	5°C	25°C	5°C	25°C
2,4-D	3.37	3.21	0.75	0.79
Picloram	3.47	3.39	0.88	0.91

[a]Reproduced from Canadian Journal of Soil Science, 53, 429, 1973, by permission of the Agricultural Institute of Canada.

The data were examined in the light of the generalized equilibrium theory proposed by Fava and Eyring (1956). The rate constants and other kinetic parameters for the adsorption and desorption processes were computed from the following equation (Haque and Sexton, 1968):

$$\frac{d\phi}{dt} = 2K' \ (1 - \phi) \ \text{Sinh} \ \{b \ (1 - \phi)\} = Y$$

where ϕ is the fraction adsorbed (amount adsorbed at time t divided by the amount adsorbed at equilibrium), K' is the rate constant for adsorption and b is a constant that yields a measure of the surface stressing energy due to loading with molecules. The advantages of the above approach are (a) the kinetic parameters for both the adsorption and desorption processes are computed from adsorption data alone, and (b) computer implementation of the solution of the above equation is rapid and convenient.

A summary of the rate parameters calculated for the adsorption of 2,4-D and picloram on HA is given in Table 8 (Khan, 1973c). The values of the rate constant K' are of the order that indicated the initial rate to be controlled by the herbicide movement to the HA surface involving a physical type of adsorption. The transference rate would be dependent upon diffusion of the herbicide molecules across the water film surrounding the HA particles and on the shaking rate of suspension. Evidence was also obtained indicating that at longer times the adsorption rate becomes slow as it is controlled by the intraparticle diffusion of the herbicide molecules into the interior of the pores of HA particles. The relatively fast rate of adsorption, low values of activation energy and heat of activation suggest the physical type of adsorption.

TABLE 8. Kinetic Parameters for the Adsorption of 2,4-D and Picloram on HA[a]

Herbicide	Temperature (^{o}C)	Rate Constant K (sec^{-1})	Energy of Activation ΔE (K cal/mol)	Heat of Activation $\Delta H^{\ddagger}$ (K cal/mol)
2,4-D	5	5.15×10^{-5}		
	25	6.53×10^{-5}	1.95	1.36
Picloram	5	8.13×10^{-5}		
	25	10.13×10^{-5}	1.81	1.22

[a]Reproduced from Canadian Journal of Soil Science, 53, 429, 1973, by permission of the Agricultural Institute of Canada.

Linuron. In recent years, considerable evidence has been accumulated to indicate that significant amounts of substituted urea herbicides can be adsorbed by various organic surfaces (Hance, 1971; Jordan and Smith, 1971). We investigated the adsorption of linuron by various cation saturated HA's at 5 and 25°C (Khan and Mazurkewich, 1974). At both temperatures the adsorption capacity of cation-saturated HA's for linuron decreased in the following order: $H^{+} > Fe^{3+} > Al^{3+} > Cu^{2+} > Ca^{2+} > Zn^{2+} > Ni^{2+}$. These observations suggest that the adsorption of linuron by humic material is affected by the cation with which HA is saturated. The nature of the cation as manifested by valence, size, and polarizability will affect certain characteristics of the adsorbent, such as geometry of the polymer chain and surface hydration. This would likely influence the adsorption capacity of the material.

From the adsorption data at 5 and 25°C, we may approximate the "isosteric" heat of adsorption ΔH (Koral, Ullman, and Eirich, 1958). Thus the values of ΔH as a function of the amount of linuron adsorbed X can be estimated from the following equation:

$$\frac{\Delta H}{R} = \left(\frac{\partial \ \ell n \ C}{\partial (1/T)}\right)_{X}$$

where R is the gas constant and T the absolute temperature. The ΔH values estimated for various samples at a specific adsorption of 10 μmol linuron/g humic preparation ranged from about -0.17 to -0.75 kcal/mol. The ΔH values estimated in this experiment are relatively small and are of the order which is consistent with a physical type of adsorption (Glasstone and Lewis, 1962).

Hance (1971) suggested that the formation of complexes with exchange cations could play a significant part in the adsorption of linuron by organic matter. In accordance with this view, for linuron molecules there could be two sites at which interaction with an exchangeable cation was most likely to occur: the oxygen of the carbonyl group and the amide nitrogen (see Table 5). By infrared spectroscopy, it should be possible to distinguish between these two binding sites. Infrared studies indicated that such a mechanism was not operative in the adsorption of linuron on cation saturated HA's used in our study (Khan and Mazurkawich, 1974). No shifts in CO (1663 cm or 1578 cm) or CN (1523 cm) stretching frequencies were observed on linuron adsorption under our experimental conditions. It appears that the adsorption of linuron on HA is more likely to be a physical type rather than adsorption due to coordination of linuron to cations on HA.

VI. CONCLUSIONS

The data presented in this chapter serve to emphasize that various herbicides differ in their relative affinities for adsorption on humic substances. Paraquat and diquat would be expected to be the most strongly bound. However, the amounts adsorbed depend upon the kind of cations associated with humic materials. Humic substances upon interaction with clay minerals may also facilitate the adsorption of herbicides in soils or sediments. Acidic and nonionic polar herbicides are adsorbed by humic substances involving mainly a physical type of adsorption. The interaction of humic substances with pesticides may have some practical implications with regard to pollution. The former may adsorb toxic compounds in such a manner as to make them unavailable to plants and animals, so that the problem may be less serious than one would expect. On the other hand, one can surmise that the quantities of toxic compounds in soils and waters are usually underestimated. Clearly, there is need for more research in order to solve such problems of great significance.

REFERENCES

1. Bellamy, L. T. 1955. The Infrared Spectra of Substituted Aromatic Compounds in Relation to the Chemical Reactivities of Their Substituents, J. Chem. Soc., 2818.

2. Best, J. A., Webber, J. W. and Weed, S. B. 1972. Competitive Adsorption of diquat^{2+}, paraquat^{2+}, and Ca^{2+} on Organic Matter and Exchange Resins. Soil Sci. 114, 444.

3. Burns, I. G. and Hayes, M. H. B. 1974. Some Physico-Chemical Principles Involved in the Adsorption of the Organic Cation Paraquat by Soil Humic Materials. Residue Rev. 52, 117.

4. Burns, I. G., Hayes, M. H. B. and Stacey, M. 1973a. Some Physico-Chemical Interactions of Paraquat with Soil Organic Materials and Model Compounds. I. Effects of Temperature, Time and Adsorbate Degradation on Paraquat Adsorption. Weed Res 13, 67.

5. Burns, I. G., Hayes, M. H. B. and Stacey, M. 1973b. Some Physico-Chemical Interactions of Paraquat with Soil Organic Materials and Model Compounds. II. Adsorption and Desorption Equilibria in Aqueous Suspensions. Weed Res. 13, 79.

6. Burns, I. G., Hayes, M. H. B., and Stacey, M. 1973c. Spectroscopic Studies on the Mechanisms of Adsorption of Paraquat by Humic Acid and Model Compounds. Pest. Sci. 4, 201.

7. Fava, A. and Eyring, H. 1956. Equilibrium and Kinetics of Detergent Adsorption - A Generalized Equilibrium Theory. J. Phys. Chem. 60, 890.

8. Foster, R. 1969. Organic Charge - Transfer Complexes. Academic, New York.

9. Glasstone, S. and Lewis D. 1962. Elements of Physical Chemistry. Van Nostrand, New York.

10. Greenland, D. J. 1965. Interaction Between Clays and Organic Compounds in Soils. Part II. Adsorption of Soil Organic Compounds and Its Effect on Soil Properties. Soils Fert. 28, 521.

11. Hamaker, J. W. and Thompson, J. M. 1972. Adsorption. In Organic Chemical in Soil Environment, Vol. 1. (C.A.I. Goring and J. W. Hamaker, Eds.). Dekker, New York, p. 49.

12. Hance, R. J. 1965. The Adsorption of Urea and Some of Its Derivatives by a Variety of Soils. Weed Res. 5, 98.

13. Hance, R. J. 1971. Complex Formation as an Adsorption Mechanism for Linuron and Atrazine. Weed Res. 11, 106.

14. Haque, R. and Lilley, S. 1972. Infrared Spectroscopic Studies of Charge-Transfer Complexes of Diquat and Paraquat. J. Agr. Food. Chem 20, 57.

15. Haque, R., Lilley, S. and Coshow, W. R. 1970. Mechanism of Adsorption of Diquat and Paraquat on Montmorillonite Surface. J. Colloid Interface Sci 33, 185.

16. Haque, R. and Sexton, R. 1968. Kinetic and Equilibrium Study of the Adsorption of 2,4-dichlorophenoxyacetic Acid on Some Surfaces. J. Colloid Interface Sci. 27, 818.

17. Hayes, M. H. B. 1970. Adsorption of Triazine Herbicides on Soil Organic Matter, Including a Short Review on Soil Organic Matter Chemistry. Residue Rev. 32, 131.

18. Homer, R. F. and Tomlinson, T. E. 1959. Redox Properties of Some Dipyridyl Quaternary Salts. Nature 184, 2012.

19. Ishiwatari, R. 1969. An Estimation of the Aromaticity of a Lake Sediment Humic Acid by Air Oxidation and Evaluation of It. Soil Sci. 107, 53.

20. Jordan, P. D. and Smith, L. W. 1971. Adsorption and Deactivation of Atrazine and Diuron by Charcoal. Weed Sci. 19, 541.

21. Khan, S. U. 1969. Interaction Between the Humic Acid Fraction of Soils and Certain Metallic Cations. Soil Sci. Soc. Amer. Proc. 33, 851.

22. Khan, S. U. 1972. Adsorption of Pesticides by Humic Substances. A review. Environ. Lett. 3, 1.

23. Khan, S. U. 1973a. Interaction of Humic Substances with Bipyridylium Herbicides. Can. J. Soil Sci. 53, 199.

24. Khan, S. U. 1973b. Interaction of Bipyridylium Herbicides with Organo-Clay Complex. J. Soil Sci. 24, 244.

25. Khan, S. U. 1973c. Equilibrium and Kinetic Studies of the Adsorption of 2,4-D and Picloram on Humic Acid. Can. J. Soil Sci. 53, 429.

26. Khan, S. U. 1974a. Humic Substances Reactions Involving Bipyridylium Herbicides in Soil and Aquatic Environments. Residue Rev. 52, 1.

27. Khan, S. U. 1974b. Adsorption of Bipyridylium Herbicides by Humic Acid. J. Environ. Qual. 3, 202.

28. Khan, S. U. and Mazurkewich, R. 1974. Adsorption of Linuron on Humic Acid. Soil Sci. 118, 339.

29. Khan, S. U. and Schnitzer, M. 1971. Sephadex Gel Filtration of Fulvic Acid: The Identification of Major Components in Two Low-Molecular Weight Fractions. Soil Sci. 112, 231.

30. Khan, S. U. and Schnitzer, M. 1972a. The Retention of Hydrophobic Organic Compounds by Humic Acid. Geochim. Cosmochim. Acta 36, 745.

31. Khan, S. U. and Schnitzer, M. 1972b. Permanganate Oxidation of Humic Acids, Fulvic Acids and Humins Extracted from Ah Horizons of a Black Chernozem, A Black Solod and a Black Solonetz Soil. Can. J. Soil Sci. 52, 43.

32. Kodama, H. and Schnitzer, M. 1969. Thermal Analysis of a Fulvic Acid - Montmorillonite Complex. Proc. Int. Clay Conf. Tokyo 1, 765.

33. Kononova, M. M. 1966. Soil Organic Matter. Pergamon, New York.

34. Koral, J., Ullman, R. and Eirich, F. R. 1974. The Adsorption of Polyvinyl Acetate. J. Phys. Chem. 62, 541.

35. Lamar, W. L. 1968. Evaluation of Organic Color and Ion in Natural Surface Waters. U. S. Geol. Survey Prof. Paper 600-D, D24.

36. McCall, H. G., Bovey, R. W., McCully, M. G. and Merkle, M. G. 1972. Adsorption and Desorption of Picloram, Trifluralin, and Paraquat by Ionic and Non-Ionic Exchange Resins. Weed Sci. 20, 250.

37. Neyroud, J. A. and Schnitzer, M. 1974a. The Exhaustive Alkaline Cupric Oxide Oxidation of Humic Acid and Fulvic Acid. Soil Sci. Soc. Amer. Proc. 37, 229.

38. Neyroud, J. A. and Schnitzer, M. 1975. The Alkaline Hydrolysis of Humic Substances. Geoderma 13, 171.

39. Ogner, G. and Schnitzer, M. 1970a. The Occurrence of Alkanes in Fulvic Acid, a Soil Humic Fraction. Geochim. Cosmochim. Acta 34, 921.

40. Ogner, G. and Schnitzer, M. 1970b. Humic Substances. Fulvic Acid-Dialkyl Phthalate Complexes and Their Role in Pollution. Science 170, 317.

41. Ong, H. L. and Bisque, R. E. 1968. Coagulation of Humic Colloids by Metal Ions. Soil Sci. 106, 220.

42. Orlov, D. S. and Glebova, G. I. 1972. Electron-Microscope Investigation of Humic Acids. Sov. Soil Sci. (Eng. transl.) 4, 445.

43. Rashid, M. A. and King, L. H. 1969. Molecular Weight Distribution Measurements on Humic and Fulvic Acid Fractions from Marine Clays on the Scotian Shelf. Geochim. Cosmochim. Acta 33, 147.

44. Schnitzer, M. and Khan, S. U. 1972. Humic Substances in the Environment. Dekker, New York.

45. Schnitzer, M. and Kodama, H. 1966. Montmorillonite: Effect of pH on Its Adsorption of a Soil Humic Compound. Science 153, 70.

46. Schnitzer, M. and Ogner, G. 1970. The Occurrence of Fatty Acids in Fulvic Acid, a Soil Humic Fraction. Israel J. Chem. 8, 505.

47. Schnitzer, M. and Skinner, S. I. M. 1965. Organo-Metallic Interaction in Soils. 4. Carboxyl and Hydroxyl Groups on Organic Matter and Metal Retention. Soil Sci. 99, 278.

48. Schnitzer, M. and Skinner, S. I. M. 1974a. The Low-Temperature Oxidation of Humic Substances. Can. J. Chem. 52, 1072.

49. Schnitzer, M. and Skinner, S. I. M. 1974b. The Peracetic Acid Oxidation of Humic Substances. Soil Sci. 118, 322.

50. Stevenson, F. J. 1972a. Role and Function of Humus in Soil with Emphasis on Adsorption of Herbicides and Chelation of Micronutrients. Bioscience 22, 643.

51. Stevenson, F. J. 1972b. Organic Matter Reactions Involving Herbicides in Soil. J. Environ. Qual. 1, 333.

52. Van Dijk, H. 1971. Cation binding of Humic Acid. Geoderma 5, 53.

53. Weber, J. B. 1972. Interaction of Organic Pesticides with Particulate Matter in Aquatic and Soil Systems. Adv. Chem. Ser. 11, 55.

54. White, B. G. 1969. Bipyridylium Quaternary Salts and Related Compounds. Part III. Weak Intermolecular Charge-Transfer Complexes of Biological Interest Occurring in Solution and Involving Paraquat. Trans. Faraday Soc. 65, 2000.

Fate of Carcinogens in Aquatic Environments

JOACHIM BORNEFF
Department of Hygiene
University Mainz
Mainz, Germany

I. INTRODUCTION

As early as 1915, Yamagiwa and Ichikawa showed that coal tar could produce cancer in animals. Polycyclic aromatic hydrocarbons (PAH) were isolated as a result of a subsequent research for the responsible agent. These PAH were shown to include compounds of varying carcinogenic potency. Skin and subcutaneous applications of only a few micrograms of the stronger PAH carcinogens produced fatal tumors in laboratory animals. Oral application requires higher doses and solubilization by means of detergents, fats, or oils.

The PAH poses a similar risk to man. This is true, for instance, for occupationally induced cancer of the skin. To date, there is no data available that states the necessary minimum dose of a carcinogen, and furthermore, there is no proof of the effectiveness of oral intakes of such a substance. Nevertheless, it is the responsibility of preventive medicine to take precautionary measures by limiting both suspect and proved carcinogenic content of food. The carcinogenic content of foodstuffs is not clear cut as it is with food additives, and legal control is still a problem. Estimation of man's yearly PAH intake amounts to 3 to 4 mg with fruits, vegetables, and bread; to 0.1 mg with fats and oils of vegetable origin; and to 0.05 mg with smoked meat or fish and drinking water.

Although water is not primary among the contributors of polycyclic aromatics, its control is necessary as it is an indispensable and irreplaceable foodstuff. Man's drinking water is obtained from different sources. Therefore, attention must be paid to surface waters, that is, the water of rivers and lakes, ground water, as well as the waste water and the by-products of waste water purification as sewage sludge, which can be used for soil cultivation.

Wedgwood and Cooper (1954) pioneered the study of carcinogens in water. They qualitatively detected benzo[a]pyrene and other polycyclic aromatics in industrial waste water from a gas works. Our own research, begun in 1954, indicated that PAH from natural sources and domestic sewage proved as important a source as PAH of industrial origin.

II. THE EXISTENCE OF POLYCYCLIC AROMATIC HYDROCARBONS IN WATER

At the beginning of these experiments, it was assumed that polycyclics in water were solubilized in the presence of fats, oils, detergents, or that they were in water associated with solid material (e.g., soot). The first part of our program was concerned with the study of particles. We checked, for instance, sandfilters of a lake-waterworks and discovered other PAH in addition to benzo[a]pyrene. The amount of carcinogenic substances within the mud was up to several milligrams per kilogram. In suspended particles from the Rhine River, an almost equal quantity of carcinogenic substances could be traced. However after adjustments for water volume, the strongly contaminated Rhine River carried about 200 times higher a concentration than the lake water works. Comparative tests showed that the quantity of sedimentary particles per cubic meter river water was not the only important factor. Therefore one cannot judge the water quality just by its content of suspended and flocculated material. It is essential to know the kind of sewage input. Domestic waste waters appear to be the heaviest polluting source. Navigation happens to be of minor significance, except for accidents with oil- and gas-tankers.

Further hints concerning the possible origin of carcinogenic substances come from the examination of water works sandfilter muds. These filter muds contained high concentration of algae. We fished in the Lake of Constance for phytoplankton to isolate the source of benzo[a]pyrene. We found the algae had concentrations of one-tenth of the sandfilter muds. Thus we concluded the major portion of carcinogens is attached to the remaining particles in the mud. Since dust and soil were the remaining major components in the sandfilter muds, we expanded the investigations of Blumer (1961) to study the occurrence of polycylics in soil. Besides benzo[a]pyrene, there were about 30 different polycyclic aromatic hydrocarbons in every soil sample.

Together with analyses from other authors we can state the following: (a) the absence of benzo[a]pyrene from soil is an exception, (b) uncultivated soil contains about 1 to 10 μg/kg and fertilized soil about 100 μg/kg benzo[a]pyrene, and (c) considerably higher values were found in areas exposed to strong pollution by aerosols. As an example, 1000 μg/kg benzo[a]pyrene were found close to a railroad track (not electrified) and 200,000 μg/kg in soil from an oil refinery (Borneff and Fischer, 1962; Borneff and Fischer, 1963; Mallét and Héros, 1962; Zdrzail and Picha, 1965; Shabad, 1968; Shabad, et al., 1971; Grimmer, Jakob, and Hildebrandt, 1972; Borneff et al., 1973). However not only benzo[a]pyrene is traceable in superficial soil stratums, but other PAH were found in material of borings from a depth of 170 m and about 100,000 yr old (i.e., in ground surely not changed by civilization).

Undoubtedly, city street surface runoff is one of the main sources of PAH in surface water. Road-tar-dust contains up to half a gram of carcinogenic substances per kilogram (i.e. 100 times more than suspended solids from the Rhine River). Sartor et al. (1974) confirmed the importance of the street surface runoff for surface waters. Apparently, a continuous charge of PAH comes from sewage input into a water course. The PAH occurs mainly in particles or at least are adsorbed onto particles. We concluded this from the fact that a sewage treatment plant works partly mechanically and partly biologically. According to our tests, in raw sewage there are up to 100,000 μg or more carcinogenic substances in corpuscular form within 1 m^3 water. Furthermore, we found some dissolved. Initially, we believed PAH were solubilized by detergents, but soon we realized that the high concentration of detergents necessary to solubilize PAH does not occur in rivers and lakes.

We ascertained, using improved analytical techniques, that polycyclics are dissolved even in unobjectionable ground water, which contains no detergents or mineral oils. Therefore, one can only speak in terms of low solubility.

We now feel that precipitation, which may include some aerosols are therefore often heavily contaminated and upon seepage into the ground will release traces of PAH into the ground water. Our analyses have shown that the normal concentrations for ground water ranges between 10 and 50 μg/m^3 for PAH on the basis of the following compounds: Fluoranthene, 3,4-benzofluoranthene, 3,4-benzopyrene, 1,12-benzoperylene, 11,12-benzofluoranthene, and indeno(1,2,3-c,d)-pyrene.

We should mention that PAH in tap water is not necessarily equivalent to PAH in the original well. This result was observed by testing well waters and tap water from the end of the water supply pipe. At this locus, the values were almost 10 times higher than in the well itself. Thus an increase of PAH occurred in the supply network. A contributing factor could be the paint within the water pipes. An analysis of the paint used for the pipe construction showed some milligrams per kilogram of carcinogenic material. Further tests showed that measureable amounts of PAH were observed when water was passed over the pipes for several months.

Surface waters are exposed to PAH from aerosols, road abrasions, and waste waters, so that a considerable increase of PAH concentration is possible. Generally speaking, a low contaminated river or lake water contains five times more PAH than ground water, while a medium polluted river water contains about 10 to 20 times more PAH. Domestic waste water has about 10,000 times higher concentrations of PAH.

A. Analysis of PAH Sources on the Rhine River

Several surveys for PAH in the Rhine River, from Lake Constance to the border of the Netherlands, revealed an increasing burden from 30 to 1000 μg/m^3 PAH. Some areas of the Rhine carried a load of 10,000 μg/m^3 or more. Higher values in most cases were caused by local sewage inputs. Low effluent discharges in parts of the Rhine could be attributed to the low levels observed. Such fluctuations can depend on the precipitation, since sudden and heavy rain leads to transport of soil, road abrasions and soot. Furthermore, the phenomenon of river turbulence scouring the river is an important consideration brought to light during our own pesticide analyses. For example, PAH loads ran up to 20 kg/day in the Rhine River at Mainz at low gauge. This PAH load increases to 80 kg whenever water flow doubles. (A quantity sufficient to kill about 100 mio. of test mice.)

Thus dilution of noxious substances such as PAH by precipitation does not occur in the Rhine River as is often supposed. The opposite occurs. This might be attributable to overflow at sewage treatment plants. The identification of a load of 80 kg PAH/day demonstrates that we are still far away from efficient protection of our surface waters. However, analyses in 1976 showed a remarkable decrease of the PAH-concentrations, but the monitoring must be continued to demonstrate permanent success.

Yet surface waters can have an equal quality to ground waters, if care is taken against contaminations. This was observed by our analyses in Lake Zurich, continuously performed for over 5 yr. We detected in Lake Zurich PAH concentrations in summer, fall, and winter at levels found in normal groundwater (about 30 $\mu g/m^3$). Increases in PAH were registered only during spring. It appears that during this season the water quality suffers, because of the input of runoff. Melted snow in urban areas is often contaminated with sedimented aerosols from central heatings, cars, and industry. The degree of this concentration rise in spring is still acceptable, since occasional top concentrations of up to 150 $\mu g/m^3$ do not constitute a menace to man's health.

In summary, we can state that even unobjectionable drinking water and noncontaminant Lake water contain a rather small amount of PAH. Also, civilizing influences may cause an increase of up to 10^4 times higher. This increase of PAH was produced primarily by domestic and industrial waste waters, and secondarily by road dust and combustion products.

III. THE FATE OF THE POLYCYCLIC AROMATIC HYDROCARBONS IN WATER

Polycyclic aromatic hydrocarbons in drinking water should be as low as possible for two reasons: First, carcinogenic substances are unwanted, and second, a low content of PAH indicates the absence of sewage thereby increasing the appetizing quality of water.

In principle, elimination of unwelcome substances in drinking water can be verified by direct analytical determination or partly through use of indicator substances. A classic example is the estimation of the risk of intestinal infections based on coliform test results. Hygienic quality is not completely guaranteed by negative indicator results. In this respect, the coliform test is of less importance than the coprosterol test. The determination of the total carbon content of water is very important to help identify contamination by organic substances. The polycyclic aromatic hydrocarbon analysis can be used successfully to determine sewage contamination or to verify removal of organics from sewage during drinking water purification processes. The PAH method is very sensitive. The correlation of the $KMnO_4$-test as a measure of organic substances in determining sewage in drinking water in relation to river water containing sewage is 20:1. The correlation ranges from 200:1 to 1000:1 for PAH. The PAH test is a better way of determining sewage contamination than the $KMnO_4$-test.

A condition necessary for the use of polycyclics as indicator substances is sufficient stability of these compounds within their aquatic environment. This is actually guaranteed. Apparently, PAH existed in ground water for a long time, since the microorganisms needed for biological decomposition are absent or are only present in low quantity. This is completely different from soil, where there is a continuous synthesis and decomposition of PAH occurring. Chemical destruction does not occur in ground water; thus we may assume PAH persist there for years. The PAH have stability in surface water. Ultraviolet light from the sun does degrade PAH in laboratory experiments, but such an effect proved minor in field experiments on Lake Zurich. This was shown by our latest research. Lake Zurich is a rather clear lake. We found at depths of 30 cm and 30 m a relationship of 36:26 $\mu g/m^3$ PAH as a median taken over 5 yr. If ultraviolet light had an influence, the correlation would be reversed.

Physical sedimentation is the essential factor for the elimination of polycyclics in river water and lakes. Adsorption to algae and subsequent precipitation after the algae die occurs, also. A certain amount of PAH is probably dissolved with the lipids of plankton. For this reason, unpolluted rivers, lakes, and reservoirs generally will not show high concentrations of carcinogenic substances. Therefore, rivers, lakes, and reservoirs do not require particular treatment. However, it is a different situation when contaminated river or lake waters are used to supply drinking water. In these cases, additional purification processes must be employed.

There are several reports available about the effectiveness of water treatment processes in the removal of PAH. In 1963, Graef and Nothafft studied the effect of chlorination on dissolved benzo[a]pyrene. The authors concluded that in practice one should not expect great success with a chlorine dose of approximately 0.3 mg/liter. Traktman and Manita (1966) showed similar results. Nevertheless, we ascertained during our analyses at Zurich a 50 to 60% reduction of benzo[a]pyrene caused by primary chlorination of the raw water at 1 mg/liter chlorine level. Benzo[a]pyrene is one PAH very sensitive to chlorination. Scassellati-Sforzolini et al. (1973) confirmed this with studies of PAH in demineralized water. Yet, we should emphasize that in practice the degree of removal efficiency will never reach the 100% mark. Only with very high doses of chlorine can you expect to get most effective removal. There has been some apprehension that chlorine in drinking water may have the effect of producing co-carcinogenic substances. Thus we should be cautious when offering these recommendations and wait for results of cancer tests currently being run.

In 1968, Reichert reported that chlorine dioxide (ClO_2) could decompose to a noncarcinogenic substance more than 90% of benzo-[a]pyrene present within the first 2 or 3 hr. The half-life increased when the benzo[a]pyrene concentration was lowered. ClO_2 could not oxidize benzo[a]pyrene at concentrations less than 10 $\mu g/m^3$ even with rather extended contact. These low concentrations are not objectionable from a hygienic standpoint.

Reichert (1969) showed that O_3 was a faster oxidant than ClO_2. Such an advantage is nullified by the problems associated with the use of ozone. Water must be pre-purified for the reaction with O_3 to remove carcinogenic substances, but if there exists an efficient system it is possible to get a drinking water free of PAH ($<0.1\ \mu g/m^3$). We have confirmed this result in practice in 1977.

Physical purification methods can also be used. In 1962, we did our first activated carbon analyses in a water works. Those and subsequent investigations showed that in practice activated carbon can reduce total PAH by 90%. In laboratory tests, activated carbon has been shown to remove up to 99% PAH. The difference probably is caused by the lower ability of the activated carbon in water treatment plants after it is used for a time. This happened to us in another sense, too. We found activated carbon filtration uneconomical for PAH, when the concentration of the PAH in the water supply was in the range of normal ground water concentration of 30 $\mu g/m^3$.

Scholz (1970) noted that there is insufficient practical experience available on granular activated carbon. Nevertheless, activated carbon filtration remains one of the best ways to eliminate PAH.

Sedimentation is an important physical method to remove PAH not only for sewage purification and the natural cleaning of lakes, but also for drinking water. We found that after several hours in a sedimentation basin, about 90% of the PAH was eliminated from strongly polluted river water. In accordance with the same principle, slow sand filtration efficiently removes PAH. However, a rapid sand filtration process shows less efficiency. River bank filtration effectively removes PAH, if the bank consists of very fine sands. Gravel-like sand and shingles in the river bank have negative effects.

In summary, we can say that sedimentation of surface water and sewage treatment with sedimentation and sand filtration removes about two-thirds of the PAH. Individual cases of more efficient removal may be attributed to the association of the PAH to large particles. Coagulation processes can transfer the PAH into larger particles, which can sediment or be filtered. With chlorination as a final step, we can expect elimination of dissolved carcinogenic substances. During 2 yr of pilot research on the Danube River, we observed on the average 90% removal of PAH after the coagulation process and 95% after final filtration. The removal effected a level down to that of ground water (30 μg/m^3), when an optimum initial and secondary dosing of $FeCl_3$ was used in the coagulation process in conjunction with ozonation and final activated carbon filtration.

IV. SUMMARY

The carcinogenic PAH are part of the chronically active noxious substances in drinking water. They are partly of biological origin and partly produced by pyrolytic processes. The PAH occurs in sewage in extremely high concentrations, which might serve as an indicator of fecal pollution. It is imperative that PAH should be eliminated from drinking water or reduced to the low level found in ground water (30 μg/m^3). One-third of the PAH in river water occur as particulates, another third is finely dispersed, and the remaining third is dissolved. Modern water treatment plants can eliminate two-thirds of the PAH by sedimentation, coagulation, and filtration. Residue PAH could be removed by oxidation or treatment with activated carbon. The efficiency of the different purification processes was tested by lab experiments and pilot plant tests. Control analyses at several water works assured the effectiveness of the purification treatment in practice.

In our opinion, PAH analysis is a valuable complement to conventional drinking water testing. Its application will help guarantee pulbic acceptance and safety of water. This is also recommended by WHO. A normal level for the six above-mentioned polycyclics is 50 μg/m^3; in tap water 100 μg/m^3 is acceptable. Drinking water containing more than 150 μg/m^3 might be taxed as "possibly polluted" and must be accurately checked. If there are more than 200 μg/m^3, the drinking water supply should be rejected as "objectionable." These limits take into account the standard deviation of these analyses. Initially, standards should be set high. A lower standard should be discussed as soon as a broader collection of data is available.

REFERENCES

1. Blumer, M. 1961. Science 134, 474.

2. Borneff, J., et al. 1959-1971. Kanzerogene Substanzen in Wasser and Boden, I-XXVII; Report: Arch. Hyg. esp. Zbl. Bakt. Hyg. I. Abt. Orig. B, Volumes 143 to 155.

3. Borneff, J., Farkasdi, G., Glathe, H., and Kunte, H. 1973. Zbl. Bakt. Hyg. I. Abt. Orig. B, 157, 151.

4. Graef, W. and Nothafft, G. 1963. Arch. Hyg. 147, 135.

5. Grimmer, G., Jakob, J. and Hildebrandt, A. 1972. Zschr. Krebsforsch. 78, 65.

6. Mallét, M. L. and Héros, M. 1962. Compt. Rend. Acad. Sci. 254, 958.

7. Reichert, J. 1963. Arch. Hyg. 152, 37a, 265.

8. Reichert, J. 1969. Gas Wasserfach 110, 477.

9. Sartor, J. D., et al. 1974. J. Water Pollut. Control. Fed. 46, 458.

10. Scholz, L. 1970. Vortrag 3. Arb. tag DGHM Mainz 8./10. October.

11. Scassellati-Sforzolini, G., Savino, A., Monarca, S. and Lollini, M. N. 1973. L'Igiena Mod. 66, 309.

12. Shabad, L. M. 1968. Zschr. Krebsforsch. 70, 204.

13. Shabad, L. M., Cohan, Y. L., Knitzky, A. P., Khesina, A. Y., Shcherbak, N. P. and Smirnov, G. A. 1971. J. Nat. Cancer Inst. 47, 1179.

14. Traktman, N. and Manita, M. 1966. Gig. San. 31, 21.

15. Wedgwood, P. H. and Cooper, R. L. 1954. Analyst 79, 163.

16. Yamagiwa, K. and Ichikawa, K. 1915. M. H. Med. Ges. Tokyo 15, 295.

17. Zdrazil, J. F. and Picha, D. 1965. Slevarenstvi (Czech) 13, 198.

APPENDIX (3/77)

Analysis of Polycyclic Aromatic Hydrocarbons (PAH)

1. Principle

The PAH are extracted from the water with organic solvent and separated, after a cleanup procedure, by two-dimensional thin layer chromatography. They are identified and quantified by their fluorescence in UV light. Evaluation may be accomplished either by visual comparison (semi-quantitative) (Section 5.1) or by elution of the substances and measurement of their fluorescence in solution (Section 5.2.1). Alternatively, fluorescence can be measured directly from the TLC-plate with a scanner (Section 5.2.2).

2. Apparatus

Wide necked cans, 5 - 10 liter capacity (glass-, stainless steel- or aluminum containers may be used, but no plastic materials!).
Stirrer, (3000 r.p.m.) capable of stirring 10 liters with explosion proof motor (lubricants must not come in contact with the water),
Vacuum rotary evaporator, with 1 liter evaporating flask,
Equipment for thin layer chromatography,
Separatory funnels, 1 liter capacity,
Beakers, funnels, micropipettes,
Small conical flasks, about 10 ml capacity, glass stoppered,
Drying oven,
UV-lamp with filter 365 nm,
if available: Fluorometer with scanner for direct measuring of TLC plates.

3. Reagents

All solvents should be of high purity. Cyclohexane is additionally purified by distillation or by percolation through a column of active Al_2O_3. The residue of 600 ml of the cyclohexane should not contain any of the PAH when chromatographed on thin layer.

Cyclohexane
Benzene (e.g. Benzol f. d. Fluoreszenz-Spektroskopie, Fa. Merck, Art.Nr. 1785)
n-Hexane

Methanol
Ethanol (99%)
Diethyl ether
Sodium sulphate, anhydrous
Aluminum oxide powder for TLC
Acetylated cellulose (40% acetyl) for TLC
Aluminum oxide for column chromatography

Reference substances:*

Fluoranthene
3,4-benzofluoranthene (benzo[b]fluoranthene)
11,12-benzofluoranthene (benzo[k]fluoranthene)
3,4-benzopyrene (benzo[a]pyrene)
1,12-benzoperylene (benzo[ghi]perylene)
Indeno (1,2,3-cd)pyrene

3.1 Stock solutions of the PAH

A stock solution is prepared of each of the 6 substances by dissolving 10 mg in benzene and making up to 100 ml in a volumetric flask (=100 μg/ml).

Care must be taken when weighing the substances, some of them are carcinogenic! Moreover, if pollution occurs in the laboratory, this may lead to incorrect results of subsequent analyses. The solutions should be kept in dark-glass stoppered bottles. They are stable for several years if care is taken that no solvent evaporates. This may be monitored by ultraviolet adsorption or fluorescence spectrophotometry.

3.2 Standard solution

5 ml of the stock solution of fluoranthene, and 1 ml each of the other 5 stock solutions are mixed in a small glass stoppered flask. This solution contains 50 ng of fluoranthene and 10 ng each of 3,4-benzofluoranthene, 11,12-benzofluoranthene, 3,4-benzopyrene, 1,12-benzoperylene, and Indeno (1,2,3-cd)pyrene in 1 μl.

Procedure

4.1 Extraction

5(to 10) liters of water sample are collected in a thoroughly cleaned container. In the laboratory 300(to 600) ml of purified

*e.g., Firma Ferak Berlin, E. Gründemann oHG, 1000 Berlin-West 47, Friedrichsbrunnerstr. 3-5.

cyclohexane is added and stirred with a fast stirrer for 10 min. The phases are allowed to separate (preferably over night). The cyclohexane layer is then carefully collected (with a beaker) into a 1 liter separatory funnel to remove the remaining water which has been carried over during the procedure. The cyclohexane solution is then filtered through sodium sulphate into a graduated cylinder. The amount of cyclohexane recovered should be 250(to 500) ml. The solution is concentrated to a few ml in a vacuum rotary evaporator. The concentrated extract is transferred to a small conical glass stoppered flask and the evaporating flask is rinsed 2-3 times with small amounts of cyclohexane which are added to the extract. The content of the small flask is then further concentrated in a modified rotary evaporator to about 0.5 ml. It is also possible to extract PAH from 2 x 2500 ml of water in a separatory funnel. However, any changes in the method must show a recovery that exceeds 80%.

4.2 Cleanup procedure

0.5 g of aluminum oxide, basic, activity II (according to BROCKMANN) is placed in a small column (about 6 mm in diameter and 80 to 100 mm long, tapered at the lower end) above a small plug of cotton wool. The column is rinsed with 1 ml of cyclohexane and the extract placed on the column. The conical flask is rinsed with 0.5 ml cyclohexane, which is added to the column. The eluates are discarded. Then 0.5 ml of cyclohexane and 3 ml of a cyclohexane-benzene mixture (1 + 1) are added to elute the PAH. These eluates are collected in the small conical flask and evaporated to about 0.1 ml for TLC.

If heavily polluted water is examined, only one aliquot of the extract is used for the cleanup and further determination.

When first introducing the method into the laboratory, the cleanup procedure should be tested with reference substances. This is because different materials and other circumstances may affect the elution.

4.3 Thin layer chromatography (TLC)

4.3.1 Preparation of plates

For the preparation of 5 TLC plates, 28 g of aluminum oxide powder and 12 g of acetylated cellulose are well mixed with 65 g ethanol (using a stirrer or mixer) and the plates coated with this mixture using a coating apparatus. As soon as the plates are superficially dry, they are activated for 30 min. at 130°C in a drying oven. TLC plates are stored in a desiccator.

4.3.2 Chromatography procedure

The whole extract, if clean water is being examined, or an aliquot of a polluted water is applied in small portions as one spot in one corner of the TLC plate about 1.5 cm from the two sides. The plate is developed in the first direction with a mixture of n-hexane + benzene (90 + 10 vol.) for 30 min. After drying, the plate is turned 90° and developed for 60 min. in the second direction with a mixture of methanol + ether + water (40 + 40 + 10 vol.). The plates should be protected from light during this procedure.

Suitable amounts of the standard solution are applied to the thin layer plates for reference chromatograms and developed as above.

5. Evaluation

5.1 Estimation by visual comparison

If a fluorometer is not available, amounts of PAH may be determined semiquantitatively by comparing the TLC plate with reference chromatograms under a UV-lamp. In this case a series of TLC plates, with 1-100 µl of the standard solution applied, are developed. Under the UV-lamp the spots on the sample plate are identified by their position and their fluorescence color. The amount of each substance is estimated by comparison with the reference plates. For greater ease of handling, the developed plates may be sprayed with a preserving dispersion (e.g., "Neatan" by Merck) which must have no fluorescence of its own. The thin layer can then be drawn off the glass plate and fixed on a piece of black paper. Reference chromatograms preserved in this way may be used up to two months if kept in the dark, but must be replaced earlier if used frequently because of fading of the fluorescence.

5.2 Quantitative determination by measuring the fluorescence

5.2.1 Elution of substances from the TLC plate and measuring the fluorescence in solution

The spots of the substances in question are located under the UV lamp, scraped off individually and eluted with benzene using small sintered glass filter funnels. The solution is made up to a definite volume and the fluorescence is measured at the following wavelengths:

Fluoranthene 465 nm
3,4-benzofluoranthene 454 nm
11,12-benzofluoranthene 434 nm
3,4-benzopyrene 431 nm
1,12-benzoperylene 420 nm
Indeno(1,2,3-cd)pyrene 505 nm
The excitation wavelength is 365 nm.

A calibration curve is prepared for each substance from the stock solution. The fluorescence of a series of standards with concentrations ranging from 0.005 to 0.5 μg/1 are measured. To cover the whole range, two different instrument settings (sensitivity and slit width) will have to be used for high and low concentrations. A fluorescence standard must be used for calibration of the instrument.

Due to background fluorescence, the values measured may be too high, unless the whole fluorescence spectrum is registered and appropriate correction made. On the other hand, losses occurring during chromatography and elution are different in each laboratory and will have to be determined.

5.2.2 <u>Measuring fluorescence directly from the TLC plate with a scanner</u>

Working principle of the scanner: the excitation radiation (in our case 365 nm, filtered from a Hg-lamp source) is directed onto the thin layer plate through a fixed slit. The resulting fluorescence reaches the fluorometer by a suitable optical device. The thin layer plate moves past the slit at a constant velocity and the intensity of fluorescence is registered by a recorder. The area of the registered peak corresponds to intensity of fluorescence and size of the spot, and therefore is proportional to the amount of substance in the spot.

Prepare two reference chromatograms with 2 and 5 μl of the test solution. Depending on the instrument used, mark the spots of the substances in a suitable way. Place the plate into the scanner and choose the slit length of the scanner according to the size of the spot to be measured. Adjust the plate so that the slit is just about in the middle of the spot. Now set the wavelength of the fluorometer according to the table below and open the slit of the fluorometer until fluorescence intensity is about 50%. Note the slit width. (Instrument sensitivity should be set at an optimum level and must not be changed thereafter.) Now locate the plate so that the scanner slit is just in front of the spot, start the scanner motor and register the fluorescence intensity on the recorder. Repeat the procedure for every spot on the reference plate.

Wavelength setting for the PAH:

Fluoranthene 462 nm
3,4-Benzofluoranthene 452 nm
11,12-Benzofluoranthene 431 nm
3,4-Benzopyrene 430 nm
1,12-Benzoperylene 419 nm
Indenopyrene 500 nm

Then measure the spots on the sample plate, making sure that the instrument settings for each substance is the same as for the reference plate (particularly the slit width of the fluorometer). Choose the reference chromatogram according to the fluorescence intensity of your sample. If the intensity is too high to be measured under the conditions of the reference plate (i.e., exceeds 100%) the slit width should be reduced and both sample and reference plate measured under the new conditions. If the difference is too large, a reference plate must be prepared with a higher concentration. It is not advisable, however, to have too large an amount of sample on the plate, and in cases where pollution of the water is to be expected, only one aliquot of the extract should be examined by thin layer. If necessary, a second plate with a suitable aliquot for measurement can then be prepared.

6. Calculation

The area of the recorded peak is proportional to the amount of substance present. This can be determined in different ways (as in gas-chromatography); one way is to measure the peak height, and the peak width at half the height and multiply the two. This method gives good results. Of course, an electronic integrator can be used, if available.

The amount of substance on a sample plate is: $A(ng) = \frac{B \times C}{D}$ where B = amount of substance on the reference plate (ng), C = peak area measured from the sample plate, D = peak area measured from the reference plate.

If one aliquot of the extract has been used, multiply A by the appropriate factor to find the amount contained in the whole extract. This represents the amount of PAH in 10 liters of water and dividing by 10 will represent the concentration in ng/l.

The concentrations of the 6 PAH substances are summed to give the concentration of PAH in the water. It must be borne in mind that this figure does not give "total" PAH concentration, but it is a good and easily comparable representation of the whole group.

7. Other detection methods

It must be mentioned that other methods of detection may be used. With gas chromatography a large number of PAH can be determined. For separation of the compounds a column of high efficiency must be used and careful cleanup prior to GC is necessary. Column chromatography followed by paper chromatography was widely used until TLC was introduced as a routine laboratory procedure, the latter method being less time consuming and more sensitive. The advantage of combined column and paper chromatography is that larger amounts of extract can be chromatographed and therefore substances of low concentration and weak fluorescence (compared with the main components) can still be detected.

Fate of High Molecular Weight-Chlorinated Paraffins in the Aquatic Environment

VLADIMIR ZITKO and EDMOND ARSENAULT
Environment Canada
Fisheries and Marine Service
Biological Station, St. Andrews
New Brunswick, Canada E0G 2X0

I. INTRODUCTION

Chlorinated paraffins, based on C_{10}-C_{28} paraffins and containing 10 to 70 percent, most frequently 40, 50, or 70 percent, chlorine by weight, have been produced on an industrial scale since about 1930, mainly in the United States, Great Britain, and

Germany. In 1969, the United States' production of chlorinated paraffins was 28 x 10^6 kg, and the main applications include plasticizers, fire retardants, and additives in lubricants, cutting oils, abrasive products, adhesives, and so on.

Chlorinated paraffins are prepared by free radical chlorination of various paraffinic stocks, usually in the presence of a solvent such as carbon tetrachloride (Panzel and Ballschmiter 1974). In the chlorination, the reactivity of tertiary, secondary, and primary carbon atoms is approximately 4:2:1, so that in straight chain paraffins the chlorination proceeds exclusively on the secondary carbon atoms. Chlorine substitution deactivates the neighboring carbon atoms and the next chlorine atom is likely to enter the molecule at least two carbons away from the first substitution. Idealized formulas of chlorinated paraffins based on pentacosane are presented in Table 1.

TABLE 1. Idealized Formulae of Chlorinated Paraffins

Weight percent chlorine	38	51	69
Formula	$C_{25}H_{46}Cl_6$	$C_{25}H_{42}Cl_{10}$	$C_{25}H_{31}Cl_{21}$
Formula weight	559	696	1075

Even purified paraffinic stocks always consist of mixtures of homologs, and consequently, a chlorinated paraffin preparation contains a mixture of chlorinated paraffin homologs. This is illustrated in Table 2, showing the composition of paraffins obtained by dechlorination of a commercial chlorinated paraffin preparation, Cereclor 42, containing 42 percent chlorine (Zitko, 1974a). Similar results were obtained on other United States chlorinated paraffins containing 40 to 50 percent chlorine. In contrast, German preparations are generally based on shorter chain paraffins (C_{14}–C_{17}, Panzel and Ballschmiter, 1974).

The chlorination of a paraffin will lead not only to a mixture of positional isomers, containing the same number of chlorine atoms, but also to molecules with different numbers of chlorine atoms. The distribution of molecules with different numbers of chlorine atoms in the mixture depends on the overall degree of chlorination and can be calculated from statistical considerations. An example of the actually observed distribution

TABLE 2. Composition of Paraffins Obtained by Dechlorination of Cereclor 42

Paraffin	C_{21}	C_{22}	C_{23}	C_{24}	C_{25}	C_{26}	C_{27}	C_{28}
Percent	3.6	8.8	14.7	18.6	19.5	17.2	11.5	6.0

of molecules, chlorinated to a different degree, is presented in Table 3 (Teubel et al., 1962).

The above-mentioned factors alone make the composition of chlorinated paraffin preparations very complex. The situation is further complicated by the presence of branched paraffins and possibly other hydrocarbons in the stocks used for the preparation of chlorinated paraffins. Chlorination of branched paraffins yields tertiary chloro compounds that split off hydrochloric acid relatively easily. The so-called labile chlorine can be determined by refluxing with silver nitrate in aqueous acetone. Vicinal chlorine atoms can be determined by refluxing with zinc dust in butanol, and data of Weintraub and Mottern (1965) on labile and vicinal chlorine atoms in two commercial chlorinated paraffin preparations are presented in Table 4.

The elucidation of other structural details of chlorinated paraffins is hampered by the high complexity of the preparations.

There are no specific methods for the determination of chlorinated paraffins. Nonspecific methods are based on the determination of chlorine after pyrolysis. The highest sensitivity is probably offered by a microcoulometric technique, capable of detecting approximately 2 ng of chlorine (Zitko, 1973). Panzel and Ballschmiter (1974) described the behavior of chlorinated paraffins on gas chromatography but did not state the sensitivity of this technique. Chlorinated paraffins can be separated from lipids by chromatography on alumina, and from PCBs and some chlorinated hydrocarbon pesticides by chromatography on silica (Zitko, 1973). Dechlorination by sodium bis(2-methoxyethoxy) aluminum hydride (Panzel and Ballschmiter, 1974; Zitko, 1974a) and the detection of straight chain paraffins may be used for confirmation.

The toxicity of chlorinated paraffins has not been studied in detail. The available data indicate low acute oral toxicity to rats, mice, and guinea pigs, and no accumulation was observed in rats, mice, and fish (Zitko and Arsenault, 1974). Little is

TABLE 3. Distribution of Chlorinated Paraffins in a C_{20}–C_{30} Preparation, Containing 26.1 percent Chlorine (4.3 atoms/mol)

Atoms/mol	0	1	2	3	4	5	6	6
Volume percent	1.1	3.7	11.0	14.2	24.1	25.2	12.2	8.2

TABLE 4. Labile and Vicinal Chlorine Atoms in Chlorinated Paraffins

Weight % of Chlorine in Preparation	Fraction of Chlorine Atoms (%)	
	Vicinal	Labile
57.9	57	15.8
67.0	70	3.7

known about the behavior of chlorinated paraffins in the environment. A review of preparation, chemistry, toxicity, and application of chlorinated paraffins is available (Zitko and Arsenault, 1974). This review describes the fractionation of chlorinated paraffins by solvent partitioning and presents preliminary data on degradation of chlorinated paraffins in marine sediments.

II. EXPERIMENTAL

A. Materials

Chlorinated paraffin preparations Cereclor 42 (42 percent chlorine, ICI America, Inc., Wilmington, Delaware), Clorafin 40 (40 percent chlorine, Hercules Inc., Wilmington, Delaware), and Chlorez 700 (70 percent chlorine, Dover Chemical Corporation, Dover, Ohio) were used.

B. Analysis

Chlorinated paraffins were determined microcoulometrically as described by Zitko (1973) in a Dohrmann microcoulometric system MCTS-20. From air-dried, spiked sediments, chlorinated paraffins were extracted with 50 percent ether in hexane. The sediment (1 g) was placed in a 45 x 0.7-cm glass column, and chlorinated paraffins were eluted with 10 ml of the solvent (recovery 90 percent).

C. Solvent Partitioning

Hexane solutions of chlorinated paraffins (500 to 1000 mg of chlorinated paraffins in 50 ml hexane) were equilibrated with 50 ml of either acetonitrile, dimethyl formamide, or sulfoxide. The hexane phase was washed with distilled water (2 x 50 ml), and hexane was evaporated in vacuum in a rotatory evaporator. The other phase was diluted with 50 ml distilled water and extracted with 50 ml of hexane. The hexane solution was worked up as above, and the aqueous phase was further extracted with 50 ml of ether. The ether solution was washed with water, dried over anhydrous sodium sulfate, and ether was evaporated in vacuum in a rotatory evaporator. The aqueous dimethyl sulfoxide phase was further extracted with 2 x 50 ml chloroform and 2 x 50 ml ethyl acetate, and the extracts were worked up separately as described for the ether extract.

D. Degradation of Chlorinated Paraffins in Sediments

A portion of sediment, collected in the vicinity of the Biological Station was air dried and suspended in a hexane solution of chlorinated paraffins, and hexane was evaporated in a rotatory evaporator. The dry sediment was then homogenized with a larger portion of wet sediment. Two spiked samples, one containing Cereclor 42 at 596 μg/g dry weight and one containing Chlorez 700 at 357 μg/g dry weight were prepared. The sediments were charged to 500-ml Erlenmeyer flasks (25 g/flask), 300 ml of seawater, and 10 ml of a suspension of decomposing organic matter in seawater collected near the Biological Station were added and the flasks were kept at room temperature (19 to 22°C). A set of three flasks (control sediment, Cereclor 42, and Chlorez 700 spiked sediment) was aerated, and the volume of seawater was kept constant by adding distilled water as required. Another set of three flasks was kept stoppered. Samples of sediments were taken periodically and analyzed as described previously.

III. RESULTS AND DISCUSSION

A. Fractionation of Chlorinated Paraffins by Solvent Partitioning

The partition coefficients of Cereclor 42 and Chlorez 700 are given in Table 5. Acetonitrile does not extract chlorinated paraffins from hexane very effectively. Dimethyl formamide is, on the other hand, a very good solvent for the extraction of chlorinated paraffins, and since at the same time it does not extract lipids, the partitioning between hexane and dimethyl formamide could be used to separate these groups of compounds. Dimethyl sulfoxide extracts preferentially the more chlorinated Chlorez 700, and as will be shown, fractionates chlorinated paraffin preparations.

TABLE 5. Distribution Coefficients of Chlorinated Paraffins, Hexane-Solvent

Solvent	Distribution Coefficient	
	Cereclor 42	Chlorez 700
Acetonitrile	0.62	0.75
Dimethyl formamide	0.12	0.12
Dimethyl sulfoxide	0.82	0.06

Cereclor 42 and Clorafin 40 contain approximately the same amount of chlorine, but the latter gives a much lower yield of straight-chain paraffins on dechlorination (Zitko, 1974a). The difference between these two chlorinated paraffin preparations is very likely due to a lower proportion of straight-chain chlorinated paraffins in Clorafin 40. The data in Table 6, showing weight yields of fractions obtained by partitioning of these two preparations between hexane and dimethyl sulfoxide, indicate significant differences between Cereclor 42 and Clorafin 40. In the case of Cereclor 42, 78 percent of the preparation is extractable from dimethyl sulfoxide with hexane, whereas in the case of Clorafin 40 the extractable amount is only 43 percent. In addition, Cereclor 42 does not contain material recoverable from dimethyl sulfoxide with chloroform and ethyl acetate.

TABLE 6. Fractionation of Chlorinated Paraffins by Partitioning between Hexane and Dimethyl Sulfoxide

Solvent	Fraction (wt. %)	
	Cereclor 42	Clorafin 40
Hexane	43.3	40.7
Recovered from aqueous dimethyl sulfoxide with:		
Hexane	35.1	1.9
Ether	20.1	36.3
Chloroform	--	16.6
Ethyl acetate	--	0.7
Total Recovery	98.5	96.2

The chlorine content of the fractions increases somewhat with their polarity, that is, the dimethyl sulfoxide fractions contain more chlorine than the hexane fraction. Among the dimethyl sulfoxide fractions, the chlorine content increases from hexane to chloroform, but the range of chlorine content is quite narrow (39 to 45 and 35 to 45 percent for hexane and the last recovered fraction of Cereclor 42 and Clorafin 40, respectively). The composition of straight-chain paraffins obtained by dechlorination of these fractions indicates only a very slight relative increase of longer chain paraffins in the more polar fractions. The most significant difference is the yield of straight-chain paraffins. The hexane fraction of both preparations yields 49 percent of straight-chain paraffins on dechlorination. On the other hand, the yield of straight-chain paraffins from the fractions recovered from dimethyl sulfoxide is, in order of increasing polarity, 26 and 29, and 27, 20, and 22 percent, for Cereclor 42 and Clorafin 40, respectively.

The presented data show the complexity of chlorinated paraffin preparations and indicate that the chlorine content alone does not adequately characterize the preparation.

B. Degradation of Chlorinated Paraffins in Sediments

The recovery of Cereclor 42 and Chlorez 700 from spiked sediments was 75.6 and 63.2 percent, respectively. The sediment itself contained a background level of 5.65 μg/g dry weight as chlorine, of a microcoulometrically detectable material. The concentrations of chlorinated paraffins, given in Table 7, are corrected both for recovery and for background.

TABLE 7. Degradation of Chlorinated Paraffins in Spiked Sediments

Time (days)	Concentration in Sediment (μg/g dry weight) Conditions: Aerobic		Anaerobic	
	Cereclor 42	Chlorez 700	Cereclor 42	Chlorez 700
0	596	357	596	357
10	257	76	80	41
21	147	128	194	33
28	377	72	98	50

The rate of degradation of chlorinated paraffins is higher under anaerobic than under aerobic conditions and Chlorez 700 is degraded faster than Cereclor 42. On the other hand, the data show only the disappearance of the starting material and not the concentration of more polar degradation intermediates. Under the conditions used for extraction of chlorinated paraffins from the sediments, transformation products containing polar groups would not be extracted and the actual degradation (conversion of organically bound chlorine to chloride) may be much less than the data in Table 7 indicate.

C. Pollution Potential of Chlorinated Paraffins

Chlorinated paraffins may reach the environment mainly as components of plastics in solid wastes, in waste oils, and by leaching from antifouling paints. These routes are similar to some of the routes by which PCBs enter the environment, and it may be useful to compare the behavior and properties of these two classes of compounds.

Chlorinated paraffins, available in North America, have higher molecular weight than PCBs, and consequently are less volatile and very likely also less soluble in water than PCBs. Both higher molecular weight and lower solubility mean that chlorinated paraffins would be less leachable from solid refuse than PCBs, and to a lesser degree taken up by living matter. It has been shown experimentally that fish do not accumulate orally administered Cereclor 42 and Chlorez 700 (Zitko, 1974b). In contrast, chlorinated paraffins manufactured in Germany (Panzel and Ballschmiter, 1974) have a much lower average molecular weight and, if molecular weight is the only uptake-limiting factor, these chlorinated paraffins could be taken up and accumulated in the biomass.

Chlorinated paraffins are much less thermally stable than PCBs and decompose to a large extent at relatively low temperatures (300 to 400°C), whereas a temperature of at least 800°C is required for the decomposition of PCBs. Consequently, in comparison with PCBs, little if any chlorinated paraffins would occur in emissions from incinerators. The main decomposition product of chlorinated paraffins is hydrochloric acid. Little is known about the possible formation of low molecular weight chlorinated hydrocarbons, and this should be investigated in view of the recently discovered carcinogenicity of vinyl chloride.

Chlorinated paraffins are also chemically less stable than PCBs and the main decomposition route is very likely dehydrochlorination. Some chlorinated hydrocarbons such as chlorinated alicyclic hydrocarbons may be more persistent than chlorinated straight-chain paraffins. Chlorinated aromatic hydrocarbons, particularly those with attached paraffinic chains, may be present as impurities in chlorinated hydrocarbon preparations and these also may be more persistent.

A more detailed study of the environmental behavior of chlorinated paraffins is hampered by the lack of sensitive analytical techniques. In general, the qualitative and quantitative characterization of chlorinated paraffins is much more difficult and less developed than that of PCBs. For example, gas chromatography with an electron-capture detector or gas chromatography-mass spectrometry, which are extremely sensitive tools for the investigation of PCBs, are probably not to the same extent applicable to chlorinated paraffins. High pressure liquid chromatography is a potentially promising technique for the investigation of chlorinated paraffins, but has not been used as yet.

IV. CONCLUSION

In conclusion, much remains to be learned about the environmental properties of chlorinated paraffins and their degradation products, and particular attention should be given to the lower molecular weight (C_{10}–C_{13}) preparations. The evidence obtained thus far indicates that, at comparable production volumes, the degree of environmental contamination by chlorinated paraffins will likely be lower than that by PCBs.

ACKNOWLEDGMENT

We thank Mrs. Madelyn M. Irwin for efficient assistance in literature documentation and typing of the manuscript.

REFERENCES

1. Panzel, H. and Ballschmiter, K. 1974. Z. Anal. Chem. 271, 182.

2. Teubel, J., Roesner, H., and Leschner, O. 1962. Chem. Tech. (Berlin) 14, 320.

3. Weintraub, L. and Mottern, H. O. 1965. I&EC Product Research and Development 4, 99.

4. Zitko, V. 1973. J. Chromatogr. 81, 152.

5. Zitko, V. 1974a. J. Assoc. Off. Anal. Chem. 57, 1253.

6. Zitko, V. 1974b. Bull. Environ. Contam. Toxicol. 12, 406.

7. Zitko, V. and Arsenault, E. 1974. Chlorinated Paraffins: Properties, Uses, and Pollution Potential. Fisheries and Marine Service Technical Report No. 491, 38 pp., Biological Station, St. Andrews, New Brunswick, Canada.

Human Exposure to Water Pollutants

FREDERICK C. KOPFLER, ROBERT G. MELTON,
J. L. MULLANEY, and ROBERT G. TARDIFF
U. S. Environmental Protection Agency
Health Effects Research Laboratory
Cincinnato, Ohio

I. INTRODUCTION

Man has always been aware of organic contaminants present in drinking water because of the colors, tastes, and odors that they sometimes cause. Consequently, methods such as oxidation and adsorption were developed for removing those undesirable characteristics of the water.

The carbon adsorption method was developed and recently improved (Anon., 1971; Buelow et al., 1973) to give a gross measurement of the organics present in drinking water as the carbon chloroform extract (CCE). At present the CCE is the only standard, except for phenol, to limit the amount of toxic organic matter that may reach the water consumer. It needs to be supported by other methods to protect man from the full spectrum of toxic organics that could be present in drinking water. Because of the additive or synergistic effect possible among toxicants, Ettinger (1965) stated that a toxicant in water cannot be discounted as harmless because its concentration is relatively small, without massive specific evidence in support of that view.

Surface waters used as drinking water sources may contain many organic wastes in addition to the naturally occurring organics. In 1968, it was estimated that after treatment, manufacturing and domestic wastes discharged into the nation's waters totaled 6.1 billion kg (Wallis, 1974). Industrial wastes are likely to contain refractory organic chemicals that resist biological degradation. Recently, Jolley (1973) studied the effects of chlorination on organics contained in domestic sewage and tentatively identified 17 chlorinated compounds including chlorinated purines and pyrimidines. He found that approximately 1 percent of the chlorine used in treating of domestic sewage was associated with stable chlorine reaction period. Based on this yield, approximately 1000 tons (91,000 kilos) of chlorine in the form of stable, chlorine-containing compounds are discharged annually from sewage treatment plants into the nation's waterways.

The problem of organics in drinking water is further complicated because there is some recent evidence that chlorine used to disinfect drinking water may combine with synthetic organic compounds, as well as those occurring naturally, and result in elevated levels of a variety of chlorinated organic compounds (Rook, 1974; Bellar et al., 1974). This problem is not totally limited to drinking water obtained from surface sources. Nelson and Lysyj (1968) reported an average total organic carbon concentration of 0.9 mg/1 in drinking water obtained from underground sources in several southwest and pacific coast communities. Bellar et al. (1974) also reported detectable quantities of halogenated organics in chlorinated well water containing approximately 0.5 mg/1 total organic carbon. Ground water is also subject to contamination with organics such as gasoline from underground storage tanks (McKee et al., 1972) and leaching from products of human activities.

Because of past and possible future contamination episodes of water supplies and because of improvements in methods to detect and identify individual organics in water, it has become both necessary and possible to examine closely the exposure of man to organic compounds via drinking water and to ascertain the significance of such exposure.

One object of such a study is to determine the relative risks associated with man's chronic exposure to organics in drinking water. This must be accomplished by determining the biological potency of these materials as well as the antagonistic and synergistic interactions that can occur among them. These biological investigations are necessarily supported by chemical analyses to offer explanations of the results in terms of cause and effect.

Isolation of the organics from the water is the first step in both biological and chemical analyses. Traditionally, organics have been collected from water by adsorption onto activated carbon followed by drying and subsequent solvent extraction. It has been suggested in the literature that adsorption of organic compounds onto carbon can possibly effect changes in the structure of organic compounds, and at least one such change has been reported in the literature (Ishizaki and Cookson, 1973). To collect enough organics from water for bioassay, exposure time of the carbon to the water must be several days' time in which bacterial growth in the column can degrade the adsorbed organics (Bishop et al., 1967).

Consequently, with the recognition that no one method will be perfect, other methods were sought, for the purposes of concentrating sufficient quantities of organic materials from water with the least amount of degradation. Reverse osmosis was chosen. Reverse osmosis is a membrane process developed almost entirely in the past 10 to 15 years. The process is based on the fact that if two solutions of different concentrations are separated by a semipermeable membrane, water will move across the membrane in the direction of higher concentration. The driving force is provided by this difference in concentration. Since many of the dissolved substances will not pass through the membrane, the passage of water will eventually cease when the solutions on both sides reach the same concentration. This process of osmosis can be reversed by applying pressure to the more concentrated solution. The concentration of dissolved substances will then tend to increase on that side of the membrane.

Although the process was initially developed for the purposes of desalination of water, the rejection of organics has also been extensively studied. A recent paper shows that reverse osmosis is one of the most promising techniques for removing refractory organics of intermediate to higher molecular weight from water (Edwards and Schubert, 1974). Much more study is needed to predict rejections of organics by reverse osmosis; it appears, however, rejection is related to the distribution coefficient of a given compound between the aqueous solution and the hydrated membrane. Obviously, this is complicated by the concentration of inorganic ions that continuously change the solubility of the organic compounds in solution and that a high salt concentration can affect the degree of hydration of the membrane.

Exposure to organic compounds via drinking water can be estimated by using data from total organic carbon analyses. Total organic carbon of tap water in the Cincinnati water supply is typically between 1 and 2 mg/liter as shown in Figure 1. If we assume that carbon represents 50 percent of the weight of the organic materials present, 1500 liters should contain 3 to 6 g of organic material, a sufficient quantity for chemical analysis and various toxicity studies contributing to health and hazard assessments.

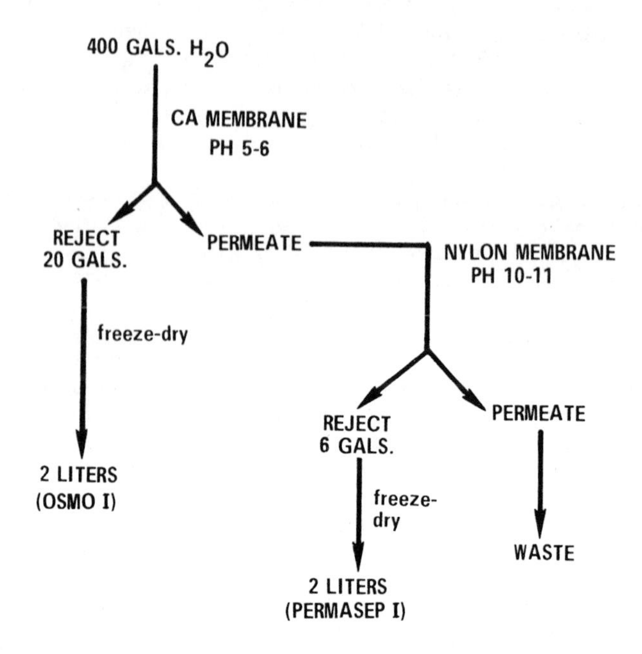

FIGURE 1. Schematic for concentration of organics from tap water.

II. REVERSE OSMOSIS CONCENTRATION OF ORGANICS FROM DRINKING WATER

The scheme presently used for isolating organics from water is shown in Figure 2. Typically, 400 gallons (1514 l) of tap water is collected into a covered 500-gallon stainless steel tank over a 2-day period. Soon after collection is begun, 140 mg of silver nitrate is added to retard bacterial growth. The collected water is adjusted to pH 5.5 with HCl. This pH is in the optimum range for stability of cellulose acetate membranes.

The water is subjected to reverse osmosis with an Osmonics Model 3319-558C unit containing a spiral would cellulose acetate membrane at a pressure of 190 lb/in^2 and a 50 percent conversion rate. Immediately before entering the reverse osmosis unit, the water passes through a copper coil immersed in a refrigerated bath maintained at 6 to 10°C. The permeate (water forced through the membrane) is collected in a second stainless steel tank while the reject is recycled into the feed tank.

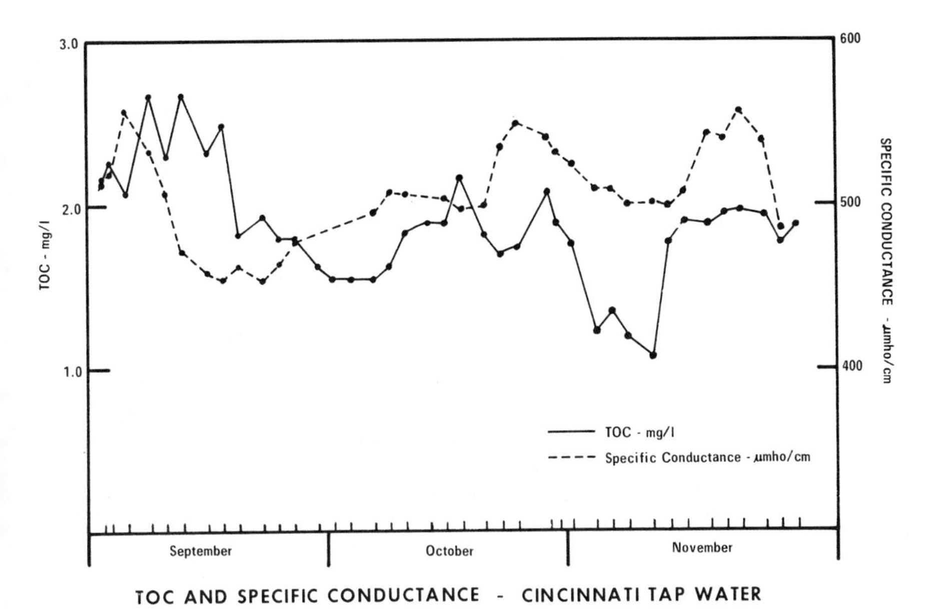

FIGURE 2. TOC and specific conductance—Cincinnati tap water.

When the volume of the feed has been reduced to 20 to 40 gallons, reverse osmosis is terminated. In experiments performed thus far the average TOC retention in this concentrate is 85 percent. The concentrate is divided into 5-liter aliquots that are each placed in a shallow stainless steel tray. These are frozen and subsequently lyophilized. The temperature of the trays is electronically controlled and is never allowed to rise above -10°C or reach complete dryness while in the lyophilizer. When the volume of a group of trays is sufficiently reduced, the contents are allowed to thaw and are then combined into a single tray, refrozen, and lyophilized. This procedure is continued until the total volume remaining is about 3 liters of liquid and solids.

The permeate collected from the cellulose acetate unit is again subjected to reverse osmosis in a similar manner using a Continental Water Conditioning Model 881-1 unit containing a DuPont B-9 Permasep permeator. This is an asymmetric aromatic polyamide membrane of hollow fiber configuration. It is easily degraded by the chlorine concentrations normally found in finished drinking water, but the chlorine is removed by the cellulose acetate membrane.

This membrane can tolerate alkaline conditions, and since most salts have been removed by the cellulose acetate membrane, the sample can be brought to pH 10.5 with sodium hydroxide. This will ionize phenols and acids to increase rejection by the membrane. To maintain a 50 percent conversion rate, a pressure of 125 lb/in^2 is maintained. The permeate is discarded and the feed solution is recycled until the volume is reduced to 2 to 3 gallons. This concentrate is frozen and lyophilized as before.

The final product obtained by lyophilizing the aqueous concentrate obtained by reverse osmosis of drinking water with the cellulose acetate membrane is filtered through a coarse, sintered-glass filter to remove the precipitated salts. These salts are again lyophilized to dryness and extracted first with pentane and then with methylene chloride.

The filtrate is extracted three times with pentane using 75 ml pentane/liter of solution for the first extraction and 50 ml pentane/liter for the other two extractions. The extracts are combined, dried with sodium sulfate, and concentrated in a Kuderna-Danish evaporator (Gunther et al., 1951).

The solution is extracted again in the same manner with methylene chloride. It is then acifidied with hydrochloric acid to pH 2 and again extracted with methylene chloride to extract acidic and phenolic compounds.

After extraction, the solvent is removed from the concentrate in a rotary evaporator at 40°C. More organics can then be obtained by passing the concentrate through a column of XAD-2 macroreticular resin and subsequent elution with ethanol. This extraction scheme is shown in Figure 3.

The concentrate obtained with the DuPont Permasep membrane is extracted in the same manner. However, since the cellulose acetate membrane removes most of the salts, little if any precipitation occurs on lyophilization and there is no need to extract salts separately.

Aliquots of each fraction are retained for chemical analysis. The remaining portions are combined for toxicity assays.

Analysis of these fractions by gas chromatography-mass spectrometry has proved the presence of some compounds not previously identified in drinking water such as those shown in Table 1. Because of tar-like residues remaining in the inlet of the gas chromatographic columns, it is apparent that much of the organic residue is nonvolatile and, as yet, remains unidentified.

The more volatile compounds present in the water, which are lost during the concentration procedure, are collected by

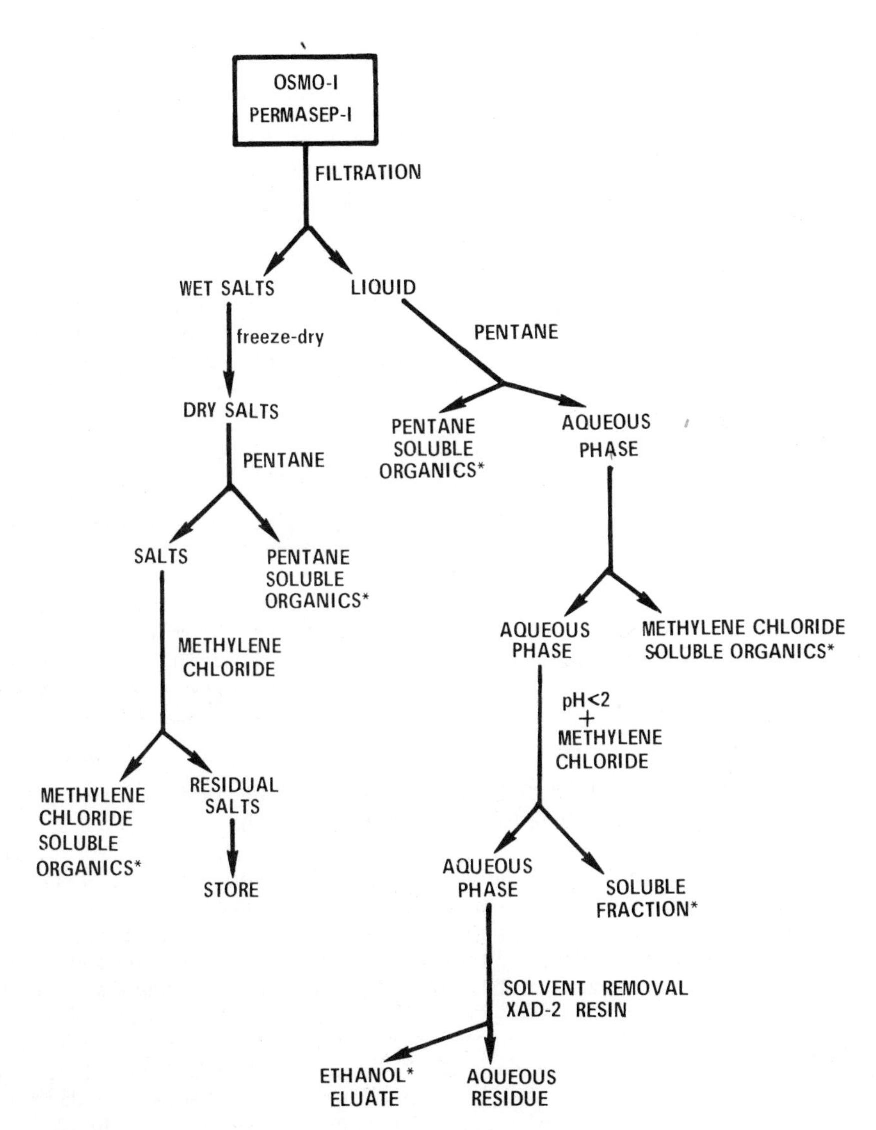

FIGURE 3. Schematic for extractions of organics from reverse osmosis concentrates of tap water.

TABLE 1. Organic Compounds Identified in Reverse Osmosis Concentrates of Cincinnati Drinking Water

Barbital	Methyl palmitate
Benzene sulfonic acid	Methyl stearate
Benzyl butyl phthalate	Methyl tetracosanoate
2-Chloroethyl methyl ether	Octadecane
Dibutyl phthalate	Pentachlorobiphenyl
Diethyl phthalate	Phenyl benzoate
Di(2-ethyl hexyl) phthalate	Phthalic anhydride
Docosane	1,1,3,3-Tetrachloroacetone
Eicosane	Tetrachlorobiphenyl
Dexadecane	Trichlorobiphenyl
Methyl docosanoate	2,4,5-Trichlorophenol

stripping with helium from a water sample onto a trap containing a porous polymer. This is accomplished with the use of the system developed by Bellar and Lichtenberg (1974), modified to contain 500 ml of sample. The trap is subsequently heated, and the volatile organics back-flushed into a gas chromatograph/mass spectrometer. Typical compounds identified in drinking water are shown in Table 2. These compounds can then be added back to the higher molecular weight organics for biological testing.

III. TOXICOLOGICAL ASSESSMENT OF TRACE ORGANICS IN WATER SUPPLIES

In the toxicological assessment of the compounds to which man is exposed in his drinking water, two general questions may be asked: First, what is the toxicity of the entire mixture with the summation of its interactive effects? And, second, what are the relatively safe levels of exposure to the individual components?

The nature of the mixtures and concentrates described above is such that the components and their relative concentrations are mostly unknown. Consequently, specific toxic effects are difficult to predict. General screening tests, however, may be employed for range finding and determining comparative potency. The tests being developed in our laboratory include (a) the LD_{50} to determine innate acute toxicity of these mixtures and concentrates to that of known compounds and other concentrates; (b) a bacterial mutagenic screen to select those samples more likely

TABLE 2. Volatile Organic Compounds Identified in Cincinnati Drinking Water

Chloromethane	Dichloroacetylene
Bromomethane	Trichloeothylene
Bromodichloromethane	Tetrachloroethylene
Dibromochloromethane	Bromotrichloroethylene
Chloroform	Cyanogen chloride
Bromoform	Methanol
Carbon tetrachloride	Ethanol
Chloroethane	2-Methyl propanal
Chloropropane	2-Butanone
1,2-Dichloroethane	3-Methyl butanal
Acetylene chloride	Toluene

to contain chemical mutagens and carcinogens; (c) *in vivo* cancer screening model employing neonatal animals to determine which samples and water supplies are more likely to be potentially tumorigenic, and (d) an *in vivo* teratologic screen to determine the potential of chemical agents in producing birth defects.

Examples of the acute toxicity of the concentrates (generated by reverse osmosis or carbon adsorption and extraction) were reported by Tardiff and Deinzer (1973). The acute toxicity of the samples was classified as "very toxic" to "extremely toxic."

Positive findings in mutagenic, carcinogenic, and teratologic screening systems initiate in-depth studies of the chemical composition and toxicity to determine the causative agents. The first step is fractionation with concurrent bioassay of the individual fractions to determine the location of the toxins. Identifying specific agents in biologically active fractions allows (a) the evaluation of the structures of the chemical agents to prioritize compounds suspected of being the etiologic agents of the toxicoses and (b) to subject these suspected toxins to specialized and specific tests such as lifetime exposure studies with organ function tests, three-generation reproductive studies, and *in vivo* mutational translocation studies.

Examples of compounds that have been identified in tap water and that may be responsible for some forms of chronic illness are the chloroethers and the chlorobenzenes. Bis(2-chloroethyl) ether and bis(2-chloroisopropyl) ether are congeners of a very potent carcinogen, bis(chloromethyl) ether. Yet, very few toxicity data are available for either compound, particularly data from lifetime exposures. Our laboratory is engaged in developing appropriate experimental models and protocols for assessing the toxicity of these agents, both singly and in combination with other environmental chemicals. The development of models applicable to the extrapolation to man is presently undergoing intensive investigation, predominantly through studies of comparative metabolism and also through investigations of comparative toxicity.

The chlorobenzenes pose a slightly different problem. Data on their individual toxicities are comprehensive (Patty, 1962; Williams, 1959), but with the exception of hexachlorobenzene, very little is known about their potential for interaction with other compounds found in man's environment. With the use of specific biochemical tools, the potential of interaction between one compound and another can be ascertained through this mechanism. Preliminary data have shown that trichlorobenzene may be potent in altering the toxicity of other foreign organic compounds.

With a dose response design, exposure levels producing either no effect or minimal effects can be determined. When the slope of the dose-response curve is taken into consideration, such data can then be extrapolated to relatively safe exposure levels.

As suggested previously, one of the main difficulties is the extrapolation of data from one species to another, or more specifically from the animal model to man. This problem is approached in two ways: (a) comparative metabolism, which attempts to define man's metabolism of the compound and compare it with the metabolism of various experimental species, and (b) comparative toxicity, which attempts to define differences in sensitivity and target organs among species and to compare differences in potency among different species. The comparative metabolism is a particularly significant approach if one considers that the toxicity of any organic compound may not be due as much to the parent molecule as to one or several of its metabolites, as for example phosphorothiolates, phosphorothionates, and carbon tetrachloride.

The magnitude of the problem is assessing the hazard and safety of organic compounds in drinking water is particularly striking when the number of compounds presently identified in only a relatively few tap water samples are considered. Table 3 lists 187 compounds that have been found in various tap waters. Some gross estimations of the concentrations of these compounds in tap water allow the comparison with the total organic carbon content of tap water. Such a comparison leads to the conclusion that these 187 compounds account for perhaps 5 or even as much as 10 percent of the compounds, by weight in tap water. Little,

TABLE 3. Organic Compounds Identified in Drinking Water in the United States (Mullaney, 1975)

1. Acenaphthene
2. Acenphthylene
3. Acetaldehyde
4. Acetic acid
5. Acetone
6. Acetophenone
7. Acetylene dichloride
8. Aldrin
9. Atrazine
10. (deethyl) Atrazine
11. Barbital
12. Behenic acid, methyl ester
13. Benzaldehyde
14. Benzene
15. Benzene sulfonic acid
16. Benzoic acid
17. Benzopyrene
18. Benzothiazole
19. Benzothiophene
20. Benzyl butyl phthalate
21. Bladex
22. Borneol
23. Bromobenzene
24. Bromochlorobenzene
25. Bromodichloromethane
26. Bromoform
27. Butanal
28. Bromophenyl phenyl ether
29. Butyl benzene
30. Butyl bormide
31. Camphor
32. e-Caprolactam
33. Carbon dioxide
34. Carbon disulfide
35. Carbon tetrachloride
36. Chloran(e)
37. Chlordene
38. Chlorobenzene
39. 1,2-bix-Chloroethoxy ethane
40. Chloroethoxy ether
41. bis-2-Chloroisopropyl ether
42. 2-Chloroethyl methyl ether
43. Chloroform
44. Chlorohydroxybenzophenone
45. bis-Chloroisopropyl ether
46. Chloromethyl ether
47. Chloromethyl ethyl ether
48. m-Chloronitrobenzene
49. 1-Chloropropene
50. 3-Chloropyridine
51. o-Cresol
52. Crotonaldehyde
53. Cyanogen chloride
54. Cycloheptanone
55. DDE
56. DDT
57. Decane
58. Dibromobenzene
59. Dibromochloromethane
60. Dibromodichloroethane
61. 2,6-Ditbutyl-p-benzoquinone
62. Dibutyl phthalate
63. 1,3-Dichlorobenzene
64. 1,4-Dichlorobenzene
65. Dichlorodifluoroethane
66. 1,2-Dichloroethane
67. 1,1-Dichloro-2-hexanone
68. 2,4-Dichlorophenol
69. 1,2-Dichloropropane
70. 1,3-Dichloropropene
71. Dieldrin
72. Di(2-ethylhexyl) adipate
73. Diethyl benzene
74. Diethyl phthalate
75. Di(2-ethylhexyl) phthalate
76. Dihexyl phthalate
77. Dihydrocarvone
78. Diisobutyl carbinol
79. Diisobutyl phthalate
80. 1,2-Dimethoxy benzene
81. 1,3-Dimethylnaphthalene
82. 2,4-Dimethylphenol
83. Dimethyl phthalate

(continued)

TABLE 3 (cont.)

84. Dimethyl sulfoxide
85. 4,6-Dinitro-2-aminophenol
86. 2,6-Dinitrotoluene
87. Dioctyl adipate
88. Diphenylhydrazine
89. Diphenyl phthalate
90. Docosane
91. n-Dodecane
92. Eicosane
93. Endrin
94. Ethanol
95. Ethylamine
96. Ethyl benzene
97. 2-Ethyl-n-hexane
98. cis 2-Ethyl-4-methyl-1,3-dioxolane
99. trans 2-Ethyl-4-methyl-1,3-dioxolane
100. o-Ethyltoluene
101. m-Ethyltoluene
102. p-Ethyltoluene
103. Geosmin
104. Heptachlor
105. Heptachlor epoxide
106. 1,2,3,4,5,7,7-Heptachloronorbornene
107. Hexachlorobenzene
108. Hexachlro-1,3-butadiene
109. Hexachlorocyclohexane
110. Hexachloroethane
111. Hexachlorophene
112. Hexadecane
113. 2-Hydroxyadiponitrile
114. Indene
115. Isoborneol
116. Isocyanic acid
117. Isodecane
118. Isophorone
119. 1-Isopropenyl-4-isopropylbenzene
120. Isopropyl benzene
121. Limonene
122. p-Menth-1-en-8-ol
123. Methane
124. Methanol
125. o-Methoxyphenol
126. Methyl benzoate
127. Methyl benzothiazole
128. Methyl biphenyl
129. 3-Methyl butanal
130. Methyl chloride
131. Methylene chloride
132. Methyl ethyl benzene
133. Methyl ethyl ketone
134. 2-Methyl-5-ethyl pyridine
135. Methylindene
136. Methyl methacrylate
137. Methyl naphthalene
138. Methyl palmitate
139. Methyl phenyl carbinol
140. 2-Methylpropanal
141. Methyl stearate
142. Methyl tetracosanoate
143. Naphthalene
144. Nitroanisole
145. Nitrobenzene
146. Nonane
147. Octadecane
148. Octane
149. Octyl chloride
150. Pentachlorobiphenyl
151. Pentachlorophenol
152. Pentachlorophenyl methyl ether
153. n-Pentadecane
154. Pentane
155. Pentanol
156. Phenyl benzoate
157. Phthalic anhydride
158. Piperidine
159. Propanol
160. Propazine
161. Propylamine
162. Propylbenzene
163. Simazine
164. 1,1,3,3-Tetrachloroacetone
165. Tetrachlorobiphenyl

(continued)

TABLE 3 (cont.)

166. 1,1,1,2-Tetrachloroethane
167. Tetrachloroethylene
168. n-Tetradecane
169. Tetramethyl benzene
170. Thiomethylbenzothiazole
171. Toluene
172. Trichlorobenzene
173. Trichlorobiphenyl
174. 1,1,2-Trichloroethane
175. 1,1,2-Trichloroethylene
176. Trichlorofluoromethane
177. 2,4,6-Trichlorophenol
178. n-Tridecane
179. Trimethyl benzene
180. 3,5,5-Trimethyl-bicyclo (4,1,0)heptene-2-one
181. 1,3,5-Trimethyl-2,4,6-trioxo-hexahydro-triazine
182. Triphenyl phosphate
183. n-Undecane
184. Vinyl benzene
185. o-Xylene
186. m-Xylene
187. p-Xylene

if any, chronic toxicity data exist for almost half of these compounds—data necessary to perform health and hazard assessments and to set drinking water standards.

REFERENCES

1. Anon. 1971. Standard Methods for the Examination of Water and Waste Water, 13th ed., American Public Health Association, New York.

2. Bellar, T. A. and Lichtenberg, J. J. 1974. J. Amer. Water Works Assoc. 66, 739-744.

3. Bellar, T. A., Lichtenberg, J. J., and Kroner, R. C. 1974. J. Amer. Water Works Assoc. 66, 703-706.

4. Bishop, D. F., Marshall, L. S., O'Farrel, T. P., Dean, R. B., O'Connor, B., Dobbs, R. A., Griggs, S. H., and Villers, R. V. 1967. J. Water Pollut. Control Fed. 39, 188-203.

5. Buelow, R. W., Carswell, J. K., and Symons, J. M. 1973. J. Amer. Water Works Assoc. 65, 57-72.

6. Edwards, V. H. and Schubert, P. F. 1974. J. Amer. Water Works Assoc. 66, 610-616.

7. Ettinger, M. B. 1965. J. Amer. Water Works Assoc. 57, 453-457.

8. Gunther, F. A., Blinn, R. C., Kolbezen, M. J., Barkley, J. H., Harris, W. D., and Simon, H. S. 1951. Anal. Chem. 23, 1835.

9. Ishizaki, C. and Cookson, J. T., Jr. 1973. J. Water Pollut. Control Fed. 45, 515-522.

10. Jolley, R. L. 1973. Oak Ridge National Laboratory Publication ORNL-TM-4290, October.

11. McKee, J. E., Laverty, F. B., and Hertel, R. M. 1972. J. Water Pollut. Control Fed. 44, 293-302.

12. Mullaney, J. L. 1975. Organic Compounds Identified in Drinking Waters in the United States; Safety/Hazard Evaluation File (update to April 1975) compiled by the

Water Quality Division, Health Effects Research Laboratory, Environmental Protection Agency, Cincinnati, Ohio.

13. Nelson, K. H. and Lysyj, I. 1968. Environ. Sci. Technol. 2, 61-62.

14. Patty, F. A., Ed. 1975. Industrial Hygiene and Toxicology (Toxicology, Vol. II, David W. Fasset and Don D. Irish, Eds.), 2nd rev. ed., Interscience, New York, pp. 1333-1340.

15. Rook, J. J. 1974. Water Treatment Exam. 23, 234-243.

16. Tardiff, R. G. and Deinzer, M. 1973. Proc. of the 15th Water Quality Conference, University of Illinois, Urbana-Champaign, p. 23-37.

17. Wallis, I. G. 1974. J. Water Pollut. Control Fed. 46, 438-457.

18. Williams, R. T. 1975. Detoxication Mechanisms, The Metabolism and Detoxication of Drugs, Toxic Substances, and Other Organic Compounds, 2nd rev. ed., John Wiley, New York, pp. 237-244.

Index